Springer Series on Biofilms
Volume 4

Series Editor: J. William Costerton
Los Angeles, USA

Hans-Curt Flemming • P. Sriyutha Murthy
R. Venkatesan • Keith Cooksey
Editors

Marine and Industrial Biofouling

Editors
Prof. Dr. Hans-Curt Flemming
Biofilm Centre
University of Duisburg-Essen
Geibelstraße 41
47057 Duisburg
Germany

Dr. R. Venkatesan
Organising Secretary RAMAT
Group Head Ocean Science
& Technology for Islands
National Institute of Ocean Technology
Pallikaranai, Chennai
India

Dr. P. Sriyutha Murthy
Ocean Science & Technology for Islands
National Institute of Ocean Technology
Ministry of Ocean Development
Velachery Tamabaram Main Road
Narayanapuram
Chennai 601 302
Tamil Nadu
India

Prof. Dr. Keith Cooksey
Department of Microbiology
Montana State University
109 Lewis Hall
PO Box 173520
Bozeman, MT 59717
USA

Series Editor
J. William Costerton
Director, Center for Biofilms
School of Dentistry
University of Southern California
925 West 34th Street
Los Angeles, CA 90089
USA

ISBN 978-3-540-69794-7 e-ISBN 978-3-540-69796-1
DOI: 10.1007/978-3-540-69796-1

Library of Congress Control Number: 2008940066

Cover design: SPi Publishing Services

Printed on acid-free paper

9 8 7 6 5 4 3 2 1 0

springer.com

Preface

This book describes the state of the art in antifouling measures using both conventional biocides and some advanced approaches. Related to biocides, the concept of the "Biocide Product Directive" of the European Union is presented as an example of an administrative instrument for curbing excessive use of environmentally undesirable products that may cause ecological damage.

Biofouling is defined as the unwanted accumulation of biological material on man-made surfaces. This definition includes biofilm-forming microorganisms such as bacteria, fungi and algae as well as fouling by macroorganisms like hydroids, barnacles, tubeworms and bivalves on submerged surfaces. The problem is site-, season- and substratum-specific and the control methods effective at a given geographical location may not hold good elsewhere. The definition is clearly operational, as not every biofilm or barnacle is equivalent to biofouling but only after the effect exceeds an arbitrarily given threshold of interference with a technical process. It is impossible to have an immaculately clean surface and the time has come for realization of the fact that we have to "live with biofilms and biofouling". It is for the plant managers to determine the tolerable threshold of interference, critical to plant operations, and select a biocidal dose and regime to keep biofilms/biofouling at bay.

The problems in technical processes that are posed by biofouling are substantial. An example is the interference with heat exchangers, where both macro- and microfouling contribute to losses in heat transfer and to increases in fluid friction resistance. In India, for example, in the next decade 15 large new power plants will be built, all using seawater as a coolant. Designers and operators will have to overcome serious fouling problems. The common concepts of biofouling control are still based on the use of biocides, which only partially and transiently mitigate the problem. With respect to macrofouling control, focussed research on the biocidal dosages required to prevent/inhibit settlement are lacking compared to research on the dosages required for killing established fouling communities. This is important as dosages required to prevent settlement are far lower than dosages required for killing established fouling communities. Oxidizing biocides are a better option than non-oxidizing biocides due to their known mode of action and toxicity, and knowledge of their by-products and degradation pathways. Cost–benefit analysis and meeting the environmental regulations for discharge are vital parameters governing the selection of biocides in power and desalination plants. Biofouling is a "surface-associated phenomenon" and control measures should concentrate on this aspect. For example, treating the entire bulk water with a biocidal concentration seems to be an economically unviable practice. If technological advancements could be achieved to deliver biocides at the surface on a continuous basis by the use of porous polymeric materials, this would ensure a cleaner surface, reduced biocidal requirements and reduced impact on the environment.

In seawater desalination by membrane filtration, a process meeting the equally important and increasing demand for freshwater, biofouling also represents the "Achilles' heel" of the technology. Again, the use of biocides is still the state of the art, but their use threatens the material properties of membranes and other equipment, as well as causing environmental problems when disposed of. The scenario for biofouling control measures in the case of the shipping industry is in a transient stage where foul-release coatings alone seem to be an effective alternative. Several alternative replacement techniques for tributyl tin self-polishing coatings are emerging, but currently none have demonstrated their performance at the field level and can be translated into a technology.

The sequence of events leading to biofouling of surfaces comprises the formation of (i) biofilms containing the initial colonizing organisms, causing serious problems, and (ii) layers of the most visibly obvious foulers that succeed them, i.e. macroalgae (Enteromorpha sp., Sargassum sp., Gracillaria sp.) and the hard-shelled foulants (barnacles, hydroids, tubeworms and bivalves). These organisms colonize submerged surfaces that already have a microbial film present. Whether there is a positive or negative effect of the microbial film on the colonization success of the macroorganisms depends on the make-up of the biofilm and the species of invertebrates involved. Various aspects of this topic are covered in several chapters of the book.

The common approach used against fouling biofilms can be compared to a "medical paradigm": the system is considered to be infected and the cure is seen to be the use of biocides. However, killing the organisms is not the solution, as the problem is usually not caused by their physiological activity but by their mere presence as a physical barrier. Reduction of the extent of fouling layers is clearly more important, but not yet generally the focus of countermeasures.

Antifouling measures are taken all over the world with very unequal levels of success. There is no such thing as a universal solution to the biofouling problem (as with the case of biocidal type, dose and regime) but there are many insights acquired from various fields stricken by biofouling that should be taken into consideration. This book is an attempt to collect some of these approaches and to provide the opportunity to learn from scientific research on biofouling, as well as from interesting approaches in various technical fields.

October 2008

H.-C. Flemming
P. Sriyutha Murthy
R. Venkatesan
K. Cooksey

Contents

Contributors

Haluk Beyenal
School of Chemical Engineering and BioEngineering, Washington State University, Pullman, Washington 99164-2710, USA
beyenal@wsu.edu

C. Compère
Ifremer-In Situ Measurements and Electronics, B.P. 70, 29280, Plouzané, France

K. E. Cooksey
Department of Microbiology, Montana State University, Bozeman, MT 59717, USA and Environmental Biotechnology Consultants, Manhattan, MT 59741, USA
umbkc@gemini.msu.montana.edu

H.-U. Dahms
Department of Biology and Coastal Marine Lab Hong Kong, University of Science and Technology, Clearwater Bay, Kowloon, Hong Kong, China and National Taiwan Ocean University, Keelung, Taiwan

L. Delauney
Ifremer-In Situ Measurements and Electronics, B.P. 70, 29280, Plouzané, France
laurent.delauney@ifremer.fr

Sergey Dobretsov
Marine Science and Fisheries Dep., Agriculture and Marine Sciences College, Sultan Qaboos University, Al-Khod 49, PO Box 123, Sultanate of Oman and Benthic Ecology, IFM-GEOMAR, Kiel University, Düsternbrooker Weg 20, 24105, Kiel, Germany
sergey_dobretsov@yahoo.com, sergey@squ.edu.om

Hans-Curt Flemming
Biofilm Centre, University of Duisburg-Essen, Geibelstrasse 41, D-47057, Duisburg, Germany and IWW Centre for Water, Moritzstrasse 26, 45476, Muelheim, Germany
hanscurtflemming@compuserve.com

Malcolm Greenhalgh
Malcolm Greenhalgh Consultancy Ltd (MGCL), Dale Cottage, Lower Park Royd Drive, Ripponden, West Yorkshire. HX6 3HR, UK
malcolm.greenhalgh@btopenworld.com

Michael G. Hadfield
Kewalo Marine Laboratory, University of Hawaii, 41 Ahui Street, Honolulu, HI 96813, USA
hadfield@hawaii.edu

Tilmann Harder
Centre for Marine Bio-Innovation, University of New South Wales, Sydney, NSW 2052, Australia
t.harder@unsw.edu.au

Zbigniew Lewandowski
Department of Civil Engineering and Center for Biofilm Engineering, Montana State University, EPS Building, Room 310, Bozeman, MT 59717, USA
ZL@erc.montana.edu

R. A. Long
Department of Biological Sciences, University of South Carolina, Columbia SC 29208, USA

K. V. K. Nair
National Institute of Ocean Technology, Ministry of Earth Sciences, Government of India, Velachery-Tambaram Main Road, Narayanapuram, Chennai 600 100, India

Brian T. Nedved
Kewalo Marine Laboratory, University of Hawaii, 41 Ahui Street, Honolulu, HI 96813, USA

P. Y. Qian
1. Department of Biology and Coastal Marine Lab Hong Kong, University of Science and Technology, Clearwater Bay, Kowloon, Hong Kong, China
boqianpy@ust.hk

Harry Ridgway
Stanford University & AquaMem Scientific Consultants, Department of Civil and Environmental Engineering, PO Box 251, Rodeo, New Mexico 88056, USA
ridgway@vtc.net

Dan Rittschof
Duke University Marine Laboratory, Nicholas School of the Environment, 135 Duke Marine Lab Road, Beaufort, NC 28516, USA
ritt@duke.edu

P. Sriyutha Murthy
Biofouling and Biofilm Processes Section, Water and Steam Chemistry Division, BARC Facilities, Indira Gandhi Center for Atomic Research Campus, Kalpakkam 603 102, India
psm_murthy@yahoo.co.in, psmurthy@igcar.gov.in

T. Subramoniam
National Institute of Ocean Technology, Ministry of Earth Sciences, Government of India, Velachery-Tambaram Main Road, Narayanapuram, Chennai 600 100, India

R. Venkatesan
National Institute of Ocean Technology, Ministry of Earth Sciences, Government of India, Velachery-Tambaram Main Road, Narayanapuram, Chennai 600 100, India

V. P. Venugopalan
Biofouling and Biofilm Processes Section, Water and Steam Chemistry Division, BARC Facilities, Indira Gandhi Center for Atomic Research Campus, Kalpakkam 603 102, India
vpv@igcar.gov.in

J. Verran
School of Biology, Chemistry and Health Science, Manchester Metropolitan University, Chester St, Manchester, M1 5GD, UK
j.verran@mmu.ac.uk

T. Vladkova
Department of Polymer Engineering, University for Chemical Technology and Metallurgy, 8 “Kliment Ohridsky” Blvd., 1756 Sofia, Bulgaria
tgv@uctm.edu

K. A. Whitehead
School of Biology, Chemistry and Health Science, Manchester Metropolitan University, Chester Street, Manchester, M1 5GD, UK
k.a.whitehead@mmu.ac.uk

B. Wigglesworth-Cooksey
Department of Microbiology, Montana State University, Bozeman, MT 59717, USA and Environmental Biotechnology Consultants, Manhattan, MT 59741, USA

Part I
Microbial Biofouling and Microbially Influenced Corrosion

Why Microorganisms Live in Biofilms and the Problem of Biofouling

Hans-Curt Flemming

Abstract Microbial biofouling is a problem of microbial biofilms. Biofouling occurs in very different industrial fields and is mostly addressed individually. However, the underlying phenomenon is much more general and in order to understand the processes causing biofouling, it is good to understand the basics of biofilm formation and development. Almost every surface can be colonized by bacteria, forming biofilms. After adhesion, the cells embed themselves in a layer of extracellular polymeric substances (EPS), highly hydrated biopolymers of microbial origin such as polysaccharides, proteins, nucleic acids and others. In this matrix they organize their life, develop complex interactions and resistance to biocides. The resulting biofilm structure is highly heterogeneous and dynamic. It is kept together by weak physicochemical interactions of extracellular polymeric substances, which have to be overcome when cleaning is attempted. The ecological advantages for the biofilm mode of life are so strong that almost all microorganisms on earth live in biofilm-like microbial aggregates rather than as single organisms.

1 Biofouling

Slime on surfaces is the usual manifestation of a phenomenon called "biofouling". It occurs in a wide range of industrial processes and in all of them it is a nuisance, sometimes a very expensive one. It is fought against in each industrial area individually and there are many "re-inventions of the wheel" and many common mistakes – although the underlying problem is always the same: microbial biofilms. Five common mistakes in conventional anti-fouling measures can be identified in most cases are:

1. *No early warning systems*: Biofouling is detected by losses in process performance or product quality – no monitoring system.

H.-C. Flemming
Biofilm Centre, University of Duisburg-Essen, Geibelstrasse 41, 47057, Duisburg, Germany
e-mail: hanscurtflemming@compuserve.com

Springer Series on Biofilms, doi: 10.1007/7142_2008_13

2. *No information on biofilm site/extent*: Sampling is performed of the water phase, which gives no information about site and extent of fouling films; sampling is not performed on surfaces.
3. *Disinfection is performed as a countermeasure*: This is not cleaning, while in most cases, the problem is caused by biomass – dead or alive. Biocides leave dead biomass on surface, providing good regrowth.
4. *No nutrient limitation is considered*: However, nutrients are potential biomass and are not reduced by biocides.
5. *No optimization of countermeasures*: Efficacy control is performed only by process or product quality – see point 1.

In very diverse industrial fields, biofouling problems all originate from the same cause: microbial biofilms. Biofilms follow common natural laws, which are important to be understood for more effective countermeasures. Basically, in biofouling the same processes occur as in biological filtration: microorganisms colonize surfaces, sequester nutrients from the water phase and convert them into metabolites and new biomass. Industrial systems frequently offer large surface areas, which invite colonization and subsequent use of biodegradable substances, leading to an extent of biofilm development that interferes with process parameters or product quality. Biofouling can be considered as a "biofilm reactor in the wrong place and at the wrong time". Therefore, detailed knowledge about biofilms is crucial for understanding and preventing biofouling as well as for successful anti-fouling measures.

The purpose of this chapter is to highlight the reasons why microorganisms form biofilms. They are the most successful form of life on earth and it is not surprising, that they cannot be eliminated easily. In many cases, microbial biofilms precede macroorganismic settlement (e.g. by larvae, barnacles and mussels), a phenomenon called macrofouling.

2 Microbial Biofilms

It is only few decades since microorganisms, sitting at the walls of microbiological liquid cultures, on rocks, sediments, in soil, on leaves, skin, teeth, implants or in wounds turned from a nuisance that could not be investigated by classical microbiological methods into a highly active field of research in which biofilms were acknowledged as the dominant form of life for microorganisms on earth (Flemming 2008). It became obvious that microorganisms on earth generally do not live as single cells and in pure cultures but do so in aggregates of mixed species. Such aggregates can consist of microcolonies as well as of patchy or confluent films on surfaces, but also as thick mats, sludge or flocks in suspension. By convention, all these phenomena are subsumed under the (somehow vague) term "biofilm" (Donlan 2002). It was just a shift of point of view that made it evident that this form of life could be found everywhere. In fact, biofilms are the first form of life recorded on earth, dating back 3.5 billion years (Schopf et al. 1983), and the most successful one. Biofilms are found even in extreme environments, such as the walls of pores in glaciers, in hot vents, under pressure of 1,000 bar at the bottom of the ocean, in ultra-pure water as well as highly salty solutions, and on electrodes active through the entire range of thermodynamic water stability. Biofilms occur as endolithic populations in minerals, on the walls of disinfectant concentrate pipes or even in highly radioactive environments such as nuclear power plants. The surface of almost all living organisms is colonized by biofilms, which provide in many cases a protective and supportive flora (e.g. skin flora), while in other cases they cause transient, acute, chronic and even fatal diseases. Biofilms are substantially involved in the biogeochemical cycles of carbon, oxygen, hydrogen, nitrogen, sulphur, phosphorus and many metals (Ehrlich 2002). Enhancing mineral weathering processes by microbial leaching, they mobilized metal ions that were vital for further evolution. In biofilms, photosynthetic organisms evolved from originally anaerobic conditions on earth, providing oxygen as a "waste gas" from photosynthesis to the atmosphere of this planet and restricting the space for living of anaerobic organisms, which first dominated life on earth, to oxygen-depleted areas. Predation among biofilm organisms is thought to have led to endosymbionts and, eventually, to the evolution of eukaryotic organisms and the concept of infection.

One of the reasons for the late acknowledgement of biofilms is certainly the insufficient suitability of conventional microbiological methods. The introduction of fluorescence microscopy and confocal laser scanning microscopy, micro-electrodes, advanced chemical analysis with particular respect to protein analysis, and, most powerfully, molecular biology has allowed biofilm biology to be revealed in much greater detail. As a consequence, the literature in this field has virtually exploded with at least 100,000 publications on biofilms currently. The advance of knowledge is immense and fast, and this brief chapter can only superficially cover it. From a life science point of view, the most exciting aspect is that microorganisms today cannot be viewed as blind little individuals that compete as much as they can, but as complex communities with division of labour and many aspects of multicellular life (Flemming 2008). This is certainly a new understanding of microbiology with big consequences for biotechnology, medicine and handling of microbial problems in technical processes.

The biofilm mode of life provides a range of advantages to the single cell planktonic mode of life. One of the biggest advantages is the fact that the cells can develop stable interactions, resulting in synergistic microconsortia. An example is the close association of ammonia oxidizing and nitrite oxidizing bacteria. The ammonia oxidizers produce nitrite, an inhibitory end product that is comfortably used as substrate by the nitrite oxidizers. This process occurs in the environment and has been employed in nitrification steps in waste water treatment for a long time and with great success. There are many other examples of orchestrated degradation of substrates by cascades of organisms.

3 Extracellular Polymeric Substances

A characteristic feature of biofilm organisms is that they are kept together and attached to surfaces by means of their extracellular polymeric substances (EPS, Flemming and Leis 2002). An example is shown in Fig. 1, which is a scanning electron micrograph

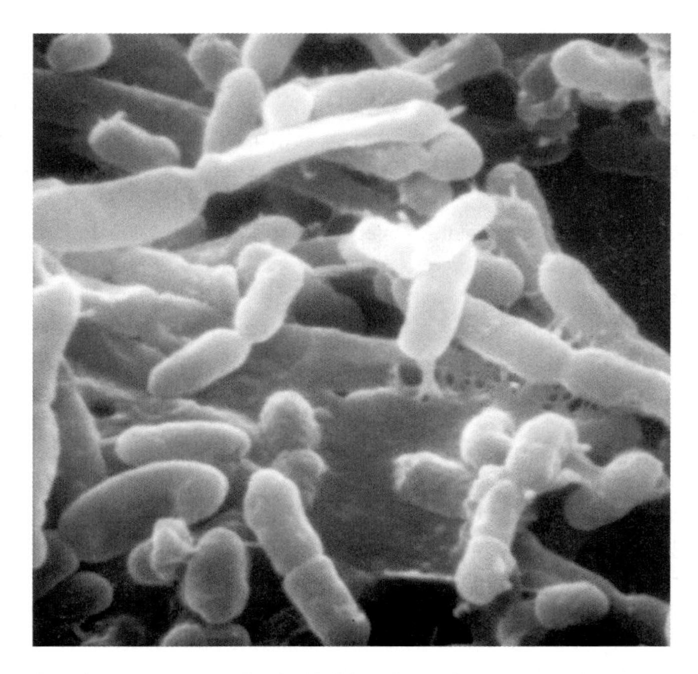

Fig. 1 Scanning electron micrograph of a biofilm of *Pseudomonas putida* on a mineral surface. EPS (dehydrated for SEM sample preparation) are surrounding the cells, keeping them together and on the surface

of *Pseudomonas putida* on a mineral surface. The sheet-like material that surrounds the cells is EPS, dehydrated by sample preparation for SEM observation.

The EPS determine the immediate conditions of life of biofilm cells living in this microenvironment by affecting porosity, density, water content, charge, sorption properties, hydrophobicity and mechanical stability – all belonging to the parameters on which the conditions of life in a biofilm depend (Branda et al. 2005). This section represents a recent synopsis of the actual state of understanding of the role of EPS (Flemming et al. 2007).

EPS are biopolymers of microbial origin in which biofilm microorganisms are embedded. In fact, the biopolymers are produced by archaea, bacteria and eukaryotic microbes. Contrary to common belief, they are certainly more than only polysaccharides. Additionally, they comprise a wide variety of proteins, glycoproteins, glycolipids and in some cases surprising amounts of extracellular DNA (e-DNA). In environmental biofilms, polysaccharides are frequently only a minor component. All EPS biopolymers are highly hydrated and form a matrix, which keeps the biofilm cells together and retains water. This matrix interacts with the environment, e.g. by attaching biofilms to surfaces and by its sorption properties, which allows for sequestering dissolved and particulate substances from the environment providing nutrients for biofilm organisms. The EPS influence predator–prey interactions, as demonstrated in a system of a predatory ciliate and yeast cells. Grazing led to an increase in biofilm mass and viability with EPS as preferred food source.

Curli as proteinaceous fibrils have gained more interest beyond infection as curli-like fibrils have also been found to play an important role in natural biofilms produced by a variety of different microorganisms. An abundance of amyloid adhesions in natural biofilms has been found, which may contribute considerably to their mechanical properties. Strengthening of biofilm structure is crucial for the stability of the "house" and the continuation of synergistic interactions based on spatial proximity of various biofilm organisms.

Cellulose has been found to be a constituent EPS component in amoebae, algae and bacteria. In agrobacteria, cellulose is involved in attachment and it seems as if cellulose plays an underestimated role in environmental EPS. It is formed by a variety of organisms and influences biofilm structure. Cellulose is also important in infectious processes when co-expressed with curli fimbriae in *Escherichia coli* (Wang et al. 2007).

Biofilms are also an ideal place for exchanging genetic material and maintaining a large and well-accessible gene pool. Horizontal gene transfer is facilitated as the cells are maintained in close proximity to each other, not fully immobilized, and can exchange genetic information. Significantly higher rates of conjugation in bacterial biofilms compared to planktonic populations have been reported (Hausner and Wuertz 1999).

The EPS matrix is not only composed of a variety of components but, in addition, these are able to interact. One example is the retention of extracellular proteins such as lipase by alginate. Such mechanisms are crucial for preventing the wash-out of enzymes, keeping them close to the cells that produced them and allowing for effective degradation of polymeric and particulate material. This leads to the concept of an "activated matrix". Activation is made even more dynamic and versatile by the excretion of membrane vesicles (MVs). These highly ordered nanostructures act as "parcels" containing enzymes and nucleic acids, sent into the depth of the EPS matrix. Such vesicles, along with phages and viruses (which are of similar size), can serve as carriers for genetic material and thereby enhance gene exchange. Through their chemistry, the MVs may bind extraneous components; their enzymes may help degrade polymers, providing nutrients or inimical agents and thereby inactivating them. Furthermore, they seem to be part of the "biological warfare" within biofilms, occurring as predatory vesicles containing lytic enzymes. This biological warfare is also long-range as, in common with other matrix material, MVs are shed from the biofilm. In this respect, vesicles are "missiles" delivering, among others, virulence factors and cell-to-cell signals (Schooling and Beveridge 2006).

The composition, architecture and function of the EPSmatrix reveal a very complex, dynamic and biologically exciting view. First of all, the matrix is a network providing sufficient mechanical stability to maintain spatial arrangement for microconsortia over a longer period of time. This stability is provided by hydrophobic interactions, cross-linking by multivalent cations and entanglements of the biopolymers with e-DNA as a newly appreciated structural component. The forces that keep the biofilm matrixtogether are provided, thus by weak physicochemical interactions such as hydrogen bonds, van der Waals forces and eletrostatical interactions. They are schematically depicted in Fig. 2 (after Mayer et al. 1999).

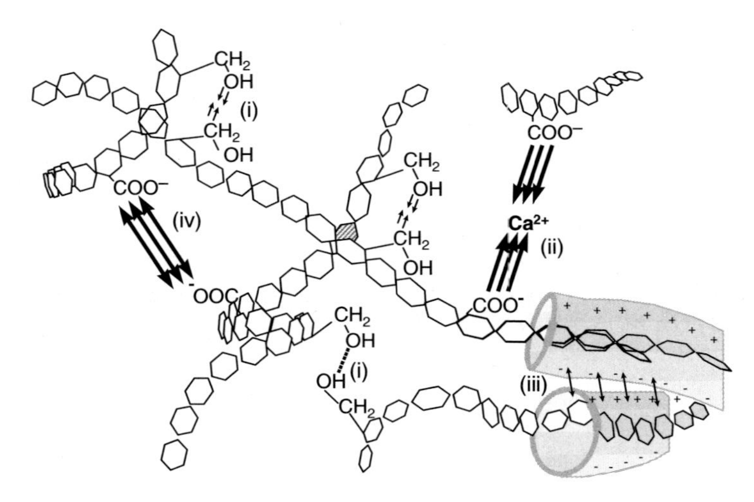

Fig. 2 Forces that keep the EPS matrix together: (**i**) hydrogen bonding, (**ii**) cation bridging, (**iii**) van der Waals forces, (**iv**) repulsive forces (after Mayer et al. 1999)

The repulsive forces are of big importance for the biofilm structure as they prevent a polymer network from collapsing. Water is equally important as it dilutes the macromolecules and limits the number of interacting groups. During desiccation, more interaction takes place and turns biofilms into practically insoluble structures (Fig. 3).

When microbial biofilms are to be removed from surfaces, as in the case of cleaning, these weak binding forces have to be overcome. Although the individual forces are low, the gross overall binding force can exceed that of covalent bonds, but it is not a directed bond. Therefore, in response to shear forces, biofilm first show characteristics of viscoelastic bodies, while when a breaking point is exceeded, they have properties of viscous liquids (Körstgens et al. 2001). Cleaning has to attempt weakening of the binding forces in order to support the efficacy of shear forces. From this point of view, it is very obvious that killing of the biofilm organisms will not contribute to cleaning unless the matrix structure is affected.

In conclusion, it seems as if "slime" has been very much underestimated and it turns out that the EPS matrix is considerably more than simply the glue for biofilms. Rather, it is a highly sophisticated system that gives the biofilm mode of life particular and successful features.

4 Structure of Biofilms

The biofilm matrix is highly hydrated and very heterogeneous. The morphology of a biofilm appears very variable. Figure 4 shows an artists view of various aspects of evolving and mature biofilms, as developed from many recent findings in biofilm research.

Fig. 3 Desiccated biofilm. The cohesive forces and the surface adhesion forces increase. Curling of biofilms occurs and sand grains from mortar are ripped out, contributing to microbially influenced weathering

The figure reveals structural aspects that make life in biofilms even more attractive. The porous architecture allows for convectional flow through the depth of the biofilm, while within the EPS matrix only diffusional transport is possible. Organisms

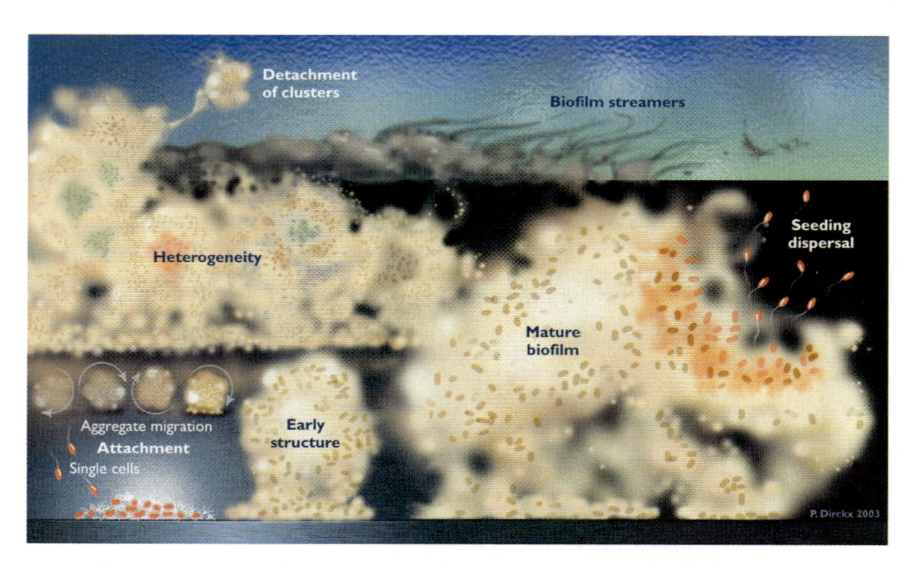

Fig. 4 Structure and processes in a biofilm (permission of Peggy Dirkx, Center for Biofilm Engineering, Montana State University, Bozeman, MT)

at the bottom of the biofilm, thus, can get access to nutrients without competition from those at the interface to the bulk water phase. Strong gradients can occur in biofilms, e.g. by actively respiring aerobic heterotrophic organisms, which consume oxygen faster than it can diffuse through the matrix. This generates anaerobichabitats just below highly active aerobic colonies in distances of less than 50 μm. Other gradients, such as pH-value, redox potential and ionic strength are known within biofilms. The result is complex interactions and a functionally structured system. The ecological relevance of this heterogeneity has inspired Watnick and Kolter (2000) to describe the biofilm as a "City of Microbes".

Another feature of biofilm cells is the increased tolerance to biocides, compared to planktonic cells (Schulte et al. 2005). It must be taken into consideration that biofilms have existed for billions of years and have survived all kinds of adverse conditions. Therefore, many different mechanisms have evolved for resistance, and they are far from being fully understood (Lewis 2001). The fact is that resistance genes can be exchanged and that biofilms have been observed even in disinfection concentrate pipes. The resistance of biofilms is particularly problematic in medicine where contaminations of implants, catheters or bones result in long-term infections, which in many cases can only be overcome by radical measures such as exchange of implants and removal of bone parts. In drinking water systems, biofilms can harbour hygienically relevant organisms that may even proliferate if nutrients are provided. Even enhanced application of disinfectants such as chlorine will not eradicate such biofilms.

5 Ecological Advantages of the Biofilm Mode of Life

From the above highlighted context, it is obvious that microorganisms gain clear advantages from the biofilm mode of life. This has been summarized very well by Costerton (2007), a biofilm pioneer. The ecological advantages of the biofilm mode of life are quite a few more and can be summarized as follows:

- Formation of stable microconsortia
- Biodiversity: gradients create different habitats
- Gene pool and facilitated genetic exchange
- Retention of extracellular enzymes in the matrix
- Access to particulate biodegradable matter by colonization
- Recycling of nutrients because lysed cells are retained in the biofilm
- Protection against biocides and other stress
- High population density: threshold concentration of signalling molecules is easily reached, facilitating intercellular communication

These are good reasons explaining the preference for the biofilm mode of life of most microorganisms on earth.

References

Branda SS, Vik A, Friedman L, Kolter R (2005) Biofilms: the matrix revisited. Trends Microbiol 13:20–26

Costerton JW (2007) The biofilm primer. Springer, Berlin Heidelberg New York

Donlan RM (2002) Biofilms: microbial life on surfaces. Emerg Infect Dis 8:881–890

Ehrlich HL (2002) Geomicrobiology, 4th edn. Marcel Dekker, New York

Flemming H-C (2008) Biofilms. In: Encyclopedia of life sciences. Wiley, Chichester, http://www.els.net/, doi: 10.1002/9780470015902.a0000342

Flemming H-C, Leis A (2002) Sorption properties of biofilms. In: Bitton G (ed.) Encyclopedia of environmental microbiology, vol 5. Wiley-Interscience, New York, pp. 2958–2967

Flemming H-C, Neu TR, Wozniak D (2007) The EPS matrix: the "House of biofilm cells". J Bacteriol 189:7945–7947

Hausner M, Wuertz S (1999) High rates of conjugation in bacterial biofilms as determined by quantitative in-situ analysis. Appl Environ Microbiol 65:3710–3713

Körstgens V, Wingender J, Flemming HC, Borchard W (2001) Influence of calcium ion concentration on the mechanical properties of a model biofilm of *Pseudomonas aeruginosa*. Water Sci Technol 43 (6) 49–57

Lewis K (2001) Riddle of biofilm resistance. Antimicrob Agents Chemother 45:999–1007

Mayer C, Moritz R, Kirschner C, Borchard W, Maibaum R, Wingender J, Flemming HC (1999) The role of intermolecular interactions: studies on model systems for bacterial biofilms. Int J Biol Macromol 26:3–16

Schooling SR, Beveridge TR (2006) Membrane vesicles: an overlooked component of the matrices of biofilms. J Bacteriol 188:5945–5947

Schopf JW, Hayes JM, Walter MR (1983) Evolution on earth's earliest ecosystems: recent progress and unsolved problems. In: Schopf JW (ed.) Earth's earliest biosphere. Princeton University Press, New Jersey, pp. 361–384

Schulte S, Wingender J, Flemming H-C (2005) Efficacy of biocides against biofilms. In: Paulus W (ed.) Directory of microbiocides for the protection of materials and processes. Kluwer, Doordrecht, pp. 90–120

Wang XM, Rochon A, Lamprokostopoulou A, Lünsdorf H, Nimtz M, Römling U (2007) Impact of biofilm matrix components on interaction of commensal *Escherichia coli* with the gastrointestinal cell line HT-29. Cell Mol Life Sci 63:352–2363

Watnick P, Kolter R (2000) Biofilms, city of microbes. J Bacteriol 182:2675–2679

The Effect of Substratum Properties on the Survival of Attached Microorganisms on Inert Surfaces

K.A. Whitehead(✉) and J. Verran

Abstract Biofilm formation is dependent on the surrounding environmental conditions and substratum parameters. Once a biofilm forms many factors may influence cell survival and resistance. Cell adhesion to a surface is a prerequisite for colonization. However, attached microorganisms may not be able to multiply, and may merely be surviving on the surface, for example at a solid–air interface, rather than forming a biofilm. Retention of attached cells is a key focus in terms of surface hygiene and biofilm control. Factors that affect this retention may differ from those affecting biofilm formed at the solid–liquid interface: the nature of the substratum, presence of organic material, vitality of the attached microorganism, and of course the surrounding environment. The majority of publications focus on the solid–liquid interface; literature addressing the solid–air interface is considerably less substantial.

1 Introduction

Microbial attachment, adhesion, retention and subsequent biofilm formation are major concerns in many settings where biofilms play a key role in ensuring the survival of microorganisms and their resistance to a range of external "attacks" for example by protozoa, environmental conditions or chemical agents. Mechanisms of resistance to these external forces are diverse.

The literature concerning biofilms and resistance is significant, and the fact that biofilms demonstrate significantly enhanced resistance is well recognized. This paper focuses on the survival of attached cells rather than on biofilm. Donlan and Costerton (2002) define a biofilm as "a microbially derived sessile community characterized by cells that are irreversibly attached to a substratum or interface or to each other and that are embedded in a matrix of extracellular polymeric substances

K.A. Whitehead
School of Biology, Chemistry and Health Science, Manchester Metropolitan University, Chester St, Manchester, M1 5GD UK
e-mail: k.a.whitehead@mmu.ac.uk

Springer Series on Biofilms, doi: 10.1007/7142_2008_23

that they themselves have produced. The cells in the biofilm may also exhibit an altered phenotype with respect to growth rate and gene transcription". Attached cells on the other hand may be surrounded by preformed extracellular polymeric substances (EPS), but will not produce more unless appropriate environmental conditions are present. Attached cells rather than biofilm are therefore found at the solid–air as well as the solid–liquid interface. Intermittent exposure of the substratum to moisture, for example during cleaning of hygienic surfaces or external surface exposure to rain, or at a meniscus (Fig. 1) generates a solid-liquid–air interface, at which fouling is apparent.

Adhesion is a prerequisite for colonization. Microorganisms can survive in very thin water films but attached microorganisms may not be able to multiply, particularly if there is little moisture available. Thus many of the factors affecting the survival of cells in a biofilm may not be applicable to cells retained on a surface in the absence of moisture, but in the presence of organic material. Thus external factors, such as the nature of the substratum and of the surrounding environment, will significantly affect survival and "biotransfer potential" (Verran and Boyd 2001).

In this chapter a range of examples showing the effect of surface features on cell survival and resistance will be discussed, focusing on the marine environment wherever possible/appropriate, and addressing any differences between the solid–liquid and solid–air interface.

Fig. 1 Plastic buoy that has biofilm growing on its surface, resulting in fouling of the material

2 Microbial Attachment to Surfaces

Viruses, bacteria, fungi, algae and protozoa may all be found in the marine biofilm community, (Fig. 2) along with "macroorganisms", such as barnacles and seaweeds. Many studies have attributed microbial survival and resistance of attachment microorganisms to their cellular physiology but it is now thought that there are a number of contributory physical and chemical factors involved. Physicochemical parameters will affect initial attachment. Once the cells attach, the surface chemistry will influence cell adhesion, whilst topographic features allow maximum cell-surface binding, enhancing strength of attachment and thus retention.

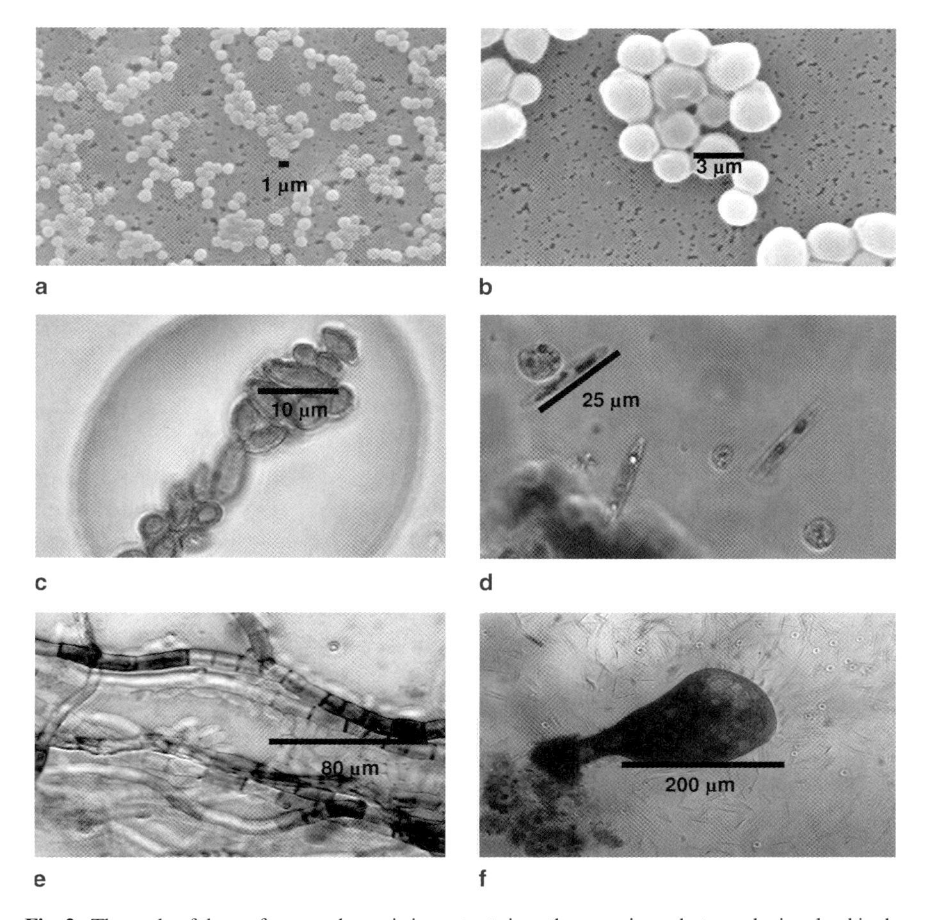

Fig. 2 The scale of the surface roughness is important since the organisms that may be involved in the formation of an environmental biofilm may vary greatly in size and shape **(a)** *Staphylococcus aureus* (bacteria) around 1 µm; **(b)** *Candida albicans* (yeast) around 3–5 µm; **(c)** *Cladsporium sp.* (fungal species) 12–18 µm; d) planktonic algae around 25 µm; **(e)** *Aureobasidum pullulans* hyphae can grow to various lengths; **(f)** *Stentor coeruleus* (cilitae) >200 µm

In an aqueous environment bacterial attachment to a surface occurs rapidly, over a few seconds to a few minutes. Moreover, the binding of microorganisms to a surface can confer advantages to cell survival, for example the attachment of cells to solid surfaces has been reported to immediately upregulate alginate synthesis in a strain of *Pseudomonas aeruginosa* (Davies et al. 1993), therefore strengthening cell–substratum binding. Bright and Fletcher (1983) also gave evidence that supported the existence of a direct substratum influence on the assimilation of amino acids in a marine *Pseudomonas* sp. thus enhancing nutrient availability to the cell.

If cells are deposited on a surface in the absence of a solid–liquid interface, for example by direct contact, then the behaviour of the passively attached cells may differ from that described above. However, in either case, since the surface of the substratum is the primary contact for cell attachment, the study of cell–surface interactions is of utmost importance.

3 Primary Adhesion to Surfaces

Much of the literature discusses microbial resistance and cell eradication once cells have attached to a surface. However, it would seem that a more proactive approach is to target the organisms in order to prevent initial cell attachment and thus subsequent retention to a surface. Primary cell adhesion to surfaces is dictated by a number of parameters. In an aqueous environment (liquid–solid), cells will first approach a surface by natural forces such as diffusion, gravitation, and Brownian motion. However, once in the vicinity of a surface, physicochemical parameters will come into play and the influence of Lifshitz–van de Waals forces, electrostatic forces and hydrogen bonding will influence the cells approach and subsequent attachment to the surface. It would seem obvious that the physicochemical, chemical and the topography have an influence on the properties of the substratum. However, the properties of the cell surface also need to be considered. The cell is a complex arrangement of different chemical species and topographies (on the micro- and nanoscale), and hence comprises islands of different physicochemical and chemical properties. Further, these properties will alter with changes in a given environment. The substratum, once in an aqueous environment will become coated with organic material, known as a conditioning film, as will cells, in addition to the presence of EPS. The EPS also plays a paramount role in primary adhesion.

At the solid–air interface in an open environment, the initial transfer of cells to a surface may occur in one fouling event where surfaces are contaminated by direct contact with the fouling material. Despite the complexities of these initial attachment scenarios it would seem logical to attempt to reduce/delay/prevent this initial cell–surface interaction in preference to managing the subsequent attached biofilm, for example by surface modification. There are a number of approaches that are directed towards this phenomenon, including the modification of surface topography (macro-, micro- and nanoscale), chemistry (self-assembled monolayers) and physicochemistry (superhydrophobic surfaces). However, it is inevitable that practically all surfaces will be colonized sooner or later (Flemming, personal communication).

4 Substratum Physicochemical Properties

Many different physicochemical interactions between microorganisms and the solid surface have been described in the literature (Boonaert and Rouxhet 2000; Simoes et al. 2007). If the physicochemical properties of a surface can be defined and controlled, then cell attachment, survival and biofilm formation can in turn be more easily managed. There is conflicting literature concerning the complex effect of surface and/or microbial physicochemical properties on microbial attachment to surfaces (Bos et al. 1999; Chen and Strevett 2001). Adhesion of vegetative cells (Sinde and Carballo 2000), bacterial spores (Husmark and Ronner 1993) and fresh-water bacteria (Pringle and Fletcher 1983) has been shown to increase with increasing surface hydrophobicity. Other organisms have also been shown to preferentially bind to hydrophobic surfaces: for example *Enteromorpha* spores (Callow et al. 2002). It has been suggested that cell attachment to hydrophobic plastics occurs very quickly (Carson and Allsopp 1980) whereas cell attachment to hydrophilic surfaces such as metallic oxides, glass and metals increases with longer exposure times (Dexter 1979).

The surface free energy of a substratum is believed to be important in initial cell attachment, but the interactions involved are complex. Biofilm formation coincides with increased inorganic positively charged elements at the surface (Carlen et al. 2001), but positive substratum surface charge has also been shown to impede bacterial surface growth despite initially promoting adhesion (Gottenbos et al. 2001). A maximum detachment rate for marine biofilms or bacteria has been demonstrated for surface free energies of 20–27 mN m^{-1} (Becker 1998; Pereni et al. 2006).

The surface energy distribution on substrata will be dependent on the surface structure and will be affected by surface imperfections such as cracks or pores, and also on the conditioning layer of the substratum, which is in turn defined by the surrounding environment. It has been suggested that the differences observed between surfaces in in-vitro hydrophobicityassessments may be due to changes in the substratum characteristics that occur during the first few minutes of exposure to the surrounding fluid, where a primary film of organic molecules known as the conditioning film is adsorbed to the substratum (Pringle and Fletcher 1983). The presence of this film clearly affects microbial retention, and also contributes to cell interactions with the surface. It is likely that conditioning films may mask some substratum properties.

This interaction may not be relevant to the attachment of cells on open surfaces, where contact between the cell and the substratum may be achieved by transient wetting or transfer between surfaces involving direct contact, or airborne transmission. However, the retention of the cells will be affected by the cohesive forces and by the area of contact between the cell and the substratum, be it conditioned or otherwise.

In an aqueous environment the conditioning of a surface by smaller molecules and ions will occur before bacterial attachment, thus the film provides the linking layer between the cells and the surface. A clear understanding of all interactions is needed if a logical attempt at controlling surface fouling is to transpire. Antifouling surfaces are possible but, since each fouling environment is essentially unique, it may be that situations have to be addressed on an individual basis. The life span and cost of production for any antifouling product must also be considered alongside the expected antifouling benefit. It is unlikely that fouling can be completely prevented, but if soil is more easily removed or if fouling is delayed then clear economical, ecological or health-associated benefits may be derived.

5 Chemical Properties of Materials

The chemical properties of materials are defined by the elements that ultimately make up the molecules of a surface. The surface chemistry, i.e. the chemical properties of the materials, has been shown to directly affect microbial attachment (Verran and Whitehead 2005: Whitehead and Verran 2007) and survival. A range of inert substrata find use in environments where microbial attachment and biofilm formation are common. The chemistry of the surface inevitably affects these interactions. Thus the choice of material must be made depending on the intended properties of the surface (e.g. immersed/exposed, high cleanability/low fouling, low wear, non-toxic, low cost etc.).

5.1 Metals

Cell attachment and thus biofilm formation can occur on metals, including aluminium (Nickels et al. 1981), stainless steel (Mittelman et al. 1990) and copper (Geesey and Bremer 1990). However, some metals such as aluminium or copper are considered toxic to bacteria (Avery et al. 1996). It has been suggested that microbial resistance to some metals, for example lead acetate, can be attributed to the high lead content of disinfectants and antiseptics, whilst resistance to copper sulfate may be due to its use as an algicide (Hiramatsu et al. 1997). However, even with concerns of increased resistance of microorganisms, and the frequent necessity of moisture to enable the antimicrobial action to occur, the incorporation of a range of metals into "antibacterial" surfaces has been reported (Kielemoes et al. 2000). The location of these surfaces, whether immersed, intermittently wet or dry, will clearly affect any intended antimicrobial effect. In particular, silver and copper have received significant attention. Antimicrobial silver and/or copper reagents have been occasionally applied to the water distribution system for inactivation of pathogens (Liu et al. 1998). However, bacterial resistance against silver and other metals may lead to limitations in the efficacy of these bactericide-releasing materials (Cloete 2003).

Copper has been shown to increase the growth rate of some bacteria (Starr and Jones 1957), whilst reduced growth in response to copper has been demonstrated for microbial populations (Jonas 1989). When compared to plastics and stainless steel surfaces, copper has been shown to have inhibitory effects on various microorganisms (De Veer et al. 1994; Keevil 2001). Copper-containing alloys have also shown increased antibacterial activity when compared to stainless steel and brasses, with increasing copper content reducing cell survival time (Wilks et al. 2005). It has been suggested that for biocidal purposes the use of copper alloyed surfaces should be restricted to regularly cleaned surfaces (Kielemoes and Verstraete 2001), since accumulation of non-microbial material and potential reaction of the cleaning agent with the copper and the fouling material may interfere with the antimicrobial effect, on open as well as closed surfaces (Airey and Verran 2007). Indeed, the ability of any antimicrobial agent in a surface to affect cells in the biofilm above will depend on the ability of the agent to diffuse through the biofilm from the substratum. Conversely, any antimicrobial agent whose effect relies on direct contact will only be active against those cells at the base of the biofilm. One might speculate that the effectiveness of an antifouling surface is only predictable for a given period of time, since once conditioning of the surface begins, surface properties will change. This will result in loss of direct contact of the surface with the foulant and consequently the loss of the surface antifouling effect. This has been demonstrated in the copper pipes containing disinfectant concentrate where biofilms have been found (Exner et al. 1983). As the copper surface becomes fouled, antimicrobial properties become diminished unless regularly cleaned (Airey and Verran 2007).

5.2 *Polymers*

Synthetic polymers may contain many additive chemicals, such as antioxidants, light stabilizers, lubricants, pigments and plasticizers, added to improve the desired physical and chemical properties of the material (Brocca et al. 2002). However, these additives may leach into the surrounding environment and provide nutrient for microorganisms present: phosphorus has been shown to increase the formation of biofilms on polyvinyl chloride in phosphorus-limited water (Lehtola et al. 2002). Several studies have shown that plastic materials can support the growth of biofilms, but it has been suggested that growth in plastic pipes is usually comparable with that on iron, steel or cement (Niquette et al. 2000). However, Bachmann and Edyvean (2006) used *Aquabacterium commune* cells under continuous cultivation with stainless steel and medium density polyethylene (MDPE) surfaces and found that biofilm cell density on MDPE slides was four times greater than on stainless steel. When various pipe materials were tested with chlorine and monochloramine disinfection, it was found that cement-based materials supported fewer fixed bacteria than plastic-based materials (Momba and Makala 2004).

Again, most of these surfaces are exposed to liquid and, potentially microorganisms, at a solid–liquid interface, often in a closed system. On open surfaces, many different properties of polymers can be exploited, depending on the intended end use. The relative softness of these surfaces makes them susceptible to surface damage, which will affect surface topography, and hence fouling and cleanability (Verran et al. 2000). However, as with all surfaces, long-term studies are required to assess the effect of surface wear and the effect of fouling, e.g. by humic substances, oil or mineral particles.

5.2.1 Incorporation and Release of Antimicrobial Agents in Polymers

In attempts to prevent/reduce cell attachment and survival on surfaces, antimicrobial agents have been incorporated in and onto polymers. Clearly, the release of the biocide/metal ions will be determined by the matrix and properties of the bulk material and surrounding environment. Biocides can be encapsulated to facilitate "delayed release", thereby extending the intended antimicrobial effect (Lukaszczyk and Kluczka 1995; Coulthwaite et al. 2005). Coatings and molecules extracted from natural sources have been suggested for use to deter microbial survival on a surface. On open surfaces, incorporation of antimicrobial agents such as biocides (for example Microban) or metals (for example BioCote or Agion) are used to achieve "antibacterial" properties, but the mechanism of the biocide action (e.g. is moisture required), duration, spectrum, speed and magnitude of effect are all important determinants of eventual effectiveness at intended point of use.

There are a number of important factors that need to be considered with respect to the development of biocide-incorporated materials, including physical and environmental aspects. The effectiveness of a biocide-incorporated surface is dependent upon the ability of the biocide to be released from the bulk material into which it is incorporated. This is a delicate balance, since if the blending, dispersion and binding properties are incorrect then the biocide release rate may be too fast (shortened life span of material), too slow (not effective), or non-existent. There is always a limited lifetime to these materials since an infinite amount of biocide is not available.

The release of biocide into the environment should also be considered. The Biocidal Product Directive (European Parliament 1998) was designed to review existing substances and aimed to provide high levels of protection for humans, animals and the environment. Many antifouling paints used to reduce the attachment of living organisms to the submerged surfaces of ships, boats and aquatic structures have biocide-release mechanisms. Two common biocides in use are the triazine herbicide Irgarol 1051 (*N*-2-methylthio-4-tert-butylamino-6-cyclopropylamino-*s*-triazine), and diuron (1-(3,4-dichlorophenyl)3,3-dimethylurea), which are designed to inhibit algal photosynthesis. It has been shown that due to leaching, environmental concentrations of the compounds pose significant risks to the plant species *Apium nodiflorum* and *Chara vulgaris* (Lambert et al. 2006). With biocide-incorporated materials there are also problems encountered with the targeted organisms, e.g. increased tolerance and resistance to the active material. Resistance to many chemical compounds including benzalkonium chloride, benzisothiazolone, chloroallyltriazine-azoniadamantane, dibromodicyanobutane, methylchloro/methylisothiazolone, tetrahydrothiadiazinthione and trifluoromethyl dichlorocarbanilide has been detected (Chapman 1998).

By definition, biocides will not assist in the accumulation and removal of organic material present on the surface and in the surrounding aqueous environment. The result may be that, although micro- and macroorganisms that attach to a surface may be inhibited or killed, the transfer of organic matter to the surface will not be affected. Thus an organic material layer will gradually build up on the surface over time, potentially masking any biocide effect.

If the biocide is not uniformly dispersed in the bulk material, then there will be areas of the surface that may allow attachment of tolerant or resistant microorganisms. Once this attachment occurs, microbial colonization and thus biofilm formation can occur, potentially enveloping the surrounding areas of material that are higher in biocide concentration. Thus although biocide-releasing surfaces may be a practical solution for surfaces that are to be used in the short term, in the long term they may be of limited value, particularly at the solid–liquid interface.

5.3 *Paints*

Coatings and paints intended for use on ships and underwater components or superstructures are a complex mixture of compounds that may include binders, pigments, extenders, solvents, thinners and additives (e.g. biocides) (Watermann et al. 2005). The purpose of antifouling paints is primarily to prevent development of macrofouling, particularly barnacles. Since microorganisms on a surface can increase the attachment of other organisms, inhibition of microbial biofilm development might decrease subsequent development of barnacles on the surface (Tang and Cooney 1998). Thus it is of importance to test new formulations for the survival and resistance of macro- and microorganisms.

As with blended polymers, the complex nature of the paint and its components will affect the activity of biocide/antimicrobial used and thus the final antimicrobial activity of the paint. To provide effective antifouling properties, organic biocides such as Irgarol, are often added in conjunction with copper to control copper-resistant fouling organisms (Voulvoulis et al. 1999). It has been shown that the release rate of copper depends not only on the copper compound and its dissolution properties, but also on the character of the paint matrix (Sandberg et al. 2007). The underlying substrata may also affect the antifouling properties of paint. Work by Tang and Cooney (1998) showed that coating surfaces with a marine paint decreased the numbers of *Pseudomonas aeruginosa* on stainless steel but had little effect on numbers of cells on fibreglass or aluminium. However, when they added copper or tributyltin (TBT) to the paint the initial development of biofilms was inhibited for 72–96 h. Biofilms that formed on surfaces coated with copper or TBT-containing paint did not synthesize greater amounts of EPS, thus the biofilms may have contained copper- or TBT-resistant cells.

There have been some attempts to use naturally extracted products as antifouling agents in paints. Four bacterial isolates from a marine environment were used to produce extracts that were formulated into ten water-based paints: nine showed activity against a test panel of fouling bacteria (Burgess et al. 2003). Five of the paints were shown to inhibit the settlement of barnacle larvae, *Balanus amphitrite*, and algal spores of *Ulva lactuca*, and for their ability to inhibit the growth of *Ulva lactuca* when grown on paint containing an extract from *Pseudomonas* sp. strain (Burgess et al. 2003).

It is interesting to note that manufacturers do not need to specify ingredients of the paint that are below 1% weight, thus antifouling paints may include significant amounts of metallic and non-metallic elements (Sandberg et al. 2007). Unfortunately, some materials used in paints (such as both organotins and copper) can be toxic to non-target marine species, such as the dog-whelk (Gibbs and Bryan 1986), oysters (Axiak et al. 1995) and juvenile carp (de Boeck et al. 1995; Tang and Cooney 1998). The use of biocidal antifouling paints has been prohibited in some European countries, such as Sweden, Denmark, Germany and France (Watermann et al. 2005). It should be noted that although copper is widely used in Europe, Sweden has prohibited its use in antifouling paints on pleasure crafts in fresh water and along the Swedish coast of the Baltic Sea (Sandberg et al. 2007). However, recent investigations have shown that newly developed, "toxin-free" antifouling paints that do not contain, e.g., copper, Irgarol or TBT may still be toxic towards marine organisms (Karlsson and Eklund 2004).

On surfaces that are not submerged but are externally exposed, a solid–liquid–air interface will form as rain droplets pass over the surface. The physical washing effect, coupled with release of any intended antimicrobial properties, will thus help reduce fouling on the surface. On internal surfaces, required properties might encompass easy cleanability rather than specifically antimicrobial properties – although in hygienic environments some biocidal effect would be desirable. Thus, photocatalytic paints are finding applications. UV radiation is an effective, but temporary photochemical method for disinfection, which requires a special irradiation source within the UV (185–254 nm) band. Photocatalysis is an alternative to direct UV disinfection and antimicrobial efficacy is possible with higher wavelengths, which are naturally present in ambient solar and artificial light (Erkan et al. 2006). Large band gap semiconductors, such as titanium dioxide (TiO_2), tin oxide and zinc oxide, are suitable photocatalytic materials with their higher wavelength UV absorption (320–400 nm) (Erkan et al. 2006). Titanium dioxide doped with metals has demonstrated photocatalytic activity, leading to an increased rate of destruction of organic compounds (Vohra et al. 2005) and microorganisms (Sunada et al. 2003). There has been some work carried out on the effectiveness of nanoparticle anatase titania on the destruction of bacteria (Allen et al. 2005; Verran et al. 2007). An example of photocatalytic paint currently on the market is Aoinn®. However, *in situ* information on the effectiveness of these materials is limited. The effectiveness of the activity of photocatalytic paint on microorganisms is further complicated by the interactions of the paint components interfering with the active chemicals (Caballero et al., 2008).

6 Substratum Roughness

There are a number of engineering terms used to define surface roughness, but the R_a, (the average of the peak and valley distances measured along a centre line) is the most universally used roughness parameter for general quality control (Verran and Maryan 1997) and in microbiological publications (Verran and Boyd 2001). An important consideration when describing surface topography is that there are several scales that can be used to characterize material surfaces in terms of surface waviness, roughness and topography (Table 1). Thus the surface feature dimension should be considered alongside the dimensions of the organism of concern.

Simplistically, an increase in surface roughness will increase the retention of microorganisms on a surface (Boulange-Petermann et al. 1997; Verran and Whitehead 2005; Whitehead and Verran 2006). However, there is some debate over the phenomenon (Duddridge and Prichard 1983; Taylor and Holah 1996), which may be accounted for by a consideration of the scale of topography, the "patterning" of the features on the surface and of the testing methodology used.

Electropolishinghas shown to be advantageous in minimizing initial bacterial adhesion (Arnold et al. 2001) but, in the long-term, surface roughness has been shown not to affect the development of mature biofilms (Hunt and Parry 1998)

Table 1 Descriptions of the different scale of surface topographies

	Size of surface features	Description
Macro-topography	$R_a > 10$ μm	Will include surface finishes produced by industrial processes, e.g. the use of cutting tools (uniform spacing of surface features with a well-defined direction) or grinding processes (usually directional in character with generally of irregular spacing). Roughening of a surface will increase the area available for microbial adhesion and retention; however, if the surface roughness is greatly increased, this may result in wash out of microorganisms
Micro-topography	$Ra \sim 1$ μm	Surfaces with features of micron dimension are of importance if hygiene is of concern, e.g. in food processing
Nano-topography	$Ra < 1$ μm	Procedures such as polishing, whereby fine abrasives are used to produce a smooth shiny surface, nevertheless, all surfaces have a nanotopography. Nanotopographies are likely to have little effect on the R_a or other roughness values as usually measured, but may affect retention of organic material
Angstrom-scale topography	Surface features 1–10 nm	Angstrom-sized surface features involve the configuration and mobility or functional groups, which may be of importance for both the cell and the substratum, especially where dynamic surfaces are being investigated
Molecular topography	Molecules	The charge on surface molecules ultimately make up the overall charge on the microbial or substratum surface and will affect the initial cell–surface binding

– a property already noted in the impact of biocides. For surfaces deemed "hygienic", usually encompassing microorganisms at the surface–air interface where "surface features" are smaller than the microorganisms, topography does not affect the retention of microorganisms on a surface (Verran et al. 2001a), although the cells tend to be immobilized on the features. Work by Hilbert et al. (2003) on stainless steel that was smoothed to R_a of 0.9–0.01 μm also found that the adherence of microorganisms was not affected by differences in the surface roughness, but they did conclude that surface roughness was an important parameter for corrosion resistance of the stainless steel.

At the "macro" level it has been suggested that surface roughness may not pose a major problem in terms of bacterial adherence: because the surface features are so much larger than the bacterial cells, they can have no role in retention. However, some fungal spores, algae, protozoa and larger organisms, such as those found in marine environments and implicated in fouling, may be of significance. Thus an effect of surface roughness on attachment has been demonstrated for algal spores (Fletcher and Callow 1992) and invertebrate larvae (Crisp 1974). However, such macrofeatures may well encompass a micro- or nanotopography, which can retain smaller cells (Fig. 3). *In situ*, this means that the surface may become colonized with smaller cells such as bacteria prior to eukaryotic colonization. Not only might this provide anchorage points for the larger organisms but also for possible nutrients, thus increasing the chances of survival of the larger cells (Pickup et al. 2006). This succession of surface conditioning, micro- and macrofouling is a well-described phenomenon in immersed aquatic systems. Both viable and non-viable cells will contribute to this succession.

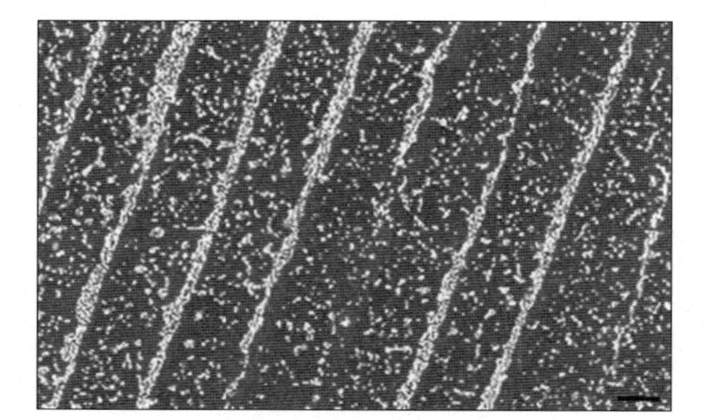

Fig. 3 Surface grooves with a macrotopography (30 μm). However, these large scale features exhibit a micro- or nanotopography in the peaks between grooves. Cells are washed from large grooves but are piled up on the top of the groove peaks R_a = 0.35 μm (image courtesy of A. Packer, MMU)

6.1 Stainless Steel: Surface Topography and Microbial Retention

Stainless steel is the most commonly used material for a number of industrial applications. Grade 316 stainless steel contains molybdenum, which increases resistance to surface pitting in aggressive environments, therefore it is widely used in the environment (Little et al. 1991). For stainless steel, different finishes will produce surfaces with differing topographies, whilst retaining low R_a values below 0.8 µm, the value used for describing "hygienic" surfaces (Flint et al. 1997). As noted above, features of appropriate dimension will retain and protect microorganisms (Fig. 4), and reduce surface cleanability and hygienic status.

On open surfaces that are regularly cleaned, biofilm formation is unlikely (Verran and Jones 2000), but in closed environments, increased retention of viable microorganisms may accelerate development of biofilm, even if more mature biofilm is unaffected by the underlying surface topography (Verran and Hissett 1999). Larger surface defects will potentially entrap accumulations of microorganisms in both open and closed systems. Work by Boyd et al. (2002) demonstrated that on stainless steel surfaces, lateral changes of 0.1 µm were sufficient to increase the strength of bacterial attachment. Such surfaces should ideally be free from defects and chemical inhomogeneity in order to minimize microbial attachment. However, Bachmann and Edyvean (2006) suggested that electropolishing of stainless steel pipes for drinking water installations was not necessary, although at joints, welds, dead ends and other features on pipelines, polishing may be necessary since microbial accumulation is more likely at these sites.

6.2 Controlling Topography to Manage Fouling

Recently, it has been noted that the shape of surface features is of importance in microbiological binding to a surface (Edwards and Rutenberg 2001; Whitehead et al. 2005, 2006). Since surface topography affects the amount and strength of attachment and retention, several groups have produced surfaces with defined topographies in order to truly assess the interactions. Callow et al. (2002) showed that when using textured surfaces consisting of valleys or pillars, the swimming spores of the

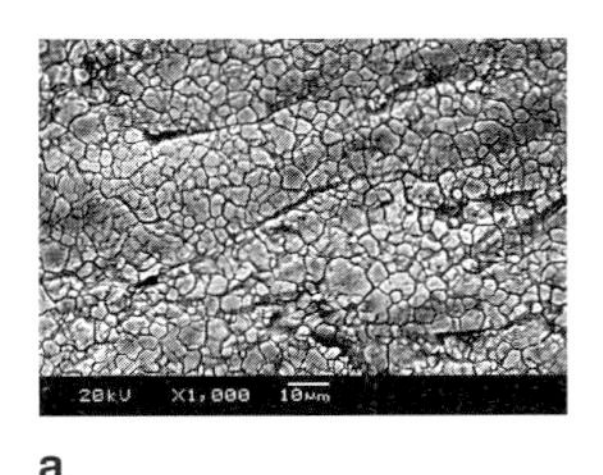

a

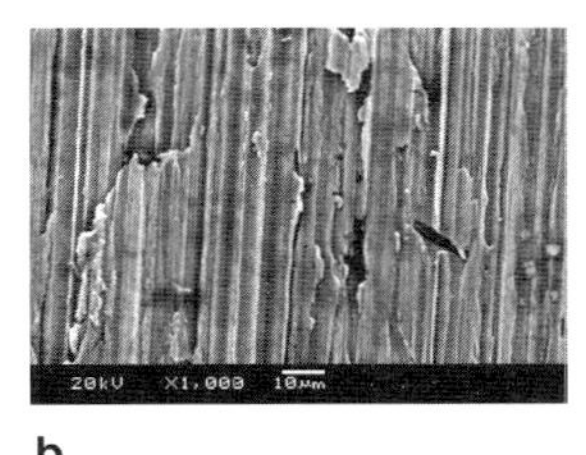

b

c

Fig. 4 SEM of a brushed finish stainless steel surface demonstrating microbial retention within linear features

green fouling alga *Enteromorpha* settled preferentially in valleys and against pillars, and that the number of spores that settled increased as the width of the valley decreased. Surface textural features of 50–100 μm have been shown to be significantly less fouled by barnacles (Andersson et al. 1999); features smaller than the diameter of the barnacle prevented attachment. The attachment of barnacle larvae has similarly been shown to be enhanced or reduced according to the scale, shape and periodicity of surface roughness (Hills and Thomason 1998; Berntsson et al. 2000). Surface roughness has also been shown to influence the settlement behaviour of fouling larvae (Howell and Behrends 2006).

On a smaller scale, a range of engineered surfaces with controlled topographical features, i.e. pits (Whitehead et al. 2004) and grooves (Packer et al. 2007), has been developed to demonstrate the effect of surface topography on cell binding. Using microbial retention assays, Whitehead et al. (2005) demonstrated that with a range of differently sized unrelated microorganisms, the dimensions of the surface feature are important with respect to the size of the cell and its subsequent retention, with maximal retention occurring when features were of a diameter comparable with the microorganisms. This observation was supported when assessing the force of adhesion of the cells on the surfaces using atomic force microscopy (Whitehead et al. 2006). Edwards and Rutenberg (2001) have further recognized that the cross-sectional shape of a groove will have a large effect on binding potential, which is especially important where flow is concerned. Likewise, the orientation of features with respect to the flow or direction of cleaning will affect retention. It should also be noted that, in order to simplify calculations, cells are treated as rigid bodies whereas actually a living cell has a flexible wall and can deform to fit surface features (Beach et al. 2002). The study of the interactions occurring between cells and substratum features of defined dimensions is thus contributing to our understanding of surface fouling at the earliest stages of biofilm formation at both the solid–liquid and solid–air interface. Wear of materials may occur on the nanometer scale (Verran and Boyd 2001). Nanoscale surface features have been shown to affect both bacterial retention (Bruinsma et al. 2002) and cell behaviour (Dalby et al. 2002; Fan et al. 2002; Curtis et al. 2004). It may be speculated that surface nanofeatures will also invariably affect organic soil retention.

7 Substratum Conditioning

The first event that occurs when a surface comes into contact with a fluid is the adsorption of molecules to the surface; the molecules attach to the surface more rapidly than the cells, and the composition of the conditioning film is dependent on the composition of the bulk fluid (Hood and Zottola 1995) and of the substratum. Retained soil in surface features may facilitate the attachment of microorganisms to the surface, provide a nutrient source for microorganisms, be indicative of poor hygiene/cleaning processes (Verran et al. 2001b), affect the susceptibility of microorganisms to sanitising agents (Holah 1995), physically protect cells retained in surface defects (Kramer 1992) or provide attachment foci for re-colonization (Storgards et al. 1999).

Considering the effect of initial surface "conditioning" on attachment, retention and survival, adsorbed proteins have been found to either increase or decrease attachment (Carballo et al. 1991; Helke et al. 1993). The specificity of adhesion–receptor interaction is more relevant at solid–liquid interfaces, where the microorganisms can move towards a more advantageous location. At the solid–air interface, the immobilized cells tend to require another surface to facilitate transfer.

The presence of organic material may result in complexation and reduction in activity of some antifouling agents. Previous investigations have shown the majority (>80%) of the total copper in natural water to be complexed to organic matter (Bruland et al. 2000). Once natural sediments bind to a surface and reduce the effect of the antifouling agent, the surface becomes freely available for cell attachment to take place. *In-situ* field measurements on ships hulls on both pleasure crafts and navy vessels have shown lower release rates compared to laboratory tests on panels, most probably as a result of biofilm formation (Valkirs et al. 2003)

8 Microbial Resistance, Tolerance and Persistence

To help prevent the development of bacterial resistance, it is essential to understand the ramifications of the use of antimicrobial surfaces and/or cleaning and disinfection products, and to maintain excellent cleaning or management/maintenance protocols. If a cell is able to survive on a surface, resisting cleaning treatment, it can then be a source for biotransfer potential. A number of research reports have expressed concern that use of biocides may contribute to development of antibiotic resistance (Levy et al. 2000; McDonnell et al. 1999). Several workers have reported that the number of mercury-resistant bacteria in soil and aquatic environments varied according to the mercury content of the environment, where in these strains heavy metal-resistance properties were associated with multiple drug resistance (Misra 1992).

There is a vast difference between the magnitude of resistance, tolerance and persistence (RTP) of microorganisms dependent on whether the cells are found in as single units or as colonies, or if the cells are in the protective matrix of a biofilm. When *Pseudomonas aeruginosa* was tested in suspension or following deposition onto metallic or polymeric surfaces to determine the effectiveness of disinfectants (Cavicide, Cidexplus, Clorox, Exspor, Lysol, Renalin and Wavicide) and non-formulated germicidal agents (glutaraldehyde, formaldehyde, peracetic acid, hydrogen peroxide, sodium hypochlorite, phenol and cupric ascorbate) it was found that cells were on average 300-fold more resistant when present on contaminated surfaces than in suspension (Sagripanti and Bonifacino 2000). Further, it was also shown that the surface to which bacteria were attached influenced the effectiveness of disinfectants.

The development of tolerance and resistance to antimicrobial agents is not the focus of this chapter. However, although different challenges face cells at a solid–air interface in comparison with biofilms at solid–liquid interfaces, in either case the potential exists for survival, development of resistance and dissemination.

9 Conclusions

The attachment of microorganisms on inert substrata is a key to the development of biofilm at solid–liquid interfaces, and also to the potential for transfer on open surfaces at solid–air interfaces. Although the means for deposition of cells at the surface in these two systems will vary, properties of the substratum such as surface chemistry, surface topography, and the presence of organic (or inorganic) material conditioning the surfaces are essentially common to both systems.

The chemical and physicochemical properties of the substratum are important in initial cell attachment and adhesion, but once biofilm has formed, the underlying substratum has little effect on development – although surface roughness can have a significant effect on cell retention, especially under conditions of flow.

Surface modification designed to produce antifouling surfaces as an independent entity needs to focus on management of initial organic material and cell deposition in order to prevent, control or delay subsequent cell retention and multiplication. Forces used in the cleaning need to overcome those interactions that are active in adhesion of primary organic material and pioneer cells.

A variety of surface modification strategies are being explored, coupled with more fundamental investigations of factors affecting interactions occurring between cells and inert substrata. A multidisciplinary approach between biologists, chemists, physicists, engineers and modellers will facilitate the development of well-engineered and designed surfaces and systems, which are economically viable and environmentally acceptable, to enable optimum control of microbial fouling of surfaces. Promising approaches include those based on superhydrophobic surfaces. At these surfaces, the interplay of surface topography and chemistry results in contact angles approaching 180°. The development of chemically modified surfaces may be advantageous, but the use of chemical species that are detrimental to the surrounding environmental should be avoided. Mass-produced generic "solutions" may not be realistic; antifouling surface design needs to be tailored to individual applications. Although initially time-consuming, this would result in successful application and long-term cost savings.

References

Airey P, Verran J (2007) Potential use of copper as a hygienic surface; problems assocviated with cumulative soiling and cleaning. J Hosp Inf 67(3):271–277

Allen NS, Edge M, Sandoval G, Verran J, Stratton J, Maltby J (2005) Photocatalytic coatings for environmental applications. Photochem Photobiol 81:279–290

Andersson M, Berntsson K, Jonsson P, Gatenholm P (1999) Microtextured surfaces: towards macrofouling resistant coatings. Biofouling 14:167–178

Arnold JW, Boothe DH, Bailey GW (2001) Parameters of treated stainless steel surfaces important for resistance to bacterial contamination. Trans ASAE 44:347–356

Avery SV, Howlett NG, Radice S (1996) Copper toxicity towards *Saccharomyces cerevisiae*: dependence on plasma membrane fatty acid composition. Appl Environ Microbiol 62:3960–3966

Axiak V, Sammut M, Chircop P, Vella A, Mintoff B (1995) Laboratory and field investigations on the effects of organotin (tributyltin) on the oyster, *Ostrea edulis*. Sci Total Environ 171:117–120

Bachmann RT, Edyvean RGJ (2006) AFM study of the colonisation of stainless steel by *Aquabacterium commune*. Int Biodeter Biodeg 58(3–4):112–118

Beach E, Tormoen G, Drelich J, Han R (2002) Pull of force measurements between rough surfaces by atomic force microscopy. J Coll Interface Sci 247:84–99

Becker K (1998) Detachment studies on microfouling in natural biofilms on substrata with different surface tensions. Int Biodeter Biodeg 41(1):93–100

Berntsson KM, Andreasson H, Jonsson PR, Larsson L, Ring K, Petronis S, Gatenholm P (2000) Reduction of barnacle recruitment on microtextured surfaces: analysis of effective topographic characteristics and evaluation of skin friction. Biofouling 16:245–261

Boonaert CJP, Rouxhet PG (2000) Surface of lactic acid bacteria: Relationships between chemical composition and physicochemical properties. Appl Environ Microbiol 66:2548–2554

Bos R, van der Mei HC, Busscher HJ (1999) Physicochemistry of initial microbial adhesive interactions – its mechanisms and methods for study. FEMS Microbiol Rev 23:179–230

Boulange-Petermann L, Rault J, Bellon-Fontaine M-N (1997) Adhesion of *Streptococcus thermophilus* to stainless steel with different surface topography and roughness. *Biofouling* 11(3):201–216

Boyd RD, Verran J, Jones MV, Bhakoo M (2002) Use of the atomic force microscope to determine the effect of substratum surface topography on bacterial adhesion. Langmuir 18(6):2343–2346

Bright J, Fletcher M (1983) Amino acid assimilation and respiration by attached and free-living populations of a marine *Pseudomonas* sp. Microbial Ecol 9:215–226

Brocca D, Arvin E, Mosbæk H (2002) Identification of organic compounds migrating from polyethylene pipelines into drinking water. Water Res 36(15):3675–3680

Bruinsma GM, Rustema-Abbing M, de Vries J, Stegenga B, van der Mei HC, van der Linden ML, Hooymans JMM, Busscher HJ (2002) Influence of wear and overwear on surface properties of etafilcon a contact lenses and adhesion of *Pseudomonas aeruginosa*. IOVS 43:3646–3653

Bruland KW, Rue EL, Donat JR, Skrabal SA, Moffett JW (2000) Intercomparison of voltammetric techniques to determine the chemical speciation of dissolved copper in a coastal seawater sample. Anal Chim Acta 405(1–2):99–113

Burgess JG, Boyd KG, Armstrong E, Jiang Z, Yan LM, Berggren M, May U, Pisacane T, Granmo A, Adams DR (2003) The development of a marine natural product-based antifouling paint. Biofouling 19(Suppl.):197–205

Callow ME, Jennings AR, Brennan AB, Seegert CE, Gibson A, Wilson L, Feinberg A, Baney R, Callow JA (2002) Microtopographic cues for settlement of zoospores of the green fouling alga *Enteromorpha*. Biofouling 18(3):237–245

Carballo J, Ferreiros CM, Criado MT (1991) Importance of experimental design in the evaluation of proteins in bacterial adherence to polymers. Med Microbiol Immunol 180:149–155

Caballero L, Whitehead KA, Allen NS, Verran J (2008) Inactivation of *Escherichia coli* on immobilized TiO_2 using fluorescent light. J Photochem Photobiol A (submitted)

Carlen A, Nikdel K, Wennerberg A, Holmberg K, Olsson J (2001) Surface characteristics and *in vitro* biofilm formation on glass ionomer and composite resin. Biomaterials 22:481–487

Carson J, Allsopp D (1980) The enumeration of marine periphytic bacteria from a temperol sampling series. In: Becker OG, Allsopp D (eds.) Biodeterioration: proceedings of the fourth international biodeterioration symposium. Pitman, London, pp. 193–198

Chapman JS (1998) Characterizing bacterial resistance to preservatives and disinfectants. Int Biodeter Biodeg 41:241–245

Chen G, Strevett KA (2001) Impact of surface thermodynamics on bacterial transport. Environ Microbiol 3(4):237–245

Cloete TE (2003) Biofouling control in industrial water systems: what we know and what we need to know. Mater Corros – Werkstoffe Korrosion 54(7):520–526

Coulthwaite L, Bayley K, Liauw C, Craig G, Verran J (2005) The effect of free and encapsulated OIT on the biodeterioration of plasticised PVC during burial in soil for 20 months. Int Biodeter Biodeg 56:86–93

Crisp DJ (1974) Factors influencing the settlement of marine invertebrate larvae. In: Grant PT, Mackie AM (eds.) Chemoreception in marine organisms. Academic, New York, pp. 165–177

Curtis ASG, Gadegaard N, Dalby MJ, Riehle MO, Wilkinson CDW, Aitchison G (2004) Cells react to nanoscale order and symmetry in their surroundings. IEEE Trans Nanobiosci 3(1):61–65

Dalby MJ, Riehle MO, Johnstone H, Affrossman S, Curtis ASG (2002) In vitro reaction of endothelial cells to polymer demixed nanotopography. Biomaterials 23:2945–2954

Davies D, Chakrabarty A, Geesey G (1993) Exopolysaccharide production in biofilms: substratum activation of alginate gene expression by *Pseudomonas aeruginosa*. Appl Environ Microbiol 59:1181–1186

de Boeck G, Nilsson G, Elofsson U, Vlaeminck A, Blust R (1995) Brain monoamine levels and energy status in common carp (*Cyprinus carpio*) after exposure to sublethal levels of copper. Aquat Toxicol 33:265–277

De Veer I, Wilke K, Ruden H (1994) Bacteria reducing properties of copper containing and non copper containing materials: II. Relationship between microbiocide effect on copper containing materials and copper ion concentration after contamination with moist and dry hands. Zentralblatt Hygiene Umweltmedizin 195:516–528

Dexter S (1979) Influence of substratum critical surface tension on bacterial adhesion – in situ studies. Coll Interface Sci 70:346–354

Donlan R, Costerton W (2002) Biofilms: survival mechanisms of clinically relevant microorganisms. Clin Micro Rev 15:167–193

Duddridge JE, Prichard AM (1983) Factors affecting the adhesion of bacteria to surfaces. In: Microbial corrosion: proceedings from the joint National Physical Laboratory and Metal Society conference. The Metals Society, London, pp. 28–35

Edwards KJ, Rutenberg AD (2001) Microbial response to surface microtopography: the role of metabolism in localised mineral dissolution. *Chem Geol* 180:19–32

Erkan A, Bakir U, Karakas G (2006) Photocatalytic microbial inactivation over Pd doped SnO2 and TiO2 thin films. J Photochem Photobiol Chem 184(3):313–321

European Parliament and Council of the European Union (1998) Directive 98/8/EC of the European Parliament and of the Council of 16 February 1998 concerning the placing of biocidal products on the market (Biocidal Products Directive). http://ec.europa.eu/environment/biocides/index.htm. Last accessed 14 July 2008

Exner M, Tuschewitzki G-J, Thofern E (1983) Investigations on wall colonization of copper pipes in a central disinfection dosing system. Zbl Bakt Hyg I Abt Orig B 177:170–181

Fan YW, Cui FZ, Hou SP, Xu QY, Chen LN, Lee I-S (2002) Culture of neural cells on silicon wafers with nanoscale topography. J Neurosci Meth 120:17–23

Fletcher M, Callow JA (1992) The settlement, attachment and establishment of marine algal spores. Br Phycol J 27:303–329

Flint SH, Brooks JD, Bremer PJ (1997) The influence of cell surface properties of thermophilic *Streptococci* on attachment to stainless steel. J Appl Microbiol 83(4):508–517

Geesey GG, Bremer PJ (1990) Applications of Fourier-transform infrared spectrometry to studies of copper corrosion under bacterial biofilms. Marine Tech Soc J 24:36–43

Gibbs P, Bryan G (1986) Reproductive failure in populations of the dog-whelk, *Nucella lapillus,* caused by imposex induced by tributyltin from antifouling paints. J Marine Biol Assoc UK 66:767–777

Gottenbos B, Grijpma DW, van der Mei HC, Feijen J, Busscher HJ (2001) Antimicrobial effects of positively charged surfaces on adhering Gram-positive and Gram-negative bacteria. J Antimicrob Chemo 48(1):7–13

Helke DM, Somers EB, Wong ACL (1993) Attachment of *Listeria monocytogenes* and *Salmonella typhimurium* to stainless steel and buta-N in the presence of milk and individual milk components. J Food Prot 56(6):479–484

Hilbert LR, Bagge-Ravn D, Kold J, Gram L (2003) Influence of surface roughness of stainless steel on microbial adhesion and corrosion resistance. Int Biodeter Biodeg 52(3):175–185

Hills JM, Thomason JC (1998) The effect of scales of surface roughness on the settlement of barnacle (*Semibalanus balanoides*) cyprids. Biofouling 12:57–69

Hiramatsu K, Hanaki H, Ino T, Yabuta K, Oguri T, Tenover F (1997) Methicillin-resistant *Staphylococcus aureus* clinical strain with reduced vancomycin susceptibility. J Antimicrob Chemo 40:135–136

Holah JT (1995) Progress report on CEN TC 216 working group 3: Disinfectant test methods for food hygiene, institutional, industrial and domestic applications. Int Biodeter Biodeg 36(3–4):355–365

Hood SK, Zottola EA (1995) Biofilms in food processing. Food Control 6(1):9–18

Howell D, Behrends B (2006) A methodology for evaluating biocide release rate, surface roughness and leach layer formation in a TBT-free, self-polishing antifouling coating. Biofouling 22(5):303–315

Hunt AP, Parry JD (1998) The effect of substratum roughness and river flow rate on the development of a freshwater biofilm community. Biofouling 12:287–303

Husmark U, Ronner U (1993) Adhesion of *Bacillus cereus* spores to different solid surfaces: cleaned or conditioned with various food agents. Biofouling 7:57–65

Jonas RB (1989) Acute copper and cupric ion toxicity in an estuarine microbial community. Appl Environ Microbiol 55(1):43–49

Karlsson J, Eklund B (2004) New biocide-free anti-fouling paints are toxic. Marine Poll Bull 9(5–6):456–464

Keevil CW (2001) Antibacterial properties of copper and brass demonstrate potential to combat toxic *E. coli* outbreaks in the food processing industry. Paper presented at the symposium on copper and health, CEPAL, Santiago, Chile

Kielemoes J, Verstraete W (2001) Influence of copper alloying of austentic stainless steel of multi species biofilm development. Lett Appl Microbiol 33:148–152

Kielemoes J, Hammes F, Verstraete W (2000) Measurement of microbial colonisation of two types of stainless steel. Environ Tech 21(7):831–843

Kramer DN (1992) Myths: cleaning, sanitation and disinfection. Dairy Food Environ Sanit 12:507–509

Lambert SJ, Thomas KV, Davy AJ (2006) Assessment of the risk posed by the antifouling booster biocides Irgarol 1051 and diuron to freshwater macrophytes. Chemosphere 63:734–743

Lehtola MJ, Miettinen IT, Martikainen PJ (2002) Biofilm formation in drinking water affected by low concentrations of phosphorus. Can J Microbiol 48:494–499

Levy S (2000) Antibiotic and antiseptic resistance: impact on public health. Pediatr Infect Dis J 19:120–122

Little B, Wagner P, Mansfield F (1991) Microbiologically induced corrosion of metals and alloys. Int Mater Rev 36:253–272

Liu ZM, Stout JE, Boldin M, Rugh J, Diven WF, Yu VL (1998) Intermittent use of copper-silver ionization for Legionella control in water distribution systems: A potential option in buildings housing individuals at low risk of infection. Clin Infect Dis 26:138–140

Lukaszczyk J, Kluczka M (1995) Modification of the release of biocides bound to carboxylic exchange resin 1. Encapsulation by the ISP method and by evaporation from aqueous suspension. Polimery 40(10):587–590

McDonnell G, Russell AD (1999) Antiseptics and disinfectants: activity, action, and resistance. Clin Microbiol Rev 12(1):14

Misra TK (1992) Bacterial resistances to inorganic mercury salts and organomercurials. Plasmid 27(1):4–16

Mittelman MW, Nivens DE, Low C, White DC (1990) Differential adhesion, activity, and carbohydrate – protein ratios of *Pseudomonas-atlantica* monocultures attaching to stainless-steel in a linear shear gradient. Microb Ecol 19(3):269–278

Momba MNB, Makala N (2004) Comparing the effect of various pipe materials on biofilm formation in chlorinated and combined chlorine-chloraminated water systems. Water SA 30(2):175–182

Nickels JS, Bobbie RJ, Lott DF, Martz RF, Benson PH, White DC (1981) Effect of manual brush cleaning on biomass and community structure of micro-fouling film formed on aluminum and titanium surfaces exposed to rapidly flowing seawater. Appl Environ Microbiol 41(6):1442–1453

Niquette P, Servais P, Savoir R (2000) Impacts of pipe materials on densities of fixed bacterial biomass in a drinking water distribution system. Water Res 34(6):1952–1956

Packer A, Kelly P, Whitehead KA, Verran J (2007) Effects of defined linear features on surface hygiene and cleanability. Paper presented at the 50th SVC annual technical conference, Kentucky

Pereni CI, Zhao Q, Liu Y, Abel E (2006) Surface free energy effect on bacterial adhesion. Coll Surf B: Biointerfaces 48:143–147

Pickup RW, Rhodes G, Bull TJ, Arnott S, Sidi-Boumedine K, Hurley M, Hermon-Taylor J (2006) *Mycobacterium avium subsp paratuberculosis* in lake catchments, in river water abstracted for domestic use, and in effluent from domestic sewage treatment works: Diverse opportunities for environmental cycling and human exposure. Appl Environ Microbiol 72(6):4067–4077

Pringle JH, Fletcher M (1983) Influence of substratum wettability on attachment of freshwater bacteria to solid surfaces. Appl Environ Microbiol 45:811–817

Sagripanti JL, Bonifacino A (2000) Resistance of *Pseudomonas aeruginosa* to liquid disinfectants on contaminated surfaces before formation of biofilms. *J AOAC Int* 83:1415–1422

Sandberg J, Wallinder IO, Leygraf C, Virta M (2007) Release and chemical speciation of copper from anti-fouling paints with different active copper compounds in artificial seawater. Mater Corros – Werkstoffe Korrosion 58(3):165–172

Simoes LC, Simoes M, Oliveira R, Vieira MJ (2007) Potential of the adhesion of bacteria isolated from drinking water to materials. J Basic Microbiol 47:174–183

Sinde E, Carballo J (2000) Attachment of *Salmonella spp*. and *Listeria monocytogenes* to stainless steel, rubber and polytetrafluorethylene: the influence of free energy and the effect of commercial sanitizers. Food Microbiol 17(4):439–447

Starr TJ, Jones ME (1957) The effect of copper on the growth of bacteria isolated from marine environments. Limnol Oceanog 2(1):33–36

Storgards E, Simola H, Sjoberg AM, Wirtanen G (1999) Hygiene of gasket materials used in food processing equipment part 2: aged materials. Food Bioprod Proc 77(C2):146–155

Sunada K, Watanabe T, Hashimoto K (2003) Bactericidal activity of copper-deposited TiO2 thin film under weak UV light illumination. Environ Sci Tech 37(20):4785–4789

Tang RJ, Cooney JJ (1998) Effects of marine paints on microbial biofilm development on three materials. J Ind Microbiol Biotech 20(5):275–280

Taylor JH, Holah JT (1996) A comparative evaluation with respect to the bacterial cleanability of a range of wall and floor surface materials used in the food industry. J Appl Bacteriol 81(3):262–266

Valkirs AO, Seligman PF, Haslbeck E, Caso JS (2003) Measurement of copper release rates from antifouling paint under laboratory and *in situ* conditions: implications for loading estimation to marine water bodies. Mar Pollut Bull 46(6):763–779

Verran J, Boyd RD (2001) The relationship between substratum surface roughness and microbiological and organic soiling: a review. Biofouling 17(1):59–71

Verran J, Hissett T (1999) The effect of substratum surface defects upon retention of biofilm formation by microorganisms in potable water. In: Keevil CW, Godfree A, Holt D, Dow C (eds.) Biofilms in the aquatic environment. The Royal Society of Chemistry, Cambridge, pp. 25–33

Verran J, Jones M (2000) Problems of biofilms in the food and beverage industry. In: Walker J, Surmann S, Jass J (eds.) Industrial biofouling detection, prevention and control. Wiley, Chichester, pp. 145–173

Verran J, Maryan CJ (1997) Retention of *Candida albicans* on acrylic resin and silicone of different surface topography. J Prosthet Dent 77(5):535–539

Verran J, Whitehead K (2005) Factors affecting microbial adhesion to stainless steel and other materials used in medical devices. Int J Art Organs 28(11):1138–1145
Verran J, Rowe DL, Cole D, Boyd RD (2000) The use of the atomic force microscope to visualise and measure wear of food contact surfaces. Int Biodeter Biodeg 46(2):99–105
Verran J, Rowe DL, Boyd RD (2001a) The effect of nanometer dimension topographical features on the hygienic surface of stainless steel. J Food Prot 64(7):1183–1187
Verran J, Boyd R, Hall K, West RH (2001b) Microbiological and chemical analyses of stainless steel and ceramics subjected to repeated soiling and cleaning treatments. J Food Prot 64(9):1377–1387
Verran J, Sandoval G, Allen NS, Edge M, Stratton J (2007) Variables affecting, the antibacterial properties of nano and pigmentary titania particles in suspension. Dyes Pigments 73:298–304
Vohra A, Goswami DY, Deshpande DA, Block SS (2005) Enhanced photocatalytic inactivation of bacterial spores on surfaces in air. J Ind Microbiol Biotech 32(8):364–370
Voulvoulis N, Scrimshaw MD, Lester JN (1999) Analytical methods for the determination of 9 antifouling paint booster biocides in estuarine water samples. Chemosphere 38(15):3503–3516
Watermann BT, Daehne B, Sievers S, Dannenberg R, Overbeke JC, Klijnstra JW, Heemken O (2005) Bioassays and selected chemical analysis of biocide-free antifouling coatings. Chemosphere 60(11):1530–1541
Whitehead KA, Verran J (2006) The effect of surface topography on the retention of microorganisms: a review. Food Bioprod Process 84:253–259
Whitehead KA, Verran J (2007) The effect of surface properties and application method on the retention of *Pseudomonas aeruginosa* on uncoated and titanium coated stainless steel. Int Biodeter Biodegradation 60:74–80
Whitehead KA, Colligon JS, Verran J (2004) The production of surfaces of defined topography and chemistry for microbial retention studies using ion beam sputtering technology. Int Biodeter Biodegradation 54:143–151
Whitehead KA, Colligon J, Verran J (2005) Retention of microbial cells in substratum surface features of micrometer and sub-micrometer dimensions. Coll Surf B – Biointerfaces 41:129–138
Whitehead KA, Rogers D, Colligon J, Wright C, Verran J (2006) Use of the atomic force microscope to determine the effect of substratum surface topography on the ease of bacterial removal. Coll Surf B – Biointerfaces 51:44–53
Wilks SA, Michels H, Keevil CW (2005) The survival of *Escherichia coli* on a range of metal surfaces. Int J Food Microbiol 105:445–454

Mechanisms of Microbially Influenced Corrosion

Z. Lewandowski (✉) and H. Beyenal

Abstract The chapter demonstrates that biofilms can influence the corrosion of metals (1) by consuming oxygen, the cathodic reactant; (2) by increasing the mass transport of the corrosion reactants and products, therefore changing the kinetics of the corrosion process; (3) by generating corrosive substances; and (4) by generating substances that serve as auxiliary cathodic reactants. These interactions do not exhaust the possible mechanisms by which biofilm microorganisms may affect the corrosion of metals; rather, they represent those few instances in which we understand the microbial reactions and their effect on the electrochemical reactions characteristic of corrosion. In addition, we can use electrochemical and chemical measurements to detect one or more products of these reactions. An important aspect of quantifying mechanisms of microbially influenced corrosion is to demonstrate how the microbial reactions interfere with the corrosion processes and, based on this, identify products of these reactions on the surfaces of corroding metals using appropriate analytical techniques. The existence of these products, associated with the increasing corrosion rate, is used as evidence that the specific mechanism of microbially influenced corrosion is active. There is no universal mechanism of MIC. Instead, many mechanisms exist and some of them have been described and quantified better than other. Therefore, it does not seem reasonable to search for universal mechanisms, but it does seem reasonable to search for evidence of specific, well-defined microbial involvement in corrosion of metals.

1 Recent Views on Microbially Influenced Corrosion

When it is suspected that a material failure was caused by microbial corrosion, it is reasonable to ask: "How do we know that the corrosion process was influenced by microorganisms?" To address this question, many research groups

Z. Lewandowski
Department of Civil Engineering and Center for Biofilm Engineering, Montana State University, Room 310, EPS Building, Bozeman, MT 59717, USA
e-mail: ZL@erc.montana.edu

Springer Series on Biofilms, doi: 10.1007/7142_2008_8

have attempted to find a fingerprint of microbially influenced corrosion(MIC), i.e., specific characteristics distinguishing microbially stimulated corrosion from ordinary galvanic corrosion (Beech et al. 2005; Javaherdashti 1999; Lee et al. 1995; Little et al. 2000, 2007; Mansfeld and Little 1991; Videla and Herrera 2005; Wang et al. 2006). Despite significant research effort, no such fingerprint characteristic of MIC has yet been found, and there are good reasons to believe that a universal mechanism of microbially stimulated corrosion does not exist (Beech et al. 2005; Flemming and Wingender 2001; Miyanaga et al. 2007; Starosvetsky et al. 2007). Instead of a universal mechanism, several mechanisms by which microorganisms affect the rates of corrosion have been described, and the diversity of these mechanisms is such that it is difficult to expect that a single unified concept can be conceived to bring them all together. From what we now understand, and what has been demonstrated by numerous researchers, accelerated corrosion of metals in the presence of microorganisms stems from microbial modifications to the chemical environment near metal surfaces (Beech et al. 2005; Geiser et al. 2002; Lee and Newman 2003; Lewandowski et al. 1997). Such modifications depend, of course, on the properties of the corroding metal and on the microbial community structure of the biofilm deposited on the metal surface (Beech and Sunner 2004; Dickinson et al. 1996b; Flemming 1995; Olesen et al. 2000b, 2001). The conclusion that there are many mechanisms of MIC, rather than a single one, is generally accepted in the literature and can be exemplified by the paper by Starosvetsky et al. (2007), who concluded that to uncover MIC in technological equipment failures requires an individual approach to each case, and that an assessment of the destructive role of the microorganisms present in the surrounding medium is possible only by analyzing and simulating the corrosion parameters found in the field (Dickinson and Lewandowski 1998). Quite succinctly, Beech et al. (2005) describe MIC as a consequence of coupled biological and abiotic electron-transfer reactions, i.e., redox reactions of metals enabled by microbial ecology. Hamilton (2003) attempted to generate a unified concept of MIC and has found common features in only some of the possible mechanisms. It is unlikely that a unified concept of MIC can be generated at all.

MIC is caused by microbial communities attached to surfaces, known as biofilms. A biofilm is composed of four compartments: (1) the surface to which the microorganisms are attached, (2) the biofilm (the microorganisms and the matrix), (3) the solution of nutrients, and (4) the gas phase (Lewandowski and Beyenal 2007). Each compartment consists of several components, and the number of components may vary depending on the type of study. For example, in some MIC studies it is convenient to distinguish four components of the surface: (1) the bulk metal, (2) the passive layers, (3) the biomineralized deposits on the surface, and (4) the corrosion products. Microorganisms can modify each of these components in a way that enhances corrosion of the metal surface. In addition, components of the other compartments of the biofilm can be modified in ways that affect the corrosion reactions as well. Modifications in the solution compartment may include the chemical composition, hydrodynamics and mass transfer rates near the metal surface; modifications in the biofilm compartment may include the microbial community structure and the composition of the extracellular polymeric substances (EPS). Each of these modifications may be complex in itself, and each may affect the corrosion reactions in many ways. The complexity and the multitude of the possible interactions among microorganisms, their metabolic reactions, the corrosion reactions and the metal, such as those shown by Coetser and Cloete (2005), are the reasons why it is unlikely that a unifying concept of MIC can be developed (Coetser and Cloete 2005).

When biofilms accumulate on metal surfaces, reactants and products of microbial metabolic reactions occurring in the space occupied by the biofilm affect the solution chemistry and the surface chemistry, and both types of modification may interfere with the electrochemical processes naturally occurring at the interface between the metal and its environment. The reactants and products of electrochemical reactions occurring at a metal surface interact with the reactants and products of microbial metabolic processes occurring in biofilms in a complex way. Some of these interactions accelerate corrosion, and some may inhibit corrosion. The interactions that accelerate corrosion, and are characteristic enough, are called mechanisms of MIC, and much of this text is devoted to quantifying the mechanisms that we now understand. To approach the task of quantifying these interactions in an organized manner, we will start by describing the reactions characterized as galvanic corrosion and then assess the effects of various metabolic reactions on these reactions. Corrosion science has developed a succinct system of quantifying various forms of corrosion, and we will use this system to quantify the effects of microbial metabolic reactions on corrosion by referring to the principles of the chemistry and electrochemistry of metals immersed in water solutions. Traditionally, the mechanisms of corrosion are quantified using thermodynamics and kinetics, and we will follow this tradition here.

The term corrosion can be defined in various ways, and there are many forms of corrosion and many materials that can corrode – both metallic and nonmetallic. Among the well-known processes of nonmetallic corrosion is the corrosion of stone and its effect on ancient artifacts. Here, we restrict the meaning of corrosion and define it as the anodic dissolution of metals. Among the many anodic reactions that may occur at the surface of a metal, the one in which the metal itself is the reactant subjected to oxidation is singled out and termed corrosion. Noble metals, such as platinum and gold, do not undergo an oxidation reaction and serve only to facilitate charge transfer between external redox species. In contrast, active metals such as iron are oxidized and this process contributes to the net anodic reaction rate, which is typically the dominant anodic process for freely corroding metals. On corroding metals, anodic reactions are coupled with cathodic reactions (reduction). In aerated water solutions, the dominant cathodic reaction is the reduction of dissolved oxygen, while in anaerobic solutions, the reduction of protons is the dominant cathodic reaction; this is typically represented as the reduction of water.

Equations (1)–(6) show the relevant half reactions, followed by the corresponding net reactions, for the corrosion of ironin aqueous media (Lewandowski et al. 1997).

Anaerobic

$$Fe \rightarrow Fe^{2+} + 2e^{-} \text{ anodic} \tag{1}$$

$$2H_2O + 2e^{-} \rightarrow H_2 + 2OH^{-} \text{ cathodic} \tag{2}$$

$$2H_2O + Fe \rightarrow Fe(OH)_2 + H_2 \text{ net} \tag{3}$$

Aerobic

In aerobic solutions, the basic anodic reaction is of course the same as the one described by (1) – dissolution of iron – but the products of the reaction, ferric ions, are hydrolyzed and further oxidized by the available oxygen, and all these reactions are summarized as follows:

$$4OH^{-} + 4Fe(OH)_2 \rightarrow 4Fe(OH)_3 + 4e^{-} \text{ anodic} \tag{4}$$

$$O_2 + 2H_2O + 4e^{-} \rightarrow 4OH^{-} \text{ cathodic} \tag{5}$$

$$4Fe(OH)_2 + O_2 + 2H_2O \rightarrow 4Fe(OH)_3 \text{ net} \tag{6}$$

These corrosion reactions can be modified by the metabolic reactions in biofilms in many ways, and we will discuss four possible modifications here:

1. Biofilms create oxygen heterogeneities near a metal surface.
2. Biofilm matrix increases mass transport resistance near a metal surface.
3. Metabolic reactions in biofilms generate corrosive substances, such as acids.
4. Metabolic reactions in biofilms generate substances that serve as cathodic reactants.

These four possible interactions do not exhaust the possible effects of microorganisms on corrosion reactions. The reason we have selected these four interactions is that they have been extensively studied, and so we know more about them than we know about other interactions. Other mechanisms, both accelerating and inhibiting corrosion, are continually proposed and studied. For obvious reasons, using biofilms to inhibit the corrosion of metals stimulates imaginations, and several authors have described such scenarios. For example, Jayaraman et al. (1999) demonstrated axenic aerobic biofilms inhibiting generalized corrosion of copper and aluminum. Similarly, in the work by Zuo et al. (2005), Al 2024 was passive in artificial seawater in the presence of a protective biofilm of *Bacillus subtilis* WB600. When antibiotics were added to the artificial seawater to kill the bacteria in the biofilm, pitting occurred within a few hours (Zuo et al. 2005). However, as summarized by Little and Ray (2002), most of the experiments on inhibiting corrosion with biofilms were done in laboratories, and when the biofilms were exposed to natural waters they failed to protect the material. Clearly, the laboratory biofilms were different from those deposited in nature. One assumption made in attempting to use biofilms to inhibit corrosion is that biofilm formation is predictable and controllable (Little and Ray 2002). This is not true. Even pure culture biofilms in laboratory are not uniform and their structure changes all the time (Lewandowski et al. 2004).

Corroding metals fall into two categories: active metals – such as iron, and passive metals – such as stainless steels. These two types of materials are affected by different types of corrosion. To demonstrate the possible microbial modifications of the corrosion reactions, we need to specify the reactions characteristic of each type of corrosion affecting these materials.

2 Corrosion of Active Metals

2.1 *Thermodynamics of Iron Corrosion*

Using the terminology accepted in electrochemical studies, a metal immersed in water is called an electrode. The potential of an electrode in an aqueous solution depends on the rates of the anodic (oxidation) and cathodic (reduction) reactions occurring at the metal surface. When these rates are at equilibrium, thermodynamics can be used to quantify the electrode potential. When these rates are not at equilibrium, thermodynamics cannot be used to find the electrode potential and it must be found empirically. Corrosion reactions are not at equilibrium, and the potentials of corroding metals cannot be predicted from thermodynamics.

To illustrate the thermodynamic principles of galvanic corrosion, we will select a set of conditions and compute the potentials of the reactions participating in the corrosion of iron. For the anodic reaction,

$$Fe^{2+} + 2e^- \rightarrow Fe \quad E^0 = -0.44\,V_{SHE} \tag{7}$$

The Nernst equation quantifies the half-cell potential for iron oxidation as

$$E = E^0 - \frac{0.059}{n}\log\left[\frac{1}{[Fe^{2+}]}\right] \tag{8}$$

Iron is a solid metal and its activity equals one. Consequently, the potential of the anodic half reaction depends on the concentration of ferrous ions in the solution and is computed as

$$E = -0.44 + (0.059/2)\log[Fe^{2+}] \tag{9}$$

Selecting the concentration of ferrous iron, $[Fe^{2+}] = 10^{-6}$ M, the potential equals $E = -0.62\ V_{SHE}$.

The cathodic reaction– the reduction of oxygen:

$$O_2 + 2H_2O + 4e^- \rightarrow 4OH^- \quad E^0 = +0.401\,V_{SHE} \tag{10}$$

The Nernst equation quantifies the half-cell potential for oxygen reduction:

$$E = E^0 - \frac{0.059}{n}\log\left[\frac{[OH^-]^4}{pO_2}\right] \tag{11}$$

The potential of this half reaction depends on the partial pressure of oxygen and on the pH.

$$E = 0.401 + 0.059/4[\log(pO_2) + 4(14 - \mathrm{pH})] \tag{12}$$

Assuming that $p(O_2) = 0.2$ atm and pH 7, the potential of the cathodic reaction is $E = 0.804\ V_{SHE}$.

If only one of these reactions were occurring on the metal surface, the metal would assume the respective potential specified for the reaction. For example, if only the cathodic reaction were taking place, the metal would have the potential +0.804 V_{SHE}, and if only the anodic reaction were taking place, the metal would have the potential –0.62 V_{SHE}. This can be demonstrated in electrochemical studies where the anode and the cathode can be separated, placed in different half-cells, and studied in isolation. However, in corrosion, both reactions occur on the same piece of metal and at the same time, and the potential of the metal can have only one value. As a result, the potential of the corroding metal is somewhere between the potential of the anodic half reaction, –0.62 V_{SHE}, and the potential of the cathodic half reaction, +0.804 V_{SHE}. The exact potential of the corroding metal depends on the kinetics (reaction rates) of the anodic and cathodic reactions, and can be measured empirically and interpreted from the theory of mixed potentials. Here, for the purpose of this simplified argument, it is enough to assume that the potential of the corroding iron is between the potentials of the anodic and cathodic half reactions, say, in the middle: $E = (-0.62 + 0.804)/2 = 0.092\ V_{SHE}$. Setting the potential of the metal between the potential of the anodic and cathodic half reactions has consequences: it sets the position of the equilibrium for each of the participating reactions. If the potential were equal to that computed for either of the half reactions, anodic or cathodic, this half reaction would be at equilibrium. If the potential of the corroding iron is between the potentials computed for the two half reactions, none of these half reactions (1)–(6) are at equilibrium and each of them proceeds in the direction that approaches the equilibrium. To quantify the consequences of this departure from the equilibrium, we can inspect the Nernst equation describing potentials of the anodic and cathodic half reactions when the potentials are shifted from their respective equilibrium potentials. If the potential of the corroding iron is 0.092 V_{SHE}, it is higher than the equilibrium potential for the anodic reaction and lower than the equilibrium potential for the cathodic reaction. As a consequence, each reaction will proceed spontaneously toward reaching the equilibrium determined by the potential of the metal, by adjusting the concentrations of the reactants and products to satisfy the equilibrium for the given potential.

The anodic reaction, $Fe^{2+} + 2e^{-} \rightarrow Fe$, must adjust its potential to +0.092 V

$$+0.092 = -0.44 + (0.059/2)\log[Fe^{2+}] \quad (13)$$

If separated from the cathodic reaction, this reaction has a potential of $E = -0.62\ V_{SHE}$. When connected to the cathodic reaction, this reaction has a new equilibrium potential, $E = +0.092\ V_{SHE}$. To reach the new equilibrium potential, the concentration of ferric ions must increase. Consequently, the reaction proceeds to the left, to increase the concentration of ferric ions in the solution. Iron dissolves in this reaction.

The cathodic reaction, $O_2 + 2H_2O + 4e^- \rightarrow 4OH^-$, must adjust its potential to +0.0092 V as well:

$$+0.092 = 0.401 + 0.059/4[\log(pO_2) + 4(14 - \mathrm{pH})] \quad (14)$$

If separated from the anodic reaction, this reaction has a potential of $E = +0.804\ V_{SHE}$. When connected to the anodic reaction, this reaction has a new equilibrium potential, $E = +0.092\ V_{SHE}$. To reach this new equilibrium potential, the reaction proceeds to the right, to decrease the partial pressure of oxygen. Oxygen is consumed in this reaction.

As a result of setting the metal potential between the equilibrium potentials for the anodic and cathodic half reactions, the anodic reaction spontaneously proceeds toward dissolution of the iron and the cathodic reaction spontaneously proceeds toward reduction of the oxygen. Both reactions proceed until one of the reactants is exhausted or until they both adjust the concentrations of their respective reactants to reach the new equilibrium at $0.092\ V_{SHE}$. The thermodynamics of the corrosion processes explains why these processes occur but of course cannot predict the anodic or cathodic reaction rates. Kinetic computations are needed to refine what was said in the section dedicated to thermodynamic considerations.

2.2 *Kinetics of Iron Corrosion*

As discussed in the previous section, the anodic and cathodic processes occurring on metal surfaces correspond to different half reactions, and the electrode potential is used to predict the directions in which these reactions will proceed. Typically, the corrosion reactions occurring on the surfaces of corroding metals are the dominant redox reactions. However, the metal can always serve as a source or sink for electrons satisfying the dissolved redox couples in the solution, and more than one redox reaction can occur on the surface. There is a possibility that more than one reaction is occurring on the metal surface at a time and that each of the reactions uses the electrode as a source or a sink for the electrons needed to reach its own equilibrium potential. The term "mixed potential" is used to describe this condition, to distinguish it from the reduction–oxidation potential in which the anodic and cathodic reactions are simply the forward and reverse parts of a single reaction. The mixed potential in which the anodic reaction is metal oxidation is termed the corrosion potential, E_{corr}.

Let us use, as an example, the reaction described by (1) as the electrode reaction – an iron electrode immersed in a solution of ferrous ions. At equilibrium, the exchange of electrical charges between the electronic conductor – the electrode – and the ionic conductors – ferrous ions in the solution – is composed of two streams of electrical charges moving in opposite directions, to and from the electrode. In the forward reaction, ferrous ions from the metal lattice are dissolved in water. In the reverse reaction, ferrous ions are reduced and deposited on the surface of the electrode as iron atoms. At equilibrium, the rates of the charge transfers across the interface are equal to each other, and there is no net current flow across the interface; the potential at equilibrium is named E_{eq}, and the current flowing in opposite directions, named the exchange current, is usually quantified as the exchange current density, i_0. Once the electrode potential departs from equilibrium and an overpotential is applied, the electrode reaction is no longer at equilibrium and a net current flows in one direction. The direction in which this net current flows is determined by the sign of the applied overpotential: a negative sign is equivalent to cathodic polarization and a positive sign is equivalent to anodic polarization. This can be summarized as follows:

From the definition of overpotential, $E = E_{eq} + \eta$, we assign cathodic polarization:

$$\eta = (E - E_{eq}) < 0 \tag{15}$$

$$\text{Anodic polarization: } \eta = (E - E_{eq}) > 0 \tag{16}$$

The magnitude of the net current is determined by the extent of the overpotential (η) and by the intrinsic properties of the system, summarized by the exchange current density, i_0. At this condition, even though the currents in the two directions are equal, in various systems these currents may have different magnitudes, depending on the material of the electrode and the type of the electrode reaction. Polarizing the electrode, i.e., applying an overpotential, favors the flow of electric charges in one direction and inhibits the flow in the opposite direction: positive polarization amplifies the anodic current and negative polarization amplifies the cathodic current. The following equation, somewhat simplified, is known as the Butler–Volmer equation, and it quantifies the net current, equal to the difference in rate of charge transfer between the anodic and cathodic directions:

$$i = i_c - i_a = i_0 \left[\exp\left(\frac{-\alpha F \eta}{RT} \right) - \exp\left(\frac{\alpha F \eta}{RT} \right) \right] \tag{17}$$

where α is the symmetry coefficient and is assumed to be equal to 0.5, and the remaining symbols – F, R, and T – have their usual meanings. When applied potential

η is negative (cathodic polarization), the first exponential expression in the Butler–Volmer equation becomes positive and the second becomes negative. As a result, the second exponential expression is, for practical reasons, negligibly small when compared with the first exponential expression; i.e., the anodic current is negligibly small when compared with the cathodic current. The opposite is true when the overpotential has a positive sign (anodic polarization). Figure 1 shows the relation between the applied potential and the current: the potentiodynamic polarization curve.

2.3 *Microbially Stimulated Modifications of the Corrosion of Iron and Active Metals*

In corrosion, the anodic and cathodic reactions are not at equilibrium but they are related to each other by two requirements:

1. The two reactions progress on the same piece of metal, and so they must have the same potential.
2. Electrons extracted in the anodic reactions are used in the cathodic reactions; therefore, the anodic and the cathodic currents must be equal.

These two requirements combined are used to quantify the thermodynamics and kinetics of the corrosion process – the corrosion potential and the corrosion current, as shown in Fig. 2.

The reactants and products of microbial metabolism in biofilms may interact with the corrosion reactions, and these interactions may affect the thermodynamics of the process, e.g., by introducing an additional cathodic reactant and thus altering the position

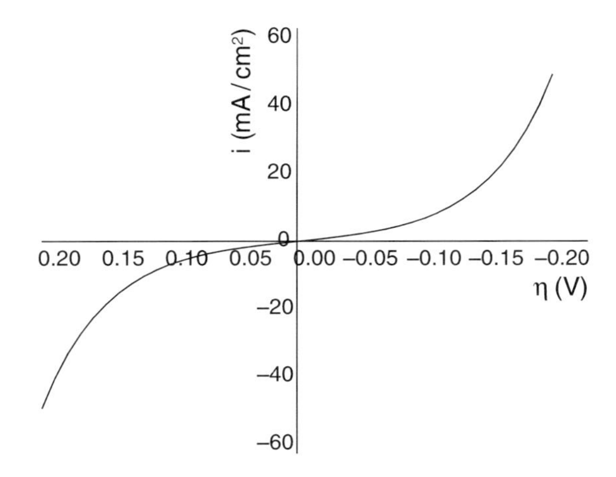

Fig. 1 Potentiodynamic polarization curve. The relationship between the overpotential (η), which varied between –0.2 and +0.2 V, and the current density (i) for $i_0 = 1$ mA cm^{-2} and $\alpha = 0.5$

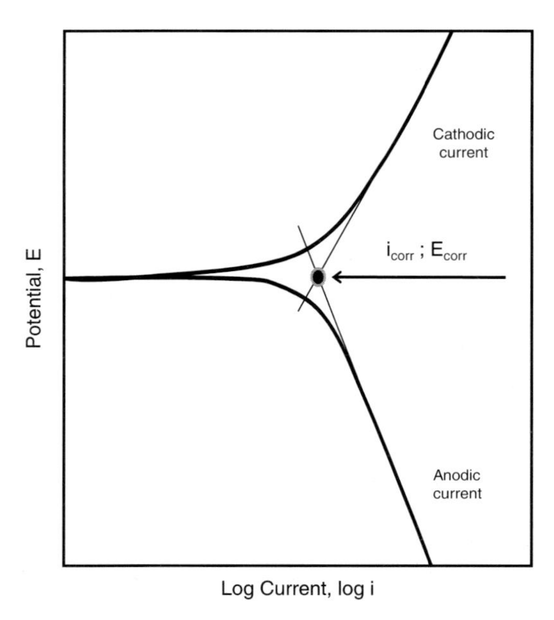

Fig. 2 The intersect of the extrapolated anodic and cathodic potentiodynamic polarization curves demonstrates the meaning of the corrosion potential, E_{corr}, and the corrosion current, i_{corr}. The corrosion potential, E_{corr}, can be measured with respect to a suitable reference electrode (see also Fig. 4, where such a measurement is run for a real sample). The corrosion current, i_{corr}, cannot be measured directly, unless the anode and the cathode are separated, but it can be estimated using other electrochemical techniques based on disturbing the potential of the electrode

of the equilibrium for the relevant reactions. These interactions may also affect the kinetics of the process, e.g., by changing the concentrations of the reactants and products of the corrosion reactions, and thus the rates of the relevant reactions. All these interactions can be presented as in Fig. 2, or a similar plot can be created for specific reactions and reactants. If such a plot is used, modifications affecting the thermodynamics change the locations of the equilibrium points on the vertical axis. For example, replacing protons with oxygen as the cathodic reactant would raise the position of the equilibrium for the cathodic reactions. This would have an effect on the kinetics of the reaction by affecting the position of the intercept between the anodic and cathodic parts of the corrosion reaction, thus affecting E_{corr} and i_{corr}. The kinetics of the participating reactions are illustrated as the slopes of the lines in Fig. 2. For example, a sudden increase in the rate of the cathodic reaction (higher cathodic current for the same potential) would be reflected by a decrease in the slope of the line representing the cathodic reaction, and thus by a change in the position of the intercept determining E_{corr} and i_{corr}. Figure 2 shows the principles of the corrosion of active metals, and the mechanisms of MIC explain the processes that can cause such changes. Despite their resistance to general corrosion, passive metals and alloys can also be affected by MIC. To evaluate the mechanisms of such effects, we will first discuss the mechanisms of the corrosion of stainless steels and other passive metals and then, as we did for active metals, discuss the possible microbial effects that can modify these mechanisms and accelerate corrosion.

3 Corrosion of Passive Metals

Passive metals and alloys show a different mechanism of corrosion than active metals do. The best known passive alloys are stainless steels, and much research has been done on MIC of these materials. The corrosion reactions for stainless steels are the same as those for iron. However, stainless steels are alloys and some of their components, when oxidized, form dense layers of oxides, passive layers, which prevent further corrosion. Passivated alloys can resist corrosion in the presence of strong oxidants that would cause corrosion on unalloyed metal. However, this protection works to a certain extent only. When a cathodic reactant polarizing the metal has a high enough oxidation potential, localized corrosion, called pitting, occurs as a result of localized damage to the passive layer. The mechanism of this process is shown in Fig. 3.

As shown in Fig. 3, when a passivating alloy is subjected to anodic polarization, i.e., to an increasing electrode potential, the corrosion current initially increases, following the increase in the polarization potential. This increase continues until the polarization potential reaches a critical value, called the passivation potential. At this potential, the alloying constituents of the metal are oxidized and form dense layers on the metal surface, which slow down the corrosion of the metal, as is demonstrated by the decreasing corrosion current. As the polarization potential increases further, the metal first reaches a passive zone, in which it is immune to the increase in the polarization potential until the polarization potential reaches a critical value, called the pitting potential. When the polarization reaches, and exceeds, the pitting potential, the

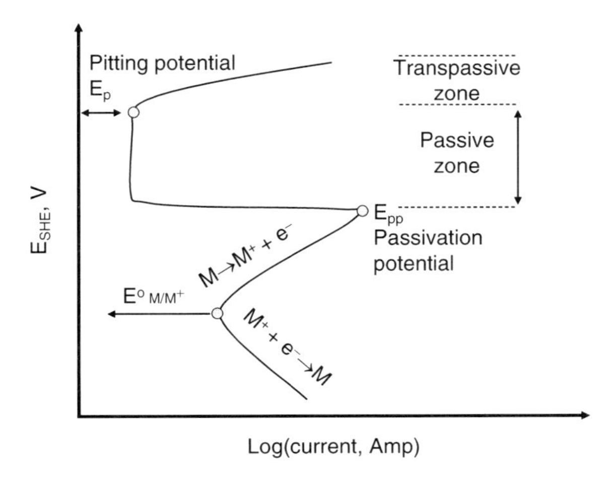

Fig. 3 Potentiodynamic polarization curve of a passive metal; thermodynamic principles of passivation and of pitting corrosion

corrosion current gradually increases, as a result of localized damage to the passive layers. The damaged areas are small compared to the surface of the metal; the damage to the surface has the form of small pits, and this type of corrosion is called pitting corrosion. Because the damaged areas become anodic sites, their small combined surface area makes the localized corrosion process particularly dangerous for the integrity of the metal. The anodic current densities can reach high values and localized damage of the material can progress much faster than it does in cases of general corrosion, in which the anodic current densities are much smaller.

4 Microbially Stimulated Modifications of the Corrosion of Passive Metals and Alloys

Passive metals and alloys, such as stainless steels, can be used within the passive zone, where the oxidation potentials of the available oxidants do not exceed the pitting potential. Microbial interference that may accelerate the corrosion of such surfaces is then necessarily related to two possible mechanisms:

1. Microbially generated oxidants (cathodic reactants) can have higher oxidation potentials than the pitting potential.
2. Microbially stimulated localized damage to the passive layers can decrease the pitting potential.

The first mechanism is related to the deposition of biomineralized manganese oxides, which can subsequently raise the potential of the passive metal above the pitting potential. The second mechanism is related to the damage of the passive metal surface by microorganisms in biofilms. We will discuss these mechanisms in more detail later in this chapter.

Figure 4 shows the relation among the corrosion potential, pitting potential, and probability of pits initiation, redrawn from Sedriks (1996). The potential (E_{corr}) of a

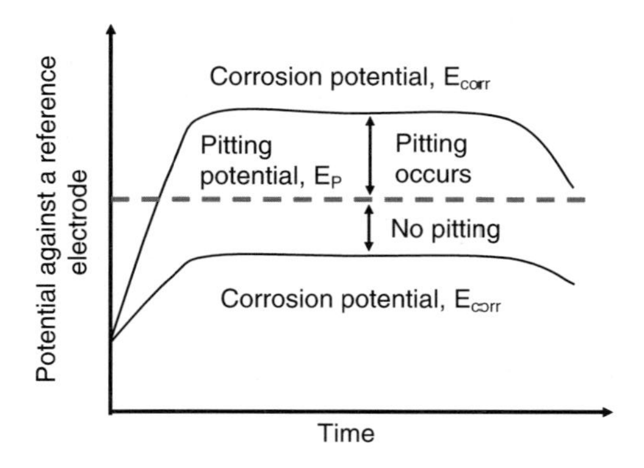

Fig. 4 When the corrosion potential, E_{corr}, reaches the pitting potential, E_p (*dashed line*), of the metal in the given solution, pits are initiated (redrawn from Sedriks, 1996)

metal such as stainless steel is measured against time. If the potential of the stainless steel is higher than the pitting potential (E_p), the stainless steel develops pits to initiate corrosion. If the potential of the stainless steel is less than the pitting potential, pits cannot develop. The pitting potential can be determined using standard electrochemical techniques described elsewhere (ASM Handbook Series 1987).

5 Mechanisms by Which Metabolic Reactions in Biofilms can Interact with Corrosion Reactions

5.1 Mechanism 1: Biofilms Create Oxygen Heterogeneities

The interaction between metabolic activity in biofilms and corrosion reactions appears to be trivial: microorganisms use the cathodic reactant, oxygen, which makes it unavailable for the corrosion reactions and, as a result, the corrosion rate decreases. If true, this mechanism would actually inhibit corrosion, and there is experimental evidence that this occurs in some situations. Hernandez et al. (1990) reported a decrease in the corrosion rate of mild steel in the presence of a uniform layer of biofilm. This decrease was attributed to the respiration of the biofilm microorganisms, resulting in a decline in oxygen concentration at the metal surface and an associated decrease in the rate of the cathodic reduction of oxygen. These authors reproduced their observations using synthetic seawater with *Pseudomonas* sp. S9 as well as with *Serratia marcescens* (Hernandez et al. 1994). We now know that to inhibit corrosion by this mechanism the biofilm must cover the surface of the metal uniformly and, in principle at least, must have uniformly distributed microbial activity. As biofilms are not uniform and microbial activity in biofilms is not uniformly distributed, this mechanism can be demonstrated in the laboratory but is unlikely to persist in a natural environment. Oxygen consumption rates and oxygen concentrations in biofilms vary from one location to another (Lewandowski et al. 1997; Lewandowski and Beyenal 2007); this leads to a more interesting interaction, mechanism 2, which increases the rate of corrosion, and there is some experimental evidence for it as well. White et al. (1985), for example, found no accumulation of iron or other metals in EPS from biofilms growing on corroding 304 stainless steel. They attributed the observed accelerated corrosion to an inhomogeneous distribution of biofilm at the metal surface, resulting in areas of differing cathodic activity, consistent with a differential aeration cell. Areas covered with biofilm exhibit lowered oxygen concentrations and become anodic, while those with less biofilm exhibit higher oxygen concentrations and become cathodic. As a result, anodic and cathodic areas are fixed at the metal surface, and this mechanism is appropriately called corrosion as a result of differential aeration cells (Ford and Mitchell 1990).

Metal corrosion through the formation of differential aeration cells results from different concentrations of oxygen occurring at different locations on the metal surface. This effect, different concentrations of oxygen at different locations on the metal surface, can be caused by the active consumption of oxygen by microorganisms in biofilms nonuniformly distributed on the metal surface, but it can also be caused by a passive mechanism in which oxygen access to some areas is physically obstructed. Placing an o-ring on a metal surface is an example of such a mechanism, but other, more subtle, scenarios are possible as well. One such scenario is based on partially covering the metal surface with a material that has nonuniformly distributed diffusivity for oxygen. The access of oxygen to some locations on the metal surface is more difficult than its access to other locations on the same surface, and differential aeration cells are formed.

These speculations lead to the question of whether depositing microbial EPS on a metal surface can cause the formation of differential aeration cells, and to a more general question: what is the role of EPS in MIC? It is well known that polysaccharides, the main constituent of EPS, can be cross-linked with metal ions. In principle, then, if EPS covers a corroding site, the metal ions can cross-link the polysaccharides and affect the position of the equilibrium between the corroding metal and its ions, thus accelerating corrosion. This mechanism is analogous to the formation of differential aeration cells, and in corrosion science both mechanisms are called differential concentration cells. The metal concentration cells do not seem to affect MIC to a large extent, at least based on the report by White et al. (1985), who found no accumulation of iron or other metals in EPS from biofilms growing on corroding 304 stainless steel. Doubt remains about the passive effect of EPS, in which it changes the access of oxygen to various locations on the metal surface. Can differential aeration cells be formed by this mechanism?

To address this question, we will first discuss the thermodynamic principles of corrosion by differential aeration cells and determine the factors that must be measured to resolve whether this mechanism is active in biofilms.

If the anodic reaction is the oxidation of iron: $\mathrm{Fe} \rightarrow \mathrm{Fe}^{2+} + 2\mathrm{e}^-$, and the cathodic reaction is the reduction of oxygen: $\mathrm{O_2} + 2\mathrm{H_2O} + 4\mathrm{e}^- \rightarrow 4\mathrm{OH}^-$, then the overall reaction describing the process is

$$2\mathrm{Fe} + \mathrm{O_2} + 2\mathrm{H_2O} \rightarrow 2\mathrm{Fe}^{2+} + 4\mathrm{OH}^- \tag{18}$$

The Nernst equation quantifying the potential for this reaction is

$$E = E^0 - \frac{0.059}{4}\log\frac{[\mathrm{Fe}^{2+}]^2[\mathrm{OH}^-]^4}{p(\mathrm{O_2})} \tag{19}$$

If the oxygen concentrations at two adjacent locations on an iron surface are different, then the cell potentials at these locations are different as well. Specifically, the location where the oxygen concentration is higher will have a higher potential (more cathodic) than the location where the oxygen concentration is lower (more anodic). The difference in potential will give rise to current flow from the anodic locations to the cathodic locations and to the establishment of a corrosion cell. This is the mechanism of differential aeration cells, and the prerequisite to this mechanism is that the concentration of oxygen vary among locations (Acuna et al. 2006; Dickinson and Lewandowski 1996; Hossain and Das 2005). Indeed, many measurements using oxygen microsensors have demonstrated that oxygen concentrations in biofilms can vary from one location to another (Lewandowski and Beyenal 2007).

This mechanism by which differential aeration cells are formed, in which a thin layer of biofilm at the surface of the substratum is discontinuous, is consistent with the current model of biofilm structure, shown in Fig. 5.

One of the most dangerous forms of localized corrosion of mild steel is tuberculation, which is the development or formation of small mounds of corrosion products. According to Herro (1991), tubercle formation originates from a differential oxygen concentration cell.

5.2 Mechanism 2: Biofilm Matrix Increases the Mass Transport Resistance near the Metal Surface, Thus Changing the Kinetics of the Corrosion Processes

Once the mechanism of differential aeration cell formation in biofilms had been demonstrated and explained, the immediately following question was whether microbial activity in biofilms is a necessary prerequisite to the formation of differential aeration cells, or perhaps, the presence of extracellular polymeric substances on the surface suffices. The idea that the presence of EPS on the surface might suffice is related to the known mechanisms of corrosion initiation based on different resistances to mass transport for oxygen at various locations on metal surfaces, similar to the initial stages of crevice formation. The possibility that the active removal of oxygen by the biofilm microorganisms might not be necessary to initiate a differential aeration cell was discussed by MIC researchers, but it was usually dismissed on the grounds that extracellular polymers are composed of 98% water and their layers on metal surfaces are only a few hundred micrometers thick, so that

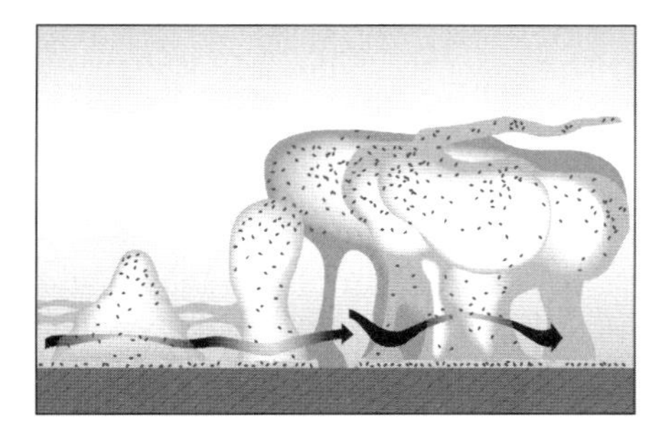

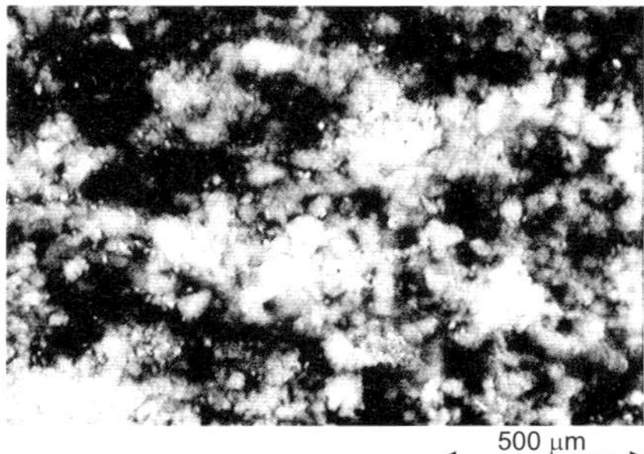

Fig. 5 Conceptual structure of biofilms (*left*) and a light microscopy image of a biofilm (*right*)

the increase in diffusion resistance expected as a result of depositing extracellular polymer could not possibly be significant. Nevertheless, the hypothesis was formulated that the deposition of extracellular polymer on a metal surface might form differential aeration cells, and an appropriate experiment was designed and executed (Roe et al. 1996). As a model of extracellular polymer, calcium alginate was used. Alginate is an extracellular biopolymer excreted by biofilm microorganisms. If alginate initiates the differential aeration cell, then the oxygen concentrations at the locations covered with alginate should be higher than those at the locations not covered with alginate. Also, pH at the locations covered with alginate should be higher than that at locations not covered with alginate. These expectations are consistent with the anodic and cathodic reactions, in which the anodic reaction decreases pH because ferrous ions hydrolyze and precipitate as hydroxides, and the cathodic reaction increases pH because it consumes protons. Two drops of sodium alginate were deposited on the surface of a corrosion coupon made of mild steel and exposed to a calcium solution which cross-linked the sodium alginate and formed a calcium alginate gel on the surface. To test whether depositing calcium alginates can initiate differential aeration cells, the variations in oxygen concentration and pH above these spots were measured using scanning microelectrodes. In addition, a scanning vibrating electrode (SVE) was used to determine the distribution of the electrical field above the surface, and it was expected that this electrode would detect the positions of the anodic and cathodic sites. The results, shown in Fig. 6

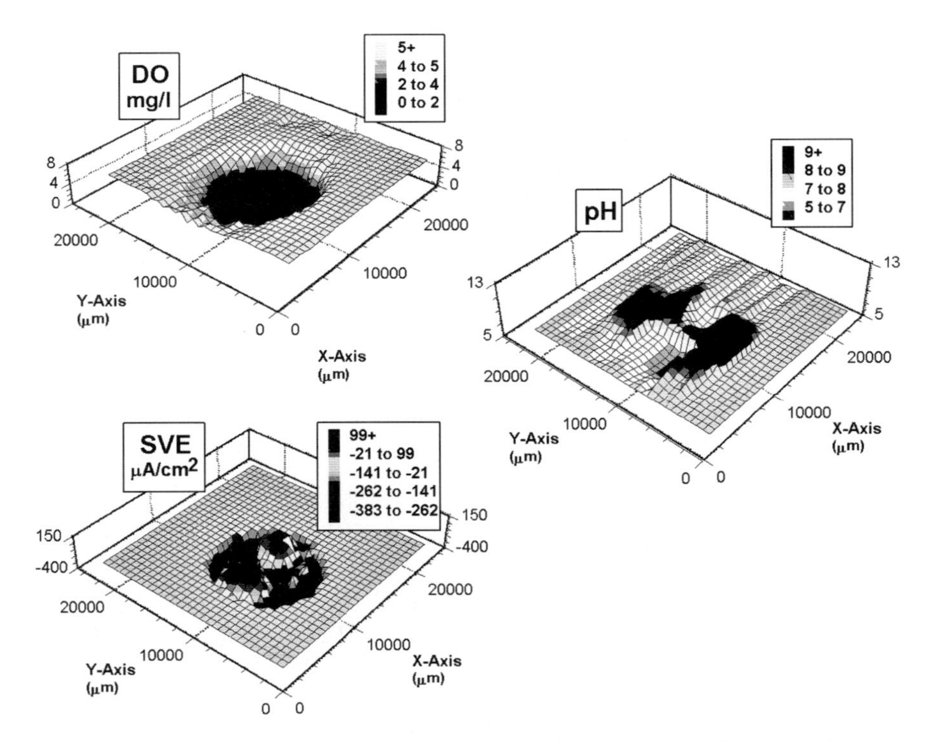

Fig. 6 Two spots of calcium alginate deposited on a surface of mild steel fix anodic sites (Roe et al. 1996)

demonstrated that the mere deposition of a thin layer of alginate on mild steel is enough to fix the anodic sites and initiate corrosion. All the characteristics of differential aeration cells were present in the system: pH was lower near the sites covered with alginate than near the sites not covered; oxygen concentration was lower near the sites not covered; and, as demonstrated by the image of the electric field distribution provided by the scanning vibrating electrode, there were anodic and cathodic sites fixed at the surface of the metal. This result, somewhat unexpected at that time, had further implications: it demonstrated that merely killing biofilm microorganisms using biocide(s) or antimicrobial(s) does not necessarily stop MIC. Once the biopolymer has been deposited on the surface, the active consumption of oxygen in the respiration reaction enhances the formation of differential aeration cells, but even without it, differential aeration cells can be formed just because EPS has been deposited on the surface. This conclusion coincides with the general notion that removing the biofilmis more important than killing the biofilm microorganisms.

Once the differential aeration cell has been established, the corrosion proceeds according to the mechanism described by (Eq. 18), which is also illustrated in Fig. 7.

5.3 *Mechanism 3: Metabolic Reactions in Biofilms Generate Corrosive Substances, Exemplified by the Sulfate-Reducing Bacteria Corrosion of Mild Steels*

The mechanism of MIC due to the formation of differential aeration cells can be called a nonspecific one, because it does not depend on the physiology of the microorganisms that deposited the extracellular polymers. There are, however, other mechanisms that are closely related to the type of microorganisms active in the biofilm and to their metabolic reactions (Beech and Gaylarde 1999; Romero et al. 2004;Videla and Herrera 2005; Xu et al. 2007). An example of such a mechanism is sulfate-reducing bacteria (SRB) corrosion (Ilhan-Sungur et al. 2007; Lee et al. 1995).

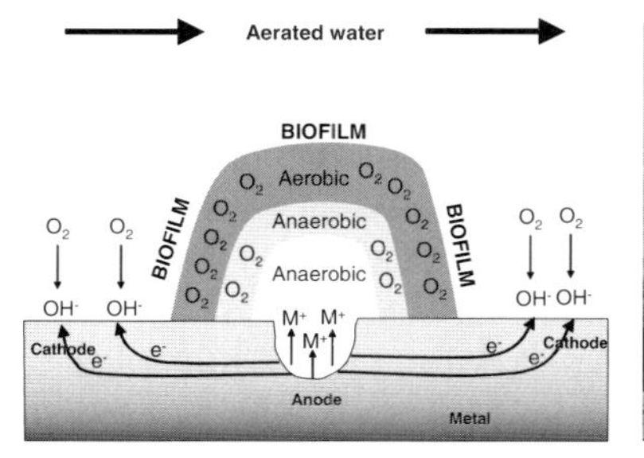

Fig. 7 (**a**) Biofilm heterogeneity results in differential aeration cells. This schematic shows pit initiation due to oxygen depletion under a biofilm (Borenstein 1994). (**b**) An anodic site and pit under the biofilm and corrosion products deposited on mild steel

The corrosion of mild steel caused by SRB is probably the most celebrated case of MIC because it provides a direct, and easy to understand, link between microbial reactions and electrochemistry (Javaherdashti 1999). Despite the progress in research, and in understanding of the process, little has been done to prevent or stop this type of corrosion once initiated, and SRB corrosion is still considered the main type of MIC. For example, Bolwell in 2006 demonstrated that engine failures in gas turbines were caused by SRB growing in the seawater lubricating oil coolers and contaminating it (Bolwell 2006). The overall progress in understanding of MIC, however, allows us to implicate other microorganisms as partners of SRB and consider more complex scenarios of MIC, in which two types of microorganisms modify the potential of the electrode in the opposite directions. For example, Rao et al. (2000) found that in the cooling water system of a nuclear test reactor iron- and manganese-oxidizing bacteria (MOB) (*Leptothrix* sp.) and SRB (*Desulfovibrio* sp.) were responsible for the corrosion of carbon steel. It is interesting to notice that these two types of microorganisms drive the redox potential in the opposite directions, thus increasing the gap between the potential of the anodic reaction and the potential of the cathodic reaction.

SRB produce hydrogen sulfide by reducing sulfate ions (Videla and Herrera 2005). According to the mechanism that was proposed by Von Wohlzogen Kuhr in 1934, SRB oxidize cathodically generated hydrogen to reduce sulfate ions to H_2S, thereby removing the product of the cathodic reaction and stimulating the progress of the reaction (Al Darbi et al. 2005). Over the years it became obvious that the mechanism must be more complex than that initially suggested, and it is now certain that the possible pathways for cathodic reactions are more complex and can, for example, include sulfides and bisulfides as cathodic reactants (Videla 2001; Videla and Herrera 2005).

Hydrogen sulfideitself can be a cathodic reactant (Antony et al. 2007; Costello 1974):

$$2H_2S + 2e^- \rightarrow H_2 + 2HS^- \tag{20}$$

Ferrous iron generated from anodic corrosion sites (21) precipitates with the metabolic product of microbial metabolism, hydrogen sulfide, forming iron sulfides, FeS_x.

$$Fe^{2+} + HS^- = FeS + H^+ \tag{21}$$

This reaction may provide protons for the cathodic reaction (Crolet 1992).

The precipitated iron sulfides form a galvanic couple with the base metal. For corrosion to occur, the iron sulfides must have electrical contact with the bare steel surface. Once contact is established, the mild steel behaves as an anode and electrons are conducted from the metal through the iron sulfide to the interface between the sulfide deposits and water, where they are used in a cathodic reaction. What exactly the cathodic reactants are is still debatable.

Surprisingly, the most notorious cases of SRB corrosion often occur in the presence of oxygen. Since the SRB are anaerobic microorganisms, this fact has been difficult to explain. Our group believes that this effect is based on mechanism 3: iron sulfides (resulting from the reaction between iron ions and sulfide and bisulfide ions) are oxidized by oxygen to elemental sulfur, a substance known to be a strong corrosion agent (Lee et al. 1995). Biofilm heterogeneity plays an important role in this process, because the central parts of microcolonies are anaerobic while the outside edges remain aerobic (Lewandowski and Beyenal 2007). This arrangement makes this mechanism possible because the oxidation of iron sulfides produces highly corrosive elemental sulfur, as illustrated by the following reaction:

$$2H_2O + 4FeS + 3O_2 \rightarrow 4S^0 + 4FeO(OH) \quad (22)$$

Hydrogen sulfide can also react with the oxidized iron to form ferrous sulfide and elemental sulfur (Schmitt 1991), thereby aggravating the situation by producing even more elemental sulfur, and closing the loop through production of the reactant in the first reaction, FeS.

$$3H_2S + 2FeO(OH) \rightarrow 2FeS + S^0 + 4H_2O \quad (23)$$

The product of these reactions – elemental sulfur – accelerates the corrosion rate. Schmitt (1991) has shown that the corrosion rate caused by elemental sulfur can reach several hundred mils per year. We have demonstrated experimentally that elemental sulphur is deposited in the biofilm during the SRB corrosion (Nielsen et al. 1993). It is also well known that sulfur disproportionation reaction that produces sulfuric acid and hydrogen sulfide is carried out by sulfur disproportionating microorganisms (Finster et al. 1998):

$$4S^0 + 4H_2O \rightarrow 3H_2S + H_2SO_4 \quad (24)$$

In summary, according to this mechanism, SRB corrosion of mild steel in the presence of oxygen is an acid corrosion:

$$\text{anodic reaction:} \quad Fe \rightarrow Fe^{2+} + 2e^- \quad (25)$$

$$\text{cathodic reaction:} \quad 2H^+ + 2e^- \rightarrow H_2 \quad (26)$$

It is worth noticing that hydrogen, the product of the cathodic reaction, can be oxidized by some species of SRB to reduce sulfate and generate hydrogen sulfide, H_2S:

$$H_2SO_4 + 4H_2 \rightarrow H_2S + 4H_2O \quad (27)$$

Hydrogen sulfide dissociates to bisulfides:

$$H_2S = H^+ + HS^- \quad (28)$$

which are used in the reaction described by (20).

Thus, this mechanism involves several loops in which reactants are consumed and regenerated, and the process continues at the expense of the energy released by the oxidation of the metal.

These reactions are linked with each other in a network of relations. To illustrate the main pathways, Fig. 8 shows the main reactions and the effect of oxygen on the SRB corrosion of mild steel.

5.4 *Mechanism 4: Metabolic Reactions in Biofilms Generate Substances That Serve as Cathodic Reactants*

One of the most puzzling aspects of MIC is the change in electrochemical properties of stainless steel that occurs as the metal surface is colonized by microorganisms in natural water. The dominant effects of colonization are a several-hundred-millivolt increase in corrosion potential (E_{corr}) to values near +350 mV versus the saturated calomel electrode (SCE) and 2–3 orders of magnitude increase in cathodic current density at potentials between approximately −300 and +300 mV_{SCE}. These effects, known as ennoblement, were first observed in the mid-1960s (Crolet 1991, 1992). Since then, numerous researchers (Braughton et al. 2001; Dickinson et al. 1997; Linhardt 2006; Washizu et al. 2004) have shown that stainless steels and other passive metals in natural waters exhibit a several-hundred-millivolt increase in corrosion potential, accompanied by an increase in cathodic current drawn, upon mild polarization. This phenomenon has been observed in a wide variety of natural and engineered environments. Washizu et al. (2004), Mattila et al. (1997), and Dexter and Gao (1988) described ennoblement in seawater (Amaya and Miyuki 1994), Dickinson et al.

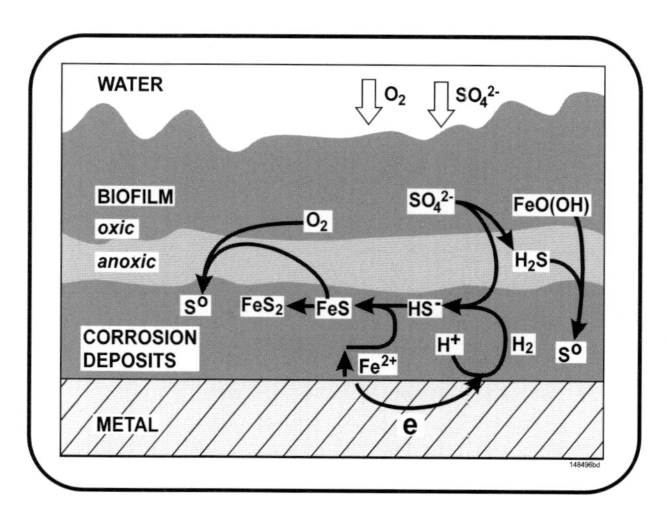

Fig. 8 Sulfate-reducing bacteria corrosion of mild steel in the presence of oxygen is an acid corrosion (Lewandowski et al. 1997)

(1996a) reported its occurrence in a freshwater stream, and Linhardt (1996) reported it in a hydroelectric power plant.

The ennoblement of stainless steels in natural waters may influence material integrity: as the corrosion potential approaches the pitting potential, the material integrity may be compromised by localized (pitting and crevice) corrosion. This sequence of events, from an increase in corrosion potential to pit initiation, is well known to material scientists, although the microbial component is new. Because the pitting potential of 316L stainless steel in seawater is around 200 mV_{SCE}, the danger of pitting initiation in such an environment is serious. There are, however, reports of microbial involvement in pitting corrosion of stainless steels immersed in fresh waters of much lower chloride concentration than that found in seawater (Hakkarainen 2003; Linhardt 2004, 2006; Olesen et al. 2001).

Temporal changes in the corrosion potential of 316L stainless steel coupons immersed in different natural water sources are illustrated by our results in Fig. 9. In all cases the potentials of 316L stainless steel coupons increased, demonstrating ennoblement of the stainless steel. Several hypotheses have been postulated to explain the mechanism of ennoblement, all suggesting that it is caused by microbial colonization of the metal surface. Mollica and Trevis (1976) attributed ennoblement to microbially produced extracellular polymeric substances. Dexter and Gao (1988) suggested that acidification of the metal–biofilm interface caused by protons derived from the metabolic reactions in the biofilm increased the potential. Chandrasekaran and Dexter (1993) proposed a combination of acidification and hydrogen peroxide production. Eashwar and Maruthamuthu (1995) believed that ennoblement was caused by microbially produced passivating siderophores. Although many authors have demonstrated

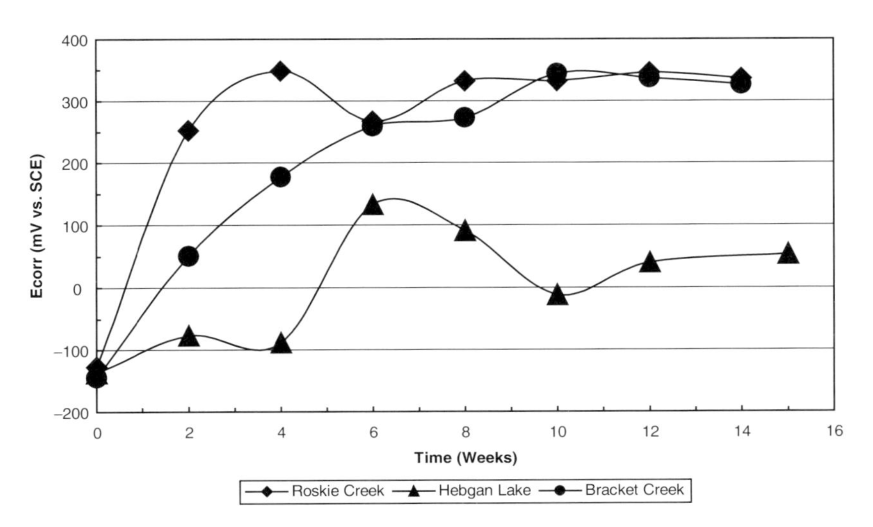

Fig. 9 Potential of 316L stainless steel coupons exposed to fresh water at three locations in Montana for 4 months. The rate and extent of ennoblement roughly correlate with the amount of biomineralized manganese recovered from the surface after 4 months (Table 1) (Braughton et al. 2001)

the relationship between ennoblement and biofilm formation, the proposed hypotheses have not been supported by convincing experimental evidence unequivocally demonstrating the mechanism of ennoblement.

We have demonstrated, in the laboratory and in the field, that stainless steels and other passive metals ennoble when colonized by MOB (Braughton et al. 2001; Dickinson et al. 1996a, 1997; Dickinson and Lewandowski 1996). While the origin of the manganese-rich material deposited on stainless steel coupons exposed to Bozeman stream water was not rigorously established, mineral-encrusted bacterial sheaths characteristic of *Leptothrix* sp. and mineralized capsules characteristic of *Siderocapsa treubii* were abundant on the surface of the ennobled stainless steel coupons, and MOB were isolated from the manganese-rich deposits (Dickinson and Lewandowski 1996). In parallel with these findings, Linhardt (1996) also demonstrated that manganese-oxidizing biofilms were responsible for pitting corrosion of stainless steel. Although biomineralization of manganese can be carried out by certain genera of the so-called iron and manganese group – *Siderocapsa*, *Leptothrix*, and *Crenothrix* – in fact the property is widely distributed in a variety of organisms, including bacteria, yeast, and fungi (Caspi et al. 1998; Francis and Tebo 2002; Tebo et al. 1997, 2004, 2005). These organisms can oxidize dissolved manganese to form highly enriched mineral–biopolymer encrustations. Deposits of manganese oxides form on submerged materials, including metal, stone, glass, and plastic, and can occur in natural waters and sediments with manganese levels as low as 10–20 ppb (Dickinson et al. 1996a, 1997; Dickinson and Lewandowski 1996).

Because biomineralized manganese oxides are in direct electrical contact with the metal, the metal exhibits the equilibrium dissolution potential of the oxides. The standard potentials (E^0) for (Eq. 29)–(Eq. 31) were calculated using the following energies of formation: ΔG_f^o Mn^{2+} = −54.5 kcal mol^{-1}, ΔG_f^o γ-MnOOH = −133.3 kcal mol^{-1}, and ΔG_f^o γ-MnO_2 = −109.1 kcal mol^{-1}.

$$\begin{gathered} MnO_{2(s)} + H^+ + e^- \rightarrow MnOOH_{(s)} \\ E^0 = +0.81\,V_{SCE} \qquad E'_{pH=7.2} = +0.383\,V_{SCE} \end{gathered} \tag{29}$$

$$\begin{gathered} MnOOH_{(s)} + 3H^+ + e^- \rightarrow Mn^{2+} + 2H_2O \\ E^0 = +1.26\,V_{SCE} \qquad E'_{pH=7.2} = +0.336\,V_{SCE} \end{gathered} \tag{30}$$

This leads to the following overall reaction:

$$\begin{gathered} MnO_{2(s)} + 4H^+ + 2e^- \rightarrow Mn^{2+} + 2H_2O \\ E^0 = +1.28\,V_{SCE} \qquad E'_{pH=7.2} = +0.360\,V_{SCE} \end{gathered} \tag{31}$$

The potentials (E') were calculated at a pH of 7.2 and [Mn^{2+}] = 10^{-6}. Dickinson et al. (1996a) demonstrated that just a 6% surface coverage by manganese oxides can increase the resting open circuit potential (OCP) of stainless steels (−200 mV$_{SCE}$) by some 500 mV, which coincides closely with the reported equilibrium potential of the oxides, +362 mV_{SCE} at a pH of 7.2 (Dickinson and Lewandowski 1996; Linhardt 1998).

The thermodynamic calculations are in good agreement with the observations as the potential of stainless steel coupons exposed to river water rises to about 360 mV, as predicted. Our results directly correlate the extent and rate of ennoblement with the amount and rate of manganese oxides deposition on metal surfaces (Braughton et al. 2001). To determine which environmental factors influence the rate of ennoblement, we exposed 316L stainless steel coupons at three locations, two creeks and a lake, for 100 days. The open circuit potential was monitored periodically, about once a week (Fig. 9). The coupons in both creeks reached a potential of +350 mV_{SCE} in 3 weeks. The coupons in the lake reached a final potential of less than +100 mV_{SCE} and the ennoblement rate was very slow. Manganese oxides were deposited on all metal coupons, and their amounts roughly correlated with the rate of ennoblement, as can be seen in Table 1.

Figure 10 shows the potentiodynamic polarization curves of nonennobled, fully ennobled, and MnO_2-plated stainless steel coupons. Both the microbial ennoblement and electroplating of MnO_2 on the metal surface shift corrosion potentials by ~300 mV in the noble direction and cause a corresponding increase in cathodic current density at modest overpotentials (around −100 mV).

In our laboratory, Dickinson and colleagues studied the effects of MOB on stainless steels and demonstrated that 3–5 % surface coverage by biofouling deposits was enough to ennoble 316L stainless steel (Dickinson et al. 1996a, 1997; Dickinson and Lewandowski 1998). Chemical examination of the deposits showed the presence of Fe(III) and Mn(IV), while epifluorescence microscopy revealed the presence of manganese- and iron-oxidizing bacteria (Dickinson and Lewandowski 1998). On the basis of these observations and other studies conducted in our laboratory, we have suggested that MOB are involved in the corrosion of stainless steels through the following mechanism (Braughton et al. 2001; Dickinson et al. 1996a, 1997; Dickinson and Lewandowski 1996; Geiser et al. 2002; Olesen et al. 2000a; Shi et al. 2002a,b): the divalent manganese (Mn^{2+}) ions are microbially oxidized to manganese oxyhydroxide, MnOOH, which is deposited on the metal surface; then the solid MnOOH is further oxidized to manganese dioxide, MnO_2. Both reactions contribute to the increase in the open circuit potential because the deposited oxides, MnOOH and MnO_2, are in electrical contact with the surface and their dissolution potential is determined by the equilibrium of the deposited minerals with the dissolved divalent manganese. The oxides deposited on the surface are reduced to divalent manganese by electrons generated at anodic sites. However, reducing the manganese oxides does not stop the ennoblement process, because the reduced products of this reaction, soluble divalent

Table 1 Amounts of biomineralized manganese recovered from the surfaces after 4 months

Source	Bracket creek	Roskie creek	Hebgen lake
Mn recovered ($\mu g/cm^2$)	9.3	33.6	1.7

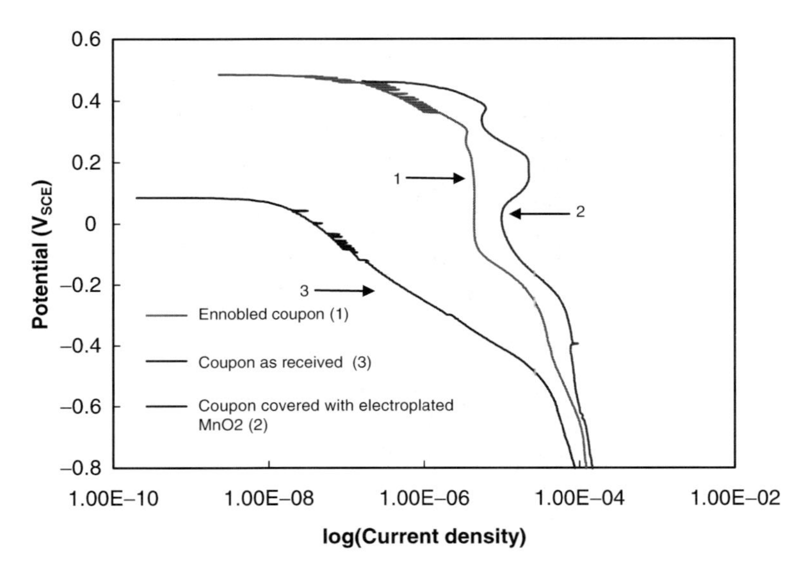

Fig. 10 Potentiodynamic polarization curves (316L stainless steel, 0.01 M Na_2SO_4, pH 8.30; scan rate: 0.167 mV s^{-1}) show typical behavior of nonennobled, fully ennobled, and MnO_2-plated stainless steel coupons. (1) Biomineralized manganese on graphite electrode, (2) electrochemically deposited manganese oxides on graphite electrode, (3) clean graphite electrode used to reduce oxygen only. Both microbial ennoblement and MnO_2 electroplating of the metal surface shift corrosion potentials by ~300 mV in the noble direction and cause a corresponding increase in cathodic current density at modest overpotentials (around −100 mV)

manganese ions, are reoxidized by the MOB attached to the metal surface. The described sequence of events, oxidation–reduction–oxidation of manganese, is a hypothetical mechanism that produces renewable cathodic reactants, MnOOH and MnO_2, and their presence on the metal surface endangers material integrity. This mechanism is illustrated in Fig. 11.

The suggested mechanism relies on the activity of MOB in biofilms deposited on metal surfaces. The biomineralization of manganese can be carried out by a variety of organisms, including bacteria, yeast, and fungi, but it is particularly associated with genera of the so-called iron and manganese group – *Siderocapsa*, *Gallionella*, *Leptothrix-Sphaerotilus*, *Crenothrix*, and *Clonothrix*. These bacteria accelerate the oxidation of dissolved iron and manganese to form highly enriched mineral–biopolymer encrustations. Deposits form on submerged materials, including metal, stone, glass, and plastic, in natural waters with manganese levels as low as 10–20 ppb.

Biomineralized manganic oxides are efficient cathodes and increase cathodic current density on stainless steel by 2–3 orders of magnitude at potentials between roughly −200 and +400 mV_{SCE}. The extent to which the elevated current density can be maintained is controlled by the electrical capacity of the mineral, which reflects both total accumulation and the conductivity of the mineral–biopolymer assemblage (only material in electrical contact with the metal will be cathodically active).

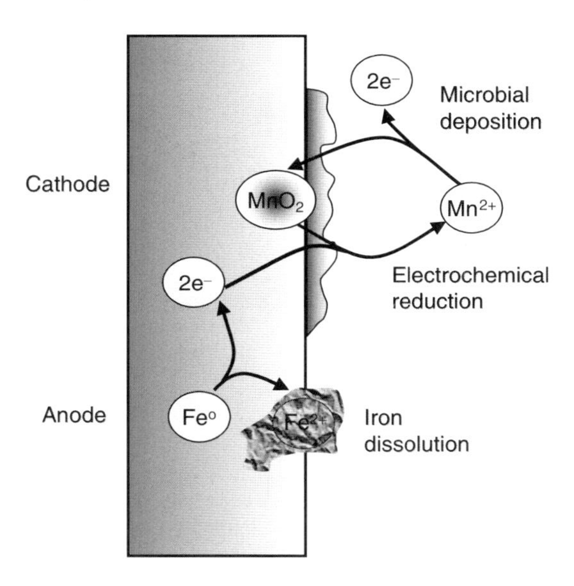

Fig. 11 Redox cycling on metal surfaces: hypothetical mechanism of microbial involvement in the corrosion of stainless steels and other passive metals (Olesen et al. 2000a)

Oxide accumulation is controlled by the biomineralization rate and by the corrosion current, in that high corrosion currents will discharge the oxide as rapidly as it is formed. It appears that this mechanism may result in redox cycling of manganese on metal surfaces, producing a renewable cathodic reactant, which agrees well with the notion that whenever biofilms accumulate on cathodic members of galvanic couples, a significant increase in the reduction current can be expected (Chandrasekaran and Dexter 1993). In conclusion, the accumulation of manganese oxides can cause pitting corrosion, as demonstrated in Fig. 12.

5.5 *Further Implications*

Our observations also suggest that MOB may be directly involved in pit initiation, in addition to the indirect effects caused by the biomineralized manganese oxides (Geiser et al. 2002). Scanning electron microscopy and atomic force microscopy images (Fig. 13) show micropits formed on 316L stainless steel ennobled by *L. discophora* SP-6. This indicates that the pits were initiated at the sites of bacterial attachment and then propagated because of the presence of manganese oxides driving the potential in the noble direction.

Our data show that the manganese oxides deposited on the surface elevate the potential, create an environment where the pits initiated by microbes can not repassivate. Because the pits are initiated at the sites of attachment, in this light, it appears that the bacteria initiate the pits and the microbially deposited manganese oxides stabilize the growth of the pits by maintaining a high potential.

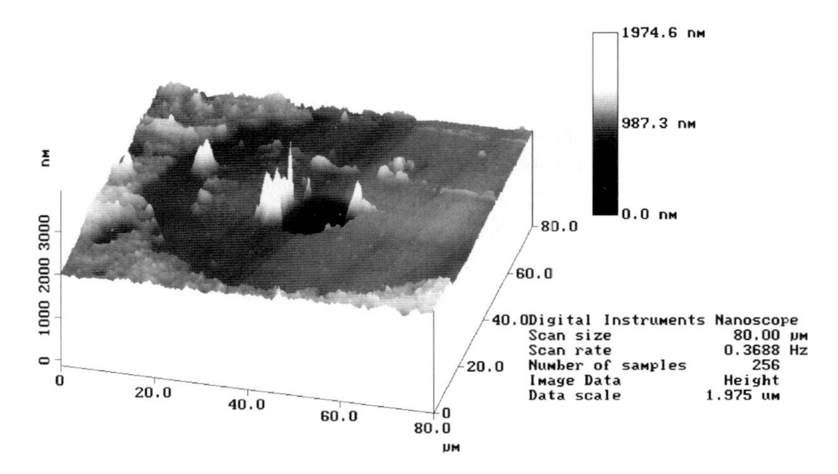

Fig. 12 Corrosion pit on a stainless steel surface covered with biomineralized manganese oxides and immersed in a 3.5% solution of NaCl

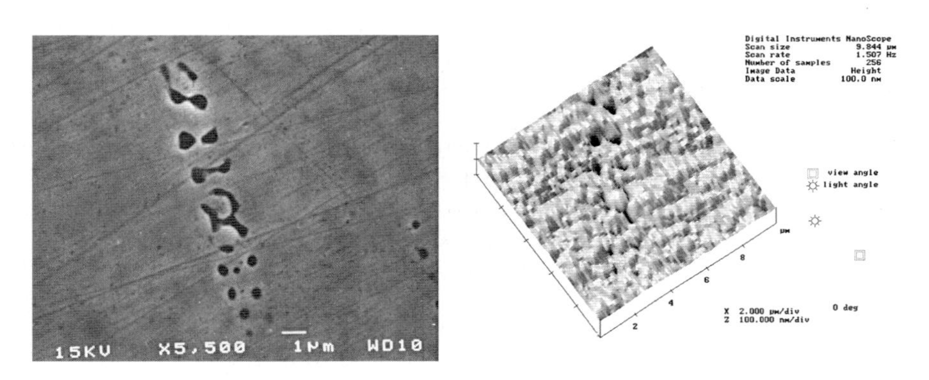

Fig. 13 Scanning electron microscopy and atomic force microscopy images of damage to a surface caused by colonization by manganese-oxidizing bacteria *L. discophora* SP-6 growing on 316L stainless steel surface. The size and shape of the indentations closely resemble the size and shape of the microorganism colonies on the surface (Geiser et al. 2002)

6 Summary and Conclusions

In conclusion, we have demonstrated that biofilms can influence the corrosion of metals (1) by metabolic reactions in the biofilms consuming oxygen, the cathodic reactant; (2) by controlling the mass transport of the corrosion reactants and products, therefore changing the kinetics of the corrosion process; (3) by generating corrosive substances; and (4) by generating substances that serve as auxiliary cathodic reactants.

These interactions do not exhaust the possible mechanisms by which biofilm microorganisms may affect the corrosion of metals; rather, they represent those few instances in which we understand the mechanism from the thermodynamic point of view. In addition, we can use electrochemical and chemical measurements to detect one or more of their products. Other mechanisms implicated in MIC involve bacteria that produce corrosive metabolites. For example *Thiobacillus thiooxidans* produces sulfuric acid and *Clostridium aceticum* produces acetic acid. These two metabolic products dissolve the passive layers of oxides deposited on the metal surface, which accelerates the cathodic reaction rate (Borenstein 1994). Other mechanisms may be initiated by hydrogen-generating microorganisms causing hydrogen embrittlement of metals or by iron-oxidizing bacteria, such as *Gallionella*. An important aspect of quantifying these mechanisms is to demonstrate exactly how they interfere with the corrosion processes. There is no universal mechanism of MIC. Instead, many mechanisms exist and some of them have been described and quantified better than others.

It does not seem reasonable to search for universal mechanisms, but it does seem reasonable to search for evidence of specific, well-defined microbial involvement in corrosion processes. For example, demonstrating the presence of elemental sulfur in the corrosion of mild steel can be considered evidence of SRB corrosion, and demonstrating the presence of manganese oxides in the corrosion of stainless steel can be considered evidence of MOB corrosion. However, even in these examples there is a possibility that some aspects of microbial participation escape our attention. The deposition of manganese oxides is easy to demonstrate on stainless steels or other passive metals because they are stable on such surfaces. However, if MOB deposit manganese oxides on mild steel where the oxides are reduced at the same rate as they are deposited, the corrosion rate may increase without the evidence of microbial participation in the process, the deposits of manganese oxides, being detectable.

Acknowledgments This work was partially supported by the United States Office of Naval Research (contract nos. N00014-99-1-0701 and N00014-06-1-0217). Beyenal was supported by Washington State University (fund no. 9904) and 3M.

References

Acuna N, Ortega-Morales BO, Valadez-Gonzalez A (2006)Biofilm colonization dynamics and its influence on the corrosion resistance of austenitic UNSS31603 stainless steel exposed to Gulf of Mexico seawater. *Mar Biotechnol* 8:62–70

Al Darbi MM, Agha K, Islam MR (2005) Comprehensive modelling of the pitting biocorrosion of steel. *Can J Chem Eng* 83:872–881

Amaya H, Miyuki H (1994) Mechanism of microbially influenced corrosion on stainless-steels in natural seawater. *Journal of the Japan Institute of Metals* 58:775–781

Antony PJ, Chongdar S, Kumar P, Raman R (2007) Corrosion of 2205 duplex stainless steel in chloride medium containing sulfate-reducing bacteria. *Electrochim Acta* 52:3985–3994

ASM Handbook Series (1987) Corrosion: Materials. ASM International

Beech IB, Gaylarde CC (1999) Recent advances in the study of biocorrosion – An overview. *Rev Microbiol* 30:177–190

Beech IB, Sunner JA, Hiraoka K (2005) Microbe–surface interactions in biofouling and biocorrosion processes. *Int Microbiol* 8:157–168

Beech WB, Sunner J (2004) Biocorrosion: Towards understanding interactions between biofilms and metals. *Curr Opin Biotechnol* 15:181–186

Bolwell R (2006) Understanding royal navy gas turbine sea water lubricating oil cooler failures when caused by microbial induced corrosion ("SRB"). *J Eng Gas Turbines Power – Trans ASME* 128:153–162

Borenstein SB (1994) microbiologically influenced corrosion handbook. Industrial Press, New York

Braughton KR, Lafond RL, Lewandowski Z (2001) The influence of environmental factors on the rate and extent of stainless steel ennoblement mediated by manganese-oxidizing biofilms. *Biofouling* 17:241–251

Caspi R, Tebo BM, Haygood MG (1998) c-Type cytochromes and manganese oxidation in *Pseudomonas putida* MnB1. *Appl Environ Microbiol* 64:3549–3555

Chandrasekaran P, Dexter SC (1993) Mechanism of potential ennoblement on passive metals by seawater biofilms (paper no. 493). CORROSION/93, NACE International, Houston, TX

Coetser SE, Cloete TE (2005) Biofouling and biocorrosion in industrial water systems. *Crit Rev Microbiol* 31:213–232

Costello JA (1974) Cathodic depolarization by sulfate-reducing bacteria. *S Afr J Sci* 70:202–204

Crolet JL (1991) From biology and corrosion to biocorrosion. In: Sequeira CAC, Tiller AK (eds.) Proceedings of the 2nd EFC workshop on Microbial Corrosion. HMSO, London, pp 50–60

Crolet JL (1992) From biology and corrosion to biocorrosion. *Oceanol Acta* 15:87–94

Dexter SC, Gao GY (1988) Effect of seawater biofilms on corrosion potential and oxygen reduction on stainless steels. *Corrosion* 44:717

Dickinson WH, Lewandowski Z (1996) Manganese biofouling and the corrosion behavior of stainless steel. *Biofouling* 10:79–93

Dickinson WH, Lewandowski Z (1998) Electrochemical concepts and techniques in the study of stainless steel ennoblement. *Biodegradation* 9:11–21

Dickinson WH, Caccavo F, Lewandowski Z (1996a) The ennoblement of stainless steel by manganic oxide biofouling. *Corros Sci* 38:1407–1422

Dickinson WH, Lewandowski Z, Geer RD (1996b) Evidence for surface changes during ennoblement of type 316L stainless steel: Dissolved oxidant and capacitance measurements. *Corrosion* 52:910–920

Dickinson WH, Caccavo F, Olesen B, Lewandowski Z (1997) Ennoblement of stainless steel by the manganese-depositing bacterium *Leptothrix discophora*. *Appl Environ Microbiol* 63:2502–2506

Eashwar M, Maruthamuthu S (1995) Mechanism of biologically produced ennoblement – Ecological perspectives and a hypothetical model. *Biofouling* 8:203–213

Finster K, Liesack W, Thamdrup B (1998) Elemental sulfur and thiosulfate disproportionation by *Desulfocapsa sulfoexigens* sp nov, a new anaerobic bacterium isolated from marine surface sediment. *Appl Environ Microbiol* 64:119–125

Flemming HC (1995) Biofouling and biocorrosion – Effects of undesired biofilms. *Chem Ing Tech* 67:1425–1430

Flemming HC, Wingender J (2001) Relevance of microbial extracellular polymeric substances (EPSs) – Part II: Technical aspects. *Water Sci Technol* 43:9–16

Ford T, Mitchell R (1990) The ecology of microbial corrosion. *Adv Microb Ecol* 11:231–262

Francis CA, Tebo BM (2002) Enzymatic manganese(II) oxidation by metabolically dormant spores of diverse *Bacillus* species. *Appl Environ Microbiol* 68:874–880

Geiser M, Avci R, Lewandowski Z (2002) Microbially initiated pitting on 316L stainless steel. *Int Biodeterior Biodegrad* 49:235–243

Hakkarainen TJ (2003) Microbiologically influenced corrosion of stainless steels – What is required for pitting? *Mater Corros/Werkstoffe Korrosion* 54:503–509

Hamilton WA (2003) Microbially influenced corrosion as a model system for the study of metal microbe interactions: A unifying electron transfer hypothesis. *Biofouling* 19:65–76

Hernandez G, Pedersen A, Thierry D, Hermansson M (1990) Bacterial effects of corrosion of steel in seawater. In: Dowling NJ, Mittelman MW, Danko JC (eds) Proceedings – Microbially influenced corrosion and biodeterioration, University of Tennessee, Knoxville

Hernandez G, Kucera V, Thierry D, Pedersen A, Hermansson M (1994) Corrosion inhibition of steel by bacteria. *Corrosion* 50:603–608

Herro HM (1991) Tubercle formation and growth on ferrous alloys (paper no. 84). NACE, Cincinnati, Ohio

Hossain MA, Das CR (2005) Kinetic and thermodynamic studies of microbial corrosion of mild steel specimen in marine environment. *J Indian Chem Soc* 82:376–378

Ilhan-Sungur E, Cansever N, Cotuk A (2007) Microbial corrosion of galvanized steel by a freshwater strain of sulphate reducing bacteria (*Desulfovibrio* sp.). *Corros Sci* 49:1097–1109

Javaherdashti RA (1999) Review of some characteristics of MIC caused by sulfate-reducing bacteria: Past, present and future. *Anti-Corros Methods Mater* 46:173–180

Jayaraman A, Ornek D, Duarte DA, Lee CC, Mansfeld FB, Wood TK (1999) Axenic aerobic biofilms inhibit corrosion of copper and aluminum. *Appl Microbiol Biotechnol* 52:787–790

Lee AK, Newman DK (2003) Microbial iron respiration: Impacts on corrosion processes. *Appl Microbiol Biotechnol* 62:134–139

Lee W, Lewandowski Z, Nielsen PH, Hamilton WA (1995) Role of sulfate-reducing bacteria in corrosion of mild-steel – A review. *Biofouling* 8:165–194

Lewandowski Z, Beyenal H (2007) Fundamentals of biofilm research. CRC

Lewandowski Z, Dickinson W, Lee W (1997) Electrochemical interactions of biofilms with metal surfaces. *Water Sci Technol* 36:295–302

Lewandowski Z, Beyenal H, Stookey D (2004) Reproducibility of biofilm processes and the meaning of steady state in biofilm reactors. *Water Sci Technol* 49:359–364

Linhardt P (1996) Failure of chromium-nickel steel in a hydroelectric power plant by manganese-oxidizing bacteria. In: Heitz E, Flemming HC, Sand W (eds.) Microbially influenced corrosion of materials. Springer Verlag, Berlin Heidelberg, pp 221–230

Linhardt P (1998) Electrochemical identification of higher oxides of manganese in corrosion relevant deposits formed by microorganisms. Electrochemical Methods in Corrosion Research VI, Pts 1 & 2. Book Series: Materials Science Forum, Vol. 289–2:1267–1274

Linhardt P (2004) Microbially influenced corrosion of stainless steel by manganese oxidizing microorganisms. *Mater Corros/Werkstoffe Korrosion* 55:158–163

Linhardt P (2006) MIC of stainless steel in freshwater and the cathodic behaviour of biomineralized Mn-oxides. *Electrochim Acta* 51:6081–6084

Little B, Ray R (2002) A perspective on corrosion inhibition by biofilms. *Corrosion* 58:424–428

Little B, Lee J, Ray R (2007) A review of 'green' strategies to prevent or mitigate microbiologically influenced corrosion. *Biofouling* 23:87–97

Little BJ, Ray RI, Pope RK (2000) Relationship between corrosion and the biological sulfur cycle: A review. *Corrosion* 56:433–443

Mansfeld F, Little BA (1991) Technical review of electrochemical techniques applied to microbiologically influenced corrosion. *Corros Sci* 32:247–272

Mattila K, Carpen L, Hakkarainen T, Salkinoja-Salonen MS (1997) Biofilm development during ennoblement of stainless steel in Baltic Sea water: A microscopic study. *Int Biodeterior Biodegrad* 40:1–10

Miyanaga K, Terashi R, Kawai H, Unno H, Tanji Y (2007) Biocidal effect of cathodic protection on bacterial viability in biofilm attached to carbon steel. *Biotechnol Bioeng* 97: 850–857

Mollica A, Trevis A (1976) Correlation entre la formation de la pellicule primaire et la modification de le cathodique sur des aciers inoxydables experimentes en eau de mer aux vitesses de 0.3 A, 5.2 m/s. Proceedings of 4th international congress on marine corrosion and fouling, Juan Les-Pins, France, 14–18 June 1976

Nielsen P, Lee WC, Morrison M, Characklis WG (1993) Corrosion of mild steel in an alternating oxic and anoxic biofilm system. *Biofouling* 7:267–284

Olesen BH, Avci R, Lewandowski Z (2000a) Manganese dioxide as a potential cathodic reactant in corrosion of stainless steels. *Corros Sci* 42:211–227

Olesen BH, Nielsen PH, Lewandowski Z (2000b) Effect of biomineralized manganese on the corrosion behavior of C1008 mild steel. *Corrosion* 56:80–89

Olesen BH, Yurt N, Lewandowski Z (2001) Effect of biomineralized manganese on pitting corrosion of type 304L stainless steel. *Mater Corros/Werkstoffe Korrosion* 52:827–832

Rao TS, Sairam TN, Viswanathan B, Nair KVK (2000) Carbon steel corrosion by iron oxidising and sulphate reducing bacteria in a freshwater cooling system. *Corros Sci* 42:1417–1431

Roe FL, Lewandowski Z, Funk T (1996) Simulating microbiologically influenced corrosion by depositing extracellular biopolymers on mild steel surfaces. *Corrosion* 52:744–752

Romero JM, Angeles-Chavez C, Amaya M (2004) Role of anaerobic and aerobic bacteria in localised corrosion: Field and laboratory morphological study. *Corros Eng Sci Technol* 39:261–264

Schmitt G (1991) Effect of elemental sulfur on corrosion in sour gas systems. *Corrosion* 47:285–308

Sedriks AJ (1996) Corrosion of stainless steel. Wiley, New York

Shi X, Avci R, Lewandowski Z (2002a) Microbially deposited manganese and iron oxides on passive metals – Their chemistry and consequences for material performance. *Corrosion* 58:728–738

Shi XM, Avci R, Lewandowski Z (2002b) Electrochemistry of passive metals modified by manganese oxides deposited by *Leptothrix discophora*: Two-step model verified by ToF-SIMS. *Corros Sci* 44:1027–1045

Starosvetsky J, Starosvetsky D, Armon R (2007) Identification of microbiologically influenced corrosion (MIC) in industrial equipment failures. *Eng Fail Anal* 14:1500–1511

Tebo BM, Ghiorse WC, van Waasbergen LG, Siering PL, Caspi R (1997) Bacterially mediated mineral formation: Insights into manganese(II) oxidation from molecular genetic and biochemical studies. Geomicrobiology: Interactions between Microbes and Minerals. Book Series: Reviews in Mineralogy, 35:225–266

Tebo BM, Bargar JR, Clement BG, Dick GJ, Murray KJ, Parker D, Verity R, Webb SM (2004) Biogenic manganese oxides: Properties and mechanisms of formation. *Annu Rev Earth Planet Sci* 32:287–328

Tebo BM, Johnson HA, McCarthy JK, Templeton AS (2005) Geomicrobiology of manganese(II) oxidation. *Trend Microbiol* 13:421–428

Videla HA (2001) Microbially induced corrosion: An updated overview (reprinted). *Int Biodeterior Biodegrad* 48:176–201

Videla HA, Herrera LK (2005) Microbiologically influenced corrosion: Looking to the future. *Int Microbiol* 8:169–180

Wang W, Wang J, Xu H, Li X (2006) Some multidisciplinary techniques used in MIC studies. *Mater Corros/Werkstoffe Korrosion* 57:531–537

Washizu N, Katada Y, Kodama T (2004) Role of H2O2 in microbially influenced ennoblement of open circuit potentials for type 316L stainless steel in seawater. *Corros Sci* 46:1291–1300

White DC, de Nivens PD, Nichols J, Mikell AT, Kerger BD, Henson JM, Geesey G, Clarke CK (1985) Role of aerobic bacteria and their extracellular polymers in facilitation of corrosion: Use of Fourier transforming infrared spectroscopy and "signature" phospholipid fatty acid analysis. In: Dexter SC (ed.) Biologically induced corrosion. NACE, Houston, p 233

Xu CM, Zhang YH, Cheng GX, Zhu WS (2007) Localized corrosion behavior of 316L stainless steel in the presence of sulfate-reducing and iron-oxidizing bacteria. *Mater Sci Eng A Struct Mater: Properties Microstruct Process* 443:235–241

Zuo RJ, Kus E, Mansfeld F, Wood TK (2005) The importance of live biofilms in corrosion protection. *Corros Sci* 47:279–287

Industrial Biofilms and their Control

P. Sriyutha Murthy (✉) and R. Venkatesan

Abstract Biofilms are considered to be ubiquitous in industrial and drinking water distribution systems. Biofilms are a major source of contribution to biofouling in industrial water systems. The problem has wide ranging effects, causing damage to materials, production losses and affecting the quality of the product. The problem of biofouling is operationally defined as biofilm development that exceeds a given threshold of interference. It is for the plant operators to keep biofilm development below the threshold of interference for effective production and to work out values for threshold limits for each of the technical systems. Industrial biofilms are quite diverse and knowledge gained with a certain type of biofilm may not be applicable to others. In recognition of this, the old concept of a universal/effective biocide is a misnomer as physical, chemical and biological parameters of source water vary from site to site and so do the interactions of biocides with these parameters. Control methods have to be tailor-made for a given technical system and cannot be extrapolated. Because of the wide-ranging complexity in industrial technical systems, understanding the biofilm processes, detection, monitoring, control and management is imperative for efficient plant operation. A successful antifouling strategy involves prevention (disinfecting regularly, not allowing a biofilm to develop beyond a given threshold), killing of organisms and cleaning of surfaces. Killing of organisms does not essentially imply cleaning as most industrial systems deploy only biocides for killing, and the cleaning process is not achieved. Cleaning is essential as dead biomass on surfaces provide a suitable surface and nutrient source for subsequent attachment of organisms. A first step in a biofilm control programme is detection and assessment of various biofilm components, like thickness of slime layer, algal and bacterial species involved, extent of extracellular polymeric substances and inorganic components. Prior to adopting a biocidal dose and regime in an industrial system, laboratory testing of biocides using side-stream monitoring devices, under dynamic conditions, should be carried out to check their effectiveness. Online monitoring strategies should be adopted and biocidal

P.S. Murthy
Biofouling and Biofilm Processes Section, Water and Steam Chemistry Division, BARC Facilities, Indira Gandhi Center for Atomic Research Campus Kalpakkam, 603 102, India
e-mails: psm_ murthy@yahoo.co.in, psmurthy@igcar.gov.in

Springer Series on Biofilms, doi: 10.1007/7142_2008_18

dosing fine-tuned to keep biofilms under control. Literature on biofilm control strategies in technical systems is rich; however, the choice of the control method often depends on cost, time constraints and the cleanliness (threshold levels) required for a technical process. Currently, there is a trend to use strong oxidizing biocides like chlorine dioxide in cooling systems and ozone in water distribution systems as low levels of chlorine have been found to be ineffective against biofilms. A number of non-oxidizing biocides are available, which are effective but the long-term effects on the environment are still unclear. New techniques for biofilm control like ultrasound, electrical fields, hydrolysis of extracellular polymeric substances and methods altering biofilm adhesion and cohesion are still in their infancy at the laboratory level and are yet to be successfully demonstrated in large industrial systems.

1 Introduction

Water drawn from natural sources is the main industrial coolant for dissipating waste heat from heat exchangers and process systems. Use of "pure" water would not eliminate biofouling problems as pure water systems still contain traces of organic carbon and, thus, also face problems due to biofouling. Apart from this, desalination plants also face biofouling problems related to accumulation of biofilms on pipe and membrane surfaces. The problems due to fouling by biofilms are more pronounced in the open and closed freshwater recirculating systems of power plants and to an extent on desalination membranes, hence problems in these systems are discussed in this chapter. However, with regard to the control of biofilms, experiences in pure water distribution systems are also discussed here as the principles and approaches are similar and they share a common goal, i.e. eliminating biofilms.

The events leading to deterioration of surfaces are:

1. Natural waters contain a large number of macromolecules released by breakdown of dead organisms. These substances adsorb onto submerged surfaces constituting a primary film (Busscher et al. 1995).
2. Initially bacteria are attracted towards this surface and are held to the substratum by weak electrostatic forces, hydrogen bonds and van der Waals interactions (Busscher et al. 1995).
3. As the bacteria grow, extracellular polymeric substances (EPS) are produced and accumulate so that the bacteria are eventually embedded in a highly hydrated matrix (Christensen and Characklis 1990; Flemming 2002). The polymeric material is largely composed of polysaccharides, proteins, nucleic acids and lipids (Flemming and Wingender 2002). It is frequently believed to represent a diffusion barrier; however, this is not the case for small molecules such as biocides as the main component of the EPS matrix is water. Therefore, the diffusion coefficients of such molecules are very close to those in free water (Christensen and Characklis 1990) unless these molecules do interact with matrix components. This effect is called diffusion–reaction limitation(Gilbert et al. 2001). The production of EPS

provides adhesion to the substrate and matrix cohesion and, thus, increases the mechanical stability of biofilms.
4. Subsequently, diatoms and microalgae colonize the substratum and the biofilm grows in thickness, further entrapping nutrients from bulk water (Flemming and Leis 2002).

Control of biofilms in industrial systems is an important component of a successful water treatment programme (Ludensky 2003). Theoretical approaches consider that the primary step in biofilm inhibition is to prevent the initial adhesion of microorganisms (Busscher et al. 1999). However, in practice, this does not work because sooner or later surfaces in technical systems will eventually be colonized. Biofilms serve as a source for production and release of microbial cells, which influences microbial levels in the water column. Codony et al. (2005) reported an interesting observation to this effect: intermittent chlorination resulted in a tenfold increase in the release of microbial cells to the water phase in the absence of biocide. Hence, it becomes more important to control biofilms. Routine monitoring procedures assess the presence of planktonic bacteria, whereas the vast majority of bacteria indigenous to aquatic environments exist attached to solid particles or industrial surfaces and go unnoticed. A particular biocide may inactivate more than one type of microorganism. With our current levels of understanding of the mechanisms of biocidal action and of microbial resistance it is pertinent to consider whether it is possible to explain why some biocides are effective while others are not. The factors that affect antimicrobial activity most are contact time, concentration, temperature, pH, the presence of organic matter and the type of microorganism. Hence, comparative assessments of different biocides are somewhat difficult. For industrial operations, system size, cleanliness, service schedules and monitoring programmes are important factors governing smooth operations.

2 Factors Influencing Biofilm Development in Industrial Systems

Industrial biofilms are quite diverse due to a wide range of contributing factors such as microbial species, temperature, nutrient availability, velocity, substratum physical and chemical characteristics, organic loading, suspended solids and general water chemistry. Therefore it is difficult to generalize about the types of biofilm that form in these systems, let alone about their control methods.

2.1 Temperature

Growth of biofilm and of species colonizing a biofilm is dependent on the operating temperatures of industrial equipment. An increase in temperature is found to favour biofilm growth. Even a small change in temperature (5°C) can cause an increase in biofilm thickness (Bott and Pinheiro 1977). In heat exchangers, raising the temperature above design value can increase the rate of corrosion, rate of chemical reactions and, for inverse solubility salts, raising the temperature might initiate deposition (Bott 1995). Heat transfer surfaces (titanium, admiralty or aluminium brass, and cupro-nickel 90:10) in an industrial heat exchanger generally experience temperatures in the range 28–45°C for auxiliary cooling systems and 60–70°C for condenser cooling, where bacterial biofilms have been shown to occur (Nebot et al. 2007).

2.2 Nutrient Availability

The basic mechanism of biofilm development involves the conversion of dissolved nutrients into accumulated biomass. Griebe and Flemming (1998) considered bio-fouling as a "biofilm reactor in the wrong place" because the same laws apply to both cases. The major factor controlling biofilm growth is nutrient availability. In industrial and drinking water systems, mass transfer of nutrients to the biofilm will tend to increase with flow velocity (Characklis 1990). The rough surfaces of bio-films also aid in increased mass transfer of nutrients by as much as threefold compared to a smooth surface (Characklis and Marshall 1990; Bott and Gunatillaka 1983). Nutrient limitation may be one way to control biofilm development without increasing disinfectant dosing in potable water distribution systems (Griebe and Flemming 1998; Chandy and Angles 2001; Flemming 2002). Adsorption of macromolecular substances increases their availability to bacteria. Industrial cooling systems offer a continuous flow of fresh water bringing in nutrients. A 400% increase in biofilm thickness was observed at a given velocity of 1.2 m s^{-1} for an increase in nutrient level from 4 mg L^{-1} to 10 mg L^{-1} (Melo and Bott 1997). Removal of organic carbon resulted in greater persistence of chlorine (Chandy and Angles, 2001).

Treatment of water to reduce the organic load is a non-viable option for power plants as once-through seawater cooling systems on an average have an intake capacity of 30 m^3 s^{-1} (for 500 MW(e) plants) and freshwater recirculating systems have a circulation rate of 80–120 m^3 h^{-1} with an intake capacity of 10 m^3 h^{-1}. However, this factor is included in this section in order to have a measure of the influence of nutrient concentration on biofilm thickness and density, which have direct implications in biocidal efficacy by reacting with the biocide dosed and neutralizing it. The method of reducing the organic load and, thus, limiting nutrients has been suggested and during the last few years has become more and more accepted in practice as a viable alternative for membrane desalination plants, as the feed water is devoid of biocides to protect the reverse osmosis membranes(Griebe and Flemming 1998; Flemming 2002). The option is viable as these plants require far less quantity of water (intake) and it may prove economical considering the consequences to membranes of biofilms and considering the treatment (organic load removal) required to meet the quality standards of permeate water, involving infrastructure facilities like coagulation chambers, activated carbon adsorption and cartridge filtration for reducing organic load.

2.3 Flow Velocity

In flowing systems, bacterial populations exist as complex, structurally heterogeneous biofilms attached to surfaces. Residence within these complex matrices provides organisms with a higher localized nutrient concentration than that found in normal waters. In the case of heat exchangers, biofilm growth can be controlled if relatively high velocities are imposed, as shear effects are likely to have an impact on biofilm development. Operating at high velocities to achieve increased shear forces also results in erosion of material surface and hence results in increased damage. An optimum shear force and temperature for minimal adhesion is yet to be worked out specifically for heat exchanger operation. Biofilms have been described as a viscoelastic material with plastic flow properties (Korstgens et al. 2001), based on their response to the modulus of elasticity and yield strength. The viscoelastic property of biofilms makes them mechanically stable and also enables them to resist detachment (Rupp et al. 2005). The EPS functions as a network of temporary junction points and yield points, which above a certain threshold results in failure of the gel system resulting in a highly viscous fluid (Korstgens et al. 2001). Hence it would be of practical importance to obtain data on the flow velocities required to either detach or induce such effects. Flow velocities of water in pure and cooling water systems govern the development of biofilms, their density and have important implications with respect to penetration of biocides.

Studies by Pujo and Bott (1991) have shown that the Reynolds number seems to have a profound effect on biofilm thickness. For a given Reynolds numberof 11,000 and fixed nutrient conditions, a velocity of 0.5 m s^{-1} generated biofilms ten times thicker than at a velocity of 2 m s^{-1} over a period of 15 days. An increase in Reynolds number increased biofilm removal (24%), but total biofilm removal was not found for all conditions (Simoes et al. 2005a) suggesting that biofilms were more mechanically stable to shear forces. Treatment of biofilms with chemicals and surfactants like cetyltrimethyl ammonium bromide (CTAB), ortho-phthalaldehyde (OPA), sodium hydroxide and sodium hypochlorite promoted weakening of biofilm mechanical stability (Simoes et al. 2005a). Similarly, velocity is also known to affect biofilm density. Experiments with unispecies *P. fluorescens* biofilms showed that an increase in velocity from 0.1 to 0.5 m s^{-1} resulted in an increase in density of biofilm from 26 kg m^{-3} to 76 kg m^{-3} (dry mass/wet volume) (Pinheiro et al. 1988). Qualitative analysis of flow effects on biofilms grown from tap water at different velocities showed that under laminar conditions biofilms were patchy and consisted of cell clusters separated by interstitial voids. In contrast, biofilms developed under turbulent flow were found to be filamentous (Stoodley et al. 1999). In flowing systems, bacteria can adapt rapidly to hydrodynamic and chemical stresses (Suci et al. 1998) and sessile cells are known to undergo complex physiological changes during the process of attachment (Sauer and Camper 2001), which reduce their susceptibility to control measures (Cloete et al. 1997; Gilbert et al. 2002).

Another factor of importance in industrial systems is shear stress on the substratum caused by flowing water. High shear forces at the substratum result in (1) increased flux of nutrients at the surface, (2) increased transport of disinfectants to the surface, (3) a greater shearing of biofilms (Percival et al. 2000) and (4) altered biofilm diversity (Rickard et al. 2004). An increase in flow velocities resulted in re-suspension of biofilms and sediments in water from pipe surfaces (laboratory study), which increased particle and turbidity counts in bulk fluid (Lehtola et al. 2006). The consequences of release of biofilm clumps from surfaces are beneficial in once-through systems where the biofilm load decreases, whereas in recirculatory and drinking water systems they pose problems of bacterial regrowth and suspension of toxic metals from the surface to bulk water. However, recent studies by Tsai (2006) showed that shear stress (0.29 N m^{-1}) and chlorination had no interaction on biofilm formation, reinstating findings of an earlier study by Peyton (1996), who observed no significant effects of flow rate on biofilm thickness. A probable reason for the observed effect in these studies is that the shear stress achieved in these studies was inadequate to remove biofilms.

It is necessary to arrive at shear stress values for biofilm removal on a variety of surfaces. Studies by Cloete et al. (2003) showed that high velocities of 3–4 m s^{-1} were required to detach biofilms from surfaces. Alternatively, fouling deposition was found to occur at a slow rate when a nominal flow velocity of (1.85 m s^{-1}) was maintained in the heat exchanger tubes (Nebot et al. 2007). Increasing the velocity regime may offer some relief from the problem of biofilms in water distribution pipelines but with respect to heat exchangers, increased velocity would increase the overall heat transfer coefficient (Bott 1995). This would mean additional surface area and increased capital costs. Further increase in velocity increases the pressure drop(i.e. pressure drop is the square of the velocity) (Bott 1995). Hence, the use of flow velocity to prevent biofilm formation is not a viable option for heat exchangers and industrial circuits because of technical problems and energy consumption. In addition, the role of velocity effects on biofilm formation is yet to be clearly understood and a clear distinction between the two contrasting schools of thought, viz: shearing effects/biofilm stability, needs to be investigated to improve our understanding of using flow velocity as a biofilm control method.

2.4 Substratum Physical and Chemical Characteristics

The type of substratum has a pronounced effect on biofilm accumulation. Smooth surfaces accumulate less biofilm mass than rough surfaces. The mechanism behind this is that individual cells are much smaller than crevices (Bott 1999) and an irregular rough surface would offer protection for cells from shear effects. However, such surface irregularities have a measurable effect only during the initial stages of biofilm development and biofilms are unavoidable in distribution systems (Veeran and Hissett 1999). When biofilm thickness exceeds roughness dimensions, roughness will no longer be of influence for biofilm accumulation; however, it will assist in better anchoring them to surfaces. Vieira et al. (1992) have shown that biofilms of *P. fluorescens* were more pronounced on aluminium plates than on brass and copper. Similarly, more biofilms were observed on polyethylene pipes than on copper pipes (Lehtola et al. 2006). This is commonly attributed to the toxic effects of copper and brass on microorganisms. However, in industrial situations, heat transfer surfaces of copper, brass and cupronickel alloys have all been shown to accumulate biofilms. Titanium heat exchanger tubes were shown to accumulate more fouling than brass tubes (Nebot et al. 2007). From the literature, it is understood that no single surface escapes fouling and that it is impossible to create smooth industrial surfaces as the surface roughness of materials used in industries is dependent on the manufacturing process. Low surface energy coatings, which are characterized by low adhesion forces of the biofilm to the surface (Busscher and van der Mei 1997), could offer some protection for structural materials like pipelines, whereas in heat exchangers chemical control methods are the only alternative.

2.5 Suspended Solids

Industrial cooling water drawn from natural sources (seawater or freshwater) contains common particulate material like sand, silt, clay or quartz and to a certain extent metal oxides resulting from the corrosion of equipment upstream. Although in industrial systems the presence of suspended particles is common, studies on their interaction with biofilms are limited. Deposition of these particles onto surfaces from suspension flows is found to occur in consecutive steps. The presence of particles in suspension influences biofilm growth by: (1) increasing the availability of nutrients to microorganisms, directly influencing their metabolism, (2) the erosion effects of particles, resulting in removal or suppression of biofilm formation and (3) the presence of biofilm enhances the capture of particulate matter from flowing systems, increasing accumulation on surfaces (Bott and Melo 1992). These mechanisms can be observed and are dependent on the shear force and size of the particles. Particulate material in flowing water influences biofilm thickness and growth. If the particle sizes are large, this results in a sloughing effect on the biofilm whereas smaller particles are known to embed within biofilms (Lowe et al. 1988).

In general, to ensure maintaining biofilms within the required threshold limits in industrial circuits, the following are necessary: operating industrial systems at velocities higher than 2–3 m s^{-1}, without additional pumping cost or erosion problems; operating at minimum (ambient) temperatures; avoiding large open sunlit areas; use of appropriate materials and surface coatings with a smooth finish; a proper biocidal and cleaning programme.

3 Problems Associated with Biofilms and Their Control in Industrial Systems

3.1 Heat Exchangers and Cooling Water Systems

In cooling water circuits, the presence of biofilms can restrict flow in pipelines (Bott 1999), decrease heat transfer in heat exchangers, increase pressure drop (Bott 1994; Characklis and Marshall 1990), enhance corrosion (Bott 1995) and alter surface roughness, which in turn can increase fluid frictional resistance resulting in decreased flow and act as a source of contamination (Camper 1993).

Two main problems encountered in heat exchanger systems due to fouling by biofilms are reduction in heat transfer (loss of thermal efficiency) and pressure drop across the heat exchangers due to flow reduction by deposits (Characklis 1990). The restrictions to flow imposed by the presence of biofilm deposits in heat exchanger surfaces increases fluid frictional resistance and, for a given throughput, the velocity will have to increase, which means additional pumping costs. In addition, the presence of biofilms may accelerate corrosion of materials in contact. Other operating costs may accrue from the presence of biofilm deposits, such as increased maintenance requirement and unplanned shutdowns for cleaning. As a result of these factors, the engineering design of heat exchangers usually incorporates allowances for fouling to accommodate a more satisfactory annual cleaning schedule.

Recirculating systems (Fig. 1) are usually located at sites where adequate water is not available for cooling purposes. In open recirculating systems, cooling water drawn from the source (usually a freshwater body) is circulated through a heat exchanger and is conveyed to a cooling tower where evaporation of some of the water results in a cooling effect and lowering of the cooling water temperature for

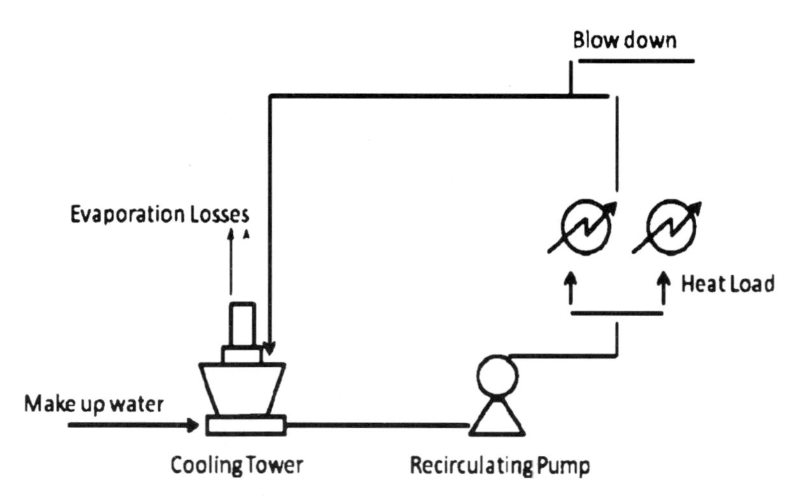

Fig. 1 General schematic of an industrial recirculatory cooling system

further recirculation. After passage through the cooling towers, the water is held in a temporary open reservoir where algal and bacterial growth occurs. In recirculatory systems, both open and closed makeup water is added to compensate for the evaporative losses as well as to maintain the quality of recirculatory water. The conductivity of recirculatory water increases due to concentration of salts on evaporation. This is usually measured as cycles of concentration. Usually, plants operate at two to three cycles of concentration, as an increase of cycles of concentration above four usually results in enhanced scaling and corrosion of equipment.

In open recirculating systems, the problems to be encountered are many as these systems are large (with a resident water of 60–80,000 m^3 for a 1,000 MW(e) power plant). Large open areas and available nutrients in the recirculating water provide adequate conditions for enhanced growth of algal species, resulting in eutrophication. This further leads to organic loading in the system as detrital matter accumulates. Further, the incoming makeup water brings in fresh nutrients that are continuously recycled in the systems.

In closed recirculating systems, the principles are as the name implies, the cooling water is conveyed through pipelines to the heat exchangers and after passage through the cooling tower is recirculated. However, even in these systems it is inevitable to have an open storage point as large volumes of water are involved. Closed recirculatory systems are not preferred as large capital investments have to be made on infrastructure. Recirculating water systems are often designed with an average flow velocity through the condenser tubes in the range of 1.8–2.4 m s^{-1}. Small heat exchangers in the process systems have lower velocities in the range of 0.3–0.6 m s^{-1}, which are prone to fouling. Water filtration devices of various types are always installed in cooling water systems fed by natural waters. These generally consist of a band screen with a coarse grid (about 1–10 cm spacing) where the flow rate is lower than 10 m^3 s^{-1} or drums for higher flow rates. Specific debris filters are also used to protect heat exchangers from clogging. An overview concerning condenser cooling circuits is given in Table 1.

Cooling towers of both open and closed recirculating systems face severe problems due to algal and bacterial growth. Cooling towers represent complex ecological niches and even different towers of identical design on a single site will generally behave quite different microbiologically (Prince et al. 2002). Conventionally, the splash-type cooling tower has been used, in which the heated discharge from the condensers is ejected through fine nozzles from the top of the cooling towers. The discharge trickles down splash bars (either concrete or wood) and collects in the cooling tower basin from where it is pumped for recirculation. The disadvantages of these splash-type towers are their extremely large size and low thermal efficiency. This led to the development of high-performance forced or induced draft cooling towers where the water trickles down through high film fills (polyvinyl chloride) to the cooling tower basin. The high film fills are comprised of corrugated parallel plates with distances of 3–5 mm between the plates. The corrugations or chevron angles result in water being broken up into fine droplets or films by the extended surfaces of the film fills. The corrugation increases the surface area and has resulted in reducing the size of cooling towers. However, these high film fills

Table 1 Biocidal regimes practiced in industrial circuits for condenser cooling

	Concentration ($mg\ L^{-1}$)	Effect	Reference
Low level targeted Cl_2	> 0.2	Effective if targeted dosing is done at inlet to heat exchangers at EDF power station France	Jenner and Khalanski (1998)
Low level Cl_2	0.2 TRC	Pilot plant device at a 550 MW plant in Spain	Nebot et al. (2007)
Low level Cl_2	0.1 TRC	Effective against planktonic cells of lake water	Nebot et al. (2007)
Discontinuous Cl_2 (30 min every 12 h)	0.5–1.0	Ineffective	Ewans et al. (1992)
Discontinuous Cl_2 (for 1 h every 8 h)	3.0	Effective at EDF Martigues-Ponteau power station on Mediterranean coast	Jenner and Khalanski (1998)
Intermittent Cl_2 (4 h on/4 h off)	0.2–0.3	Effective against biofilms at Maasvlakte power station, Rotterdam	Jenner and Khalanski (1998)
Intermittent Cl_2 (30 min on/1 h off)	1.2	Required for biofilm control on plate heat exchangers	Murthy et al. (2005)
Targeted Cl_2	1.0	Recommended by EPRI for condenser slime control	
Chlorination (30 min day^{-1})	0.5	Effective for fouling control in Netherlands – KEMA	Jenner and Khalanski (1998)
Chlorine dioxide	0.05–0.1	With residual (1 h day^{-1}) or without residuals (10–12 h day^{-1}), effective for sea-water condenser cooling in Mediterranean coast	Petrucci and Rosellini (2005)
Ozone	0.1–0.15	Killing and detaching sessile cells. Followed in Hochst unit, Germany, fed with River Main water	Jenner and Khalanski (1998)

have been prone to both inorganic and biological fouling compared to conventional low fouling, splash bar fills where algal growth is the major problem to be overcome.

In large natural cooling towers, algae tend to develop in the following regions:

– The inner surface of the shell. The wet parts that are exposed to some sunlight become covered with a cyanobacterial and algal layer. Sloughing and detachment of algae during shutdowns leads to a great input of organic matter into the system.
– In the honeycomb-like packing structures of cooling tower fills. Exposure to sunlight and the slow flow of water ($0.2\ m\ s^{-1}$) are causal factors for growth of filamentous green algae and cyanobacteria where light has access.
– In the cooling tower basins and on concrete walls and pillars of the cooling tower.

Table 2 Biocidal regimes practiced in industrial cooling towers

Regime	Concentration (mg L^{-1})	Effect	References
Discontinuous shock chlorination	2.5	Effective in killing algae; inland power station CEGB, UK	Blank (1984)
Discontinuous mass chlorination	8.0	Exposures of 6 h were effective for killing algae	Lutz and Merle (1983)
Chlorine dioxide	1.5	Elimination of filamentous algae in cooling towers	
	0.3	Requires extended time for achieving similar results	Merle and Montanat (1980)
ACTIV-OX	0.2–0.8	Chlorine dioxide treatment effective against Legionella sp. in cooling towers	Harris (1999)

Generally the walls of the cooling tower basins are not protected. The biocide dosed in the water phase is not effective as the water does not trickle through the wall in forced/induced draft towers with film technology. As a result, thick layers of cyanobacteria develop on the wall and act as source for further contamination. Some of the cooling tower water containing the biocide may come in contact with the walls. This kills the outer layers of the encrusting algae, turning the filaments white, but does not penetrate into the deep layers of horizontal filaments adhering to the walls. When the dead filaments have been washed off, the horizontal filament system is once again exposed to the flow of cooling water and growth begins again. It is important that the walls of the cooling tower basins be treated with a suitable antifouling coating or foul release coating and are subjected to periodical cleaning by high-pressure water jet and disposal of the algal debris. This will ensure smooth operation of the towers.

Chlorine has been the most common biocide used in cooling towers. Biocidal regimes practised in cooling towers are listed in Table 2. Chlorine and copper salts have been used as popular methods for controlling bacterial growth in cooling towers (Fliermans et al. 1982). Chlorine (2–4 mg L^{-1}), silver ions (0.02–0.04 mg L^{-1}) and copper ions (0.2–0.4 mg L^{-1}) have been used for treating cooling towers (Chambers et al. 1962; Cassels et al. 1995; Pedahzur et al. 1997; States et al. 1998; Kusnetsov et al. 2001; Kim et al. 2002a, b). However, the use of metal ions for biofouling control should always take into consideration the development of resistant microbial populations (Schulte et al. 2005).

Legionella sp. is an important component in natural and artificial water environments, cooling towers, plumbing systems and evaporators of large air conditioning systems, and remains a health hazard. *Legionella* sp. is known to occur in biofilms in cooling towers, showers, humidifiers (Fields et al. 2002) and hence knowledge about its response to control measures is important. These Gram-negative aerobic rods have been shown to survive at temperatures of 20–50°C and are inactivated at temperatures above 70°C (Kim et al. 2002a) and in a pH range of 5.5–8.1.The organism is known to occur in stagnant warm water bodies (Sanden et al. 1989).

This aspect is important as power plant exhaust plumes are known sources of *Legionella* deposits. *Legionella* resident within biofilms are a severe problem in cooling tower systems using freshwater.

Several disinfection methods have been tried out. In the technical context, the term "disinfection"is usually not used in the proper sense of the definition (inactivation of infecting microorganisms) but rather as getting rid of microbial problems. Chemical treatments using chlorine were the most common and widely used. Free chlorine concentrations of 1 mg L^{-1} were required for killing planktonic cells whereas a fourfold increase in concentration was required to kill sessile cells (Kim et al. 2002a). An adaptive feature exhibited by *Legionella pneumophila* associated with biofilm protozoa showed that cells were found to be less susceptible to chlorine (residual of 0.5 mg L^{-1}) (Donlan et al. 2005). Resistance by *Legionella* biofilms was also observed for the organic compound chloramine T (*N*-chloro-*p*-toluene sulfonamide), obtained by chlorinating benzene sulfonamide or *para*-toluene, on planktonic and sessile cells (Ozlem et al. 2007). In cooling systems of power plants an organic compound 1-bromo-3-chloro-5,5-dimethylhydantoin (BCDMH)containing bromine as an active ingredient has been used to control *Legionella* (Kim et al. 2002a). Effective bromine concentrations were in the range 1.0–1.5 mg L^{-1}. However, a shock dose of 3–5 mg L^{-1} of ClO_2 for a period of 1 h was required to eliminate *Legionella* from dental chair water systems (Walker et al. 1995).

3.2 *Case Study: Microbial Fouling of Cooling Tower Fills in a Power Station*

The Talcher super thermal power station (TSTPS) located in the Eastern state of Orissa, India has six units each of 500 MW(e) capacities. The plant operates on an open recirculatory mode with a residence volume of 3,600,000 m^3 h^{-1} of cooling water and a makeup water of 10,000 m^3 h^{-1}. Cooling water comes from the perennial rivers Bahmini, Trika and Singaraj, which converge to form the "Triveni Sangam" from which water is drawn and transported through underground pipelines for approximately 10 km before it reaches the recirculation system. Prior to entering the recirculation system of the plant, the water is aerated and biocide (chlorine) is added before the water is softened using alum. The pH drop after the addition of softening agent (alum) is revived by addition of lime (calcium). This results in a pH value of 8.2–8.3 in the cooling water system. The water is then clarified by removing suspended solids and reaches the pump house feeding the condenser. In the post-condenser section, the heated water from the condensers is fed into the cooling towers. The cooling towers are of forced draft type with a counter-flow direction. The water is then ejected through a fine nozzle below the demisters and falls by gravity down over the PVC fills. The water trickles down the PVC film fills through the "chevron" angle (with a flute size of 17 mm and a peak distance of 34 mm) by gravity flow. Empirical velocity across the fills is estimated to be around 0.2 m s^{-1}. The bottom of the cooling tower is of an open type for air ingress. The water is collected in a basin from where it is directed through an open channel to reach the pump house.

Severe clogging of high efficiency polyvinyl chloride film fills by deposits (Fig. 2a, b) was observed in the cooling towers (3, 4, 5 and 6) of the 4,000 MW(e) TSTPS, resulting in a loss in condenser vacuum of 40 mbar and operation of the cooling towers reaching criticality (Fig. 2c, d). The problem was found to be specific to high efficiency film fills, and was not observed in splash-type cooling towers (1 and 2) receiving the same waters. Further, the cooling towers connected in parallel and receiving the same water had different bacterial genera. Cooling towers 3 and 4 had predominantly heterotrophs and cyanobacteria (Fig. 2e), whereas iron bacteria (Fig. 2f) dominated in cooling towers 5 and 6. The problem occurred within 3 years of operation with an intermittent chlorination regime of 1.0 ± 0.1 ppm residuals for 12 h in place. The severity of the problem is reflected in the quality of the recirculating water. As a result of insufficient cooling, an increase in temperature (Fig. 3a) in the post-condenser section was observed. Reduction in flow and heat load in the condensers resulted in an increase in conductivity levels of recirculating water (Fig. 3b), further increasing the propensity of scaling in the system.

Experimental data and observations revealed the problem to be a microbially associated phenomenon. The sequence of events leading to the clogging of fills is: (i) establishment of bacterial biofilms on PVC fill surfaces due to long layoff chlorination periods and (ii) the anionic nature of the biofilms aids the entrapment of suspended, airborne particulate matter and of dissolved nutrients like the carbon, phosphate, nitrate and silicate essential for microbial growth. Estimation of bacterial loads in the cooling water during biocide dosing did not reveal significant differences between the pre- and post-condenser sections (Fig. 3c).

Chemical analysis of the high film fill deposits by X-ray photon spectroscopic (XPS) analyses showed 30–45% of silica content, which is known to precipitate, coagulate or adsorb at high concentration levels (Table 3). It is well known that naturally occurring silica can polymerize to form amorphous silica or colloidal silica under supersaturation conditions. The anionic nature of the biofilms resulted in entrapment of this compound into the matrix. The situation was noticed by plant operators when operation of the cooling towers became a concern.

The problem seems to have manifested during the layoff of biocidal dosing (during the night) when bacterial numbers multiplied. Mechanical cleaning was not performed because it is too labour-intensive, time-consuming and physically damaging to the system. Further, the towers could not be taken offline for cleaning. Based on the findings, the chlorination regime was switched over to a low-dose continuous mode (0.2 ppm residuals) and with a shock dose of 5 ppm for 15 min once a shift (8 h), coupled with increased blow-down and intake of makeup water. This resulted in slow break-down of biofilms on the fills and helped solve the problem online. Within 3 months the cooling towers had limped back to normality. The cooling towers now have improved heat transfer efficiency and are inching towards normality. Effective testing, good housekeeping during operation, proper maintenance and prompt antifouling treatment can control microbial activity in the system. The

Fig. 2 Biofouling of cooling tower fills of Talcher super thermal power station. a) a high film fill b) dry deposits on fills c) sagging of fills due to fouling load d) closer view of clogging of fills e) Cyanobacteria on cooling tower walls f) Iron bacteria on cooling tower walls

study clearly demonstrated the inefficiency of intermittent chlorination and has also shown that low-level continuous chlorination along with periodical shock chlorination is effective in breaking down biofilms.

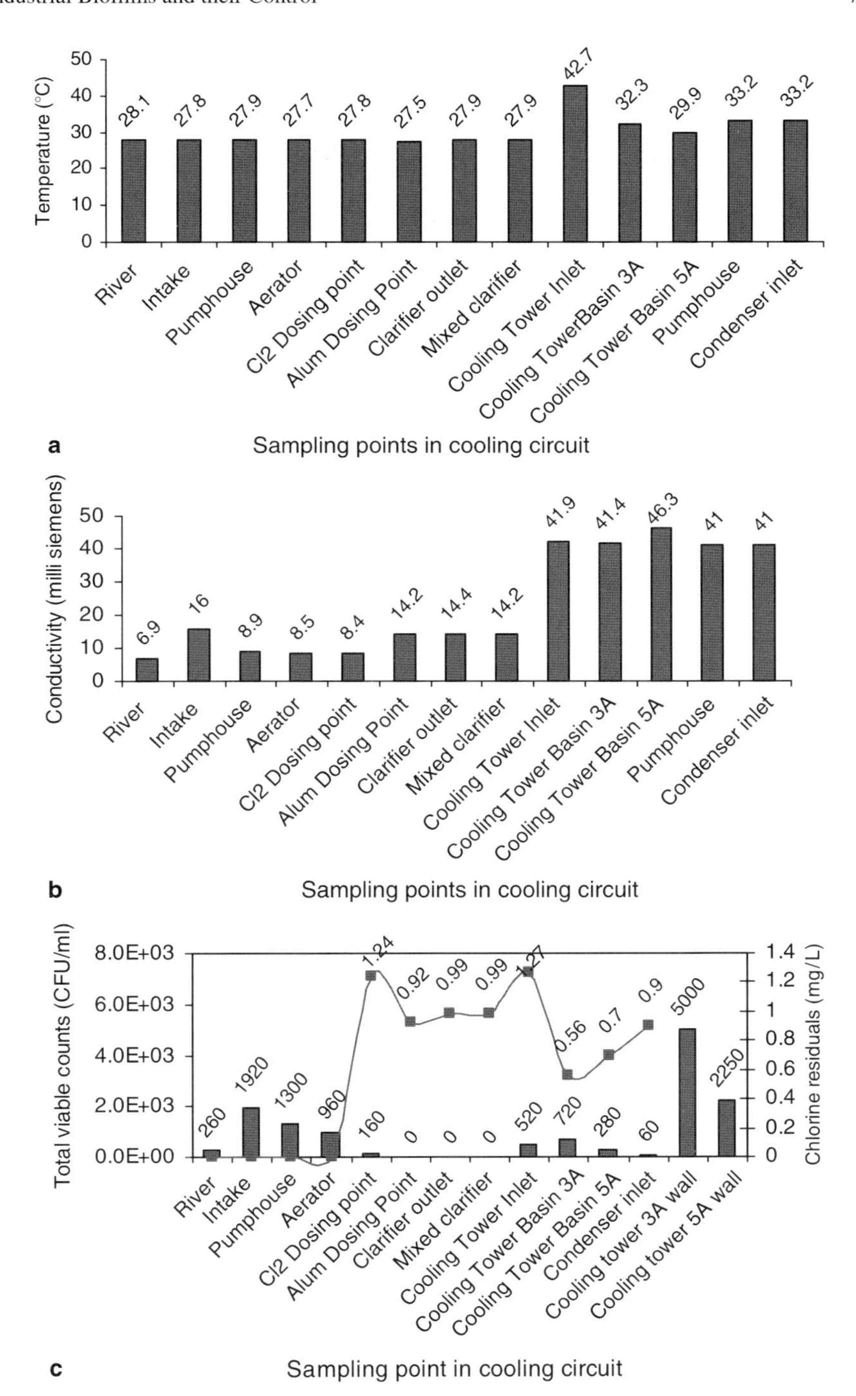

Fig. 3 Distribution of **a** temperature, **b** conductivity, **c** total viable counts and chlorine residuals in the cooling water systems of the 500 × 6 MW Talcher super thermal power station

Table 3 X-ray photon spectroscopy analysis of deposits in cooling tower fills and demisters

	Sample A	Sample B
Element	Deposits on demisters (%)	Deposits on high film fills (%)
Aluminium Al_2O_3	6.90	9.77
Calcium	4.5	3.75
Chlorine	0.0008	0.0008
Iron	–	–
Magnesium	0.63	0.70
Potassium	0.74	1.04
Silicon	31.05	41.39
Sodium	8.46	14.17
Sulfur	0.06	0.06

4 Management of Biofilms in Industrial Systems

Cooling and pure water circuits are typical ecosystems that provide an ideal environment for growth of microorganisms. The steps involved in effective management of industrial systems are: (1) detection of biofilms, (2) biocide dosing, (3) cleaning of surfaces, (4) monitoring of the effectiveness of the management strategy and (5) fine-tuning of biocidal dosing.

4.1 Detection of Biofilms

In industrial situations biofilms are visible to the naked eye as copious slime layers on surfaces. Biofilms in industrial systems are detected indirectly by symptoms noticed in the operational parameters (Flemming 2002) or failure to meet the required standards in desalination and potable water systems. The first step in detection of biofilms is sampling on surfaces, which can be a real challenge. However, water samples reveal neither the site nor extent of biofouling layers (Flemming 2002), as also demonstrated by Goysich and McCoy (1989) for cooling towers. The type of sampling method used is critical for the data to be obtained. Various methods have been used for collecting biofilm samples like sterile nylon brushes, utility knife, swabbing and stomaching for removing them from surfaces. Among the various methods, use of the stomaching procedure was found to be efficient to culture biofilm cells (Gagnon et al. 1999).

Laboratory analysis of samples involves culturing of microorganisms in biofilms and estimating the number of colony forming units. However, most of the bacteria occurring in industrial circuits cannot be cultured by standard plate methods. An European task force, with scientists from 18 different participating laboratories under the French Association des Hygienistes et Technicians Municipaux (AGHTM)

Table 4 Methods for estimating components of biofilms

Biofilm parameters	Method	References
Direct cell counting	Epifluorescence microscopy	Daley and Hobbie (1975)
Biofilm thickness	Light microscopy	Blakke and Olson (1986)
Colony forming units	Standard methods	APHA (1995)
Total living biomass	Adenosine triphosphate	Chalut et al. (1995)
	Fluorescein diacetate estimation	Rosa et al. (1998)
Total biomass	Total organic carbon	
Dry weight	Biofilm total suspended solids	APHA (1995)
Algal biomass	Chlorophyll and phaeophytin estimation	APHA (1995)
Total proteins	Protein determination	Bradford (1976)
Total sugars	Carbohydrate determination	Dubois et al. (1956)
Lipids	GC-MS	Geesey and White (1990)
Uronic acids	Uronic acid determination	Mojica et al. (2007)
Respiratory activity	CTC staining method	Schaule et al. (1993)

during the period 1996–1997, validated methods for evaluation of aqueous biofilms and recommended the use of glass beads or slides and plate counts of cells for quantifying biofilms (Keevil et al. 1999). Advances in microscopy, microfiltration membranes (nucleopore or polycarbonate) and molecular staining techniques like the Live/Dead BacLight assays are now available, which minimizes the errors in estimating viable and dead bacterial cells. A comparison of microscopic methods for biofilm examination has been reviewed by Surman et al. (1996). The use of redox dyes like CTC, which forms fluorescent and insoluble crystals after reduction, also provide a more accurate quantification of microbial numbers and activity in biofilms (Schaule et al. 1993). Several other biofilm measurement techniques or methods have been listed by Donlan (2000) and Flemming and Schaule (1996); however, in practice, results from the methods listed in Table 4 were found to be more realistic in gaining an insight into the nature and extent of the deposits.

4.2 *Biocide Dosing*

Biocide addition in industrial systems (Table 5) is the main method of controlling problems associated with microbial biofilm formation (Chen and Stewart 2000). The use of biocides is a common response to biofouling problems, resulting from a "medical paradigm" that implies that biofouling can be considered as a "technical disease" and "cured" by substances that kill the causing bacteria. However, it always should be kept in mind that killing of bacteria is not equivalent to cleaning. The complexity involved in combating biofilms in industrial systems is wide ranging, elements of which have been discussed by Flemming (2002), who has formulated

Table 5 Biocides used in industrial circuits

Oxidizing	Non-oxidizing
Bromine Chlorine Chlorine dioxide Ozone Hydrogen peroxide Para-acetic acid Bromine chloride 1-Bromo-3-chloro-5, 5-dimethylhydantoin (BCDMH)	Clamtrol: alkyl dimethyl benzyl ammonium chloride (ADBAC); Bulab 6002: poly[oxyethylene (dimethyliminio) ethylene-(dimethyliminio) ethylene dichloride]; biguanides; β-bromo-β-nitrostyrene; 2-bromo-2-nitropropano-1,3-diol (BNPD); chlorophenols; H-130; dodecyl dimethyl ammonium chloride (DDAC); 2,2-dibromo-3-nitrilo-propionamide (DBNPA); 2-dithiobisbenzamide; glutraldehyde; isothiazolone; kathion; methylenebisthiocyanate; organic sulfur and sulfones; phosphonium biocides; 2-(thiocyano-methylithio)-benzothiazole (TCMTB); thiocarbomate

a toolbox for an integrated antifouling strategy. Various devices (Robbins device, annular reactors, continuous stirred batch reactors, flow cells, mixed consortia reactors) and processes (cooling water systems, drinking water systems, model cooling towers, synthetic mediums containing high and low nutrients) have been used for assessing biocide efficacies, and have been listed extensively by Donlan (2000). Each of these systems and processes is unique and hence comparisons or extrapolation of data to other systems is very difficult. Furthermore, knowledge from these studies on the response of microorganisms to different biocides and processes is difficult to utilize in choosing a biocide type, dosing or regime. For a given industrial system and process, preliminary studies have to be carried out to arrive at the biocidal dose and concentration with respect to the environmental and hydro-biological conditions on site.

As indicated above, elimination of biofilms is an important task, and mechanical cleaning has been found to be the most satisfactory method for removing biofilms (Walker and Percival 2000) because the problems caused by biofilms in heat exchanger systems are due to their physical presence and properties. However, in most industrial systems, design and construction of equipment does not facilitate mechanical cleaning, except for tubular heat exchangers where online and offline cleaning techniques have been used. Several complexities are involved in the action of biocides in controlling biofilms, which are discussed in this section.

Good housekeeping practices (cleaning regularly) along with appropriate biocidal and surfactant or biodispersant dosings are required to keep biofilms under the threshold of interference. Biocides aid only in killing of cells, and dead biomass often accelerates the attachment process by offering a rough surface. Further, biocides increase the biodegradable organic matter(BDOC) in treated water. Instead of cleaning the system they actually increase the amount of nutrients available for growth. In general, the cleanliness and the effectiveness of the microbial control agent used should be periodically monitored using a combination of visual inspection and monitoring of differential bacterial counts like total autotrophs, heterotrophs, iron oxidizers, iron reducers, sulfate reducers, slime formers and pathogens such as *Legionella pneumophila* in both bulk water and on surfaces in order to determine the efficacy of the biocidal programme in practice.

4.2.1 Role and Action of Biocides on Microorganisms

The ideal biocide for a particular system would meet each of the following requirements: (1) active at a low concentration against a wide range of microorganisms, (2) a low order of toxicity to humans and non-target aquatic life, (3) biodegradable, (4) active in hard and soft water, (5) non-corrosive and (6) not readily inactivated in the presence of a wide range of soils.

The essential duty of the microbiocide is both to prevent primary biofilm formation and to prevent excessive growth of microorganisms, which can either induce corrosion (e.g. sulfate-reducing bacteria) or cause degradation of chemical additives (e.g. nitrifying bacteria).

4.2.2 Factors Influencing Efficacy of Biocides in Industrial Cooling Systems

In practice, the effectiveness of a biocidal programme is assessed by recovery of process parameters in industrial systems (Flemming and Schaule 1996). In turn, efficacy of biocides is determined by the Chick and Watson law (Chick et al. 1908; Watson et al. 1908):

$$\text{In}(N/N_o) = -kC^n t$$

where N/N_0 is the ratio of surviving organisms at time t, C is the disinfectant concentration, and k and n are empirical constants (n is referred to the coefficient of dilution).

The Chick and Watson law, with its concentration C multiplied by the contact time t (Ct) factor, has been the basis for all subsequent models (LeChevallier et al. 1988). Further, the efficacy of a disinfectant programme can be assessed by the recovery of process parameters (Flemming and Griebe 2000). In an industrial cooling system or water distribution system, dosing of biocides is done to prevent bacterial growth and colonization. However, experience over the years has shown that maintaining a biocide residual alone could not result in preventing microbial growth and biofilms in industrial systems. From a better understanding of the principles of microbial adhesion, the action of biocides and the quality of abstracted water, it is now becoming obvious that living with biofilms is imperative (Flemming and Griebe 2000). Biofilms are ubiquitous and cannot be totally eradicated even at a very high cost factor and for environmental safety. All systems in contact with water carry biofilms, but not all have biofouling problems. It is now being increasingly recognized that to control biofouling means to maintain biofilm development below the threshold limits so that operations are not affected (Flemming 2002).

The *Ct* values for all biocides and disinfectants are affected by a number of parameters including temperature, pH value and biocide demand as commonly caused by organic matter and protective cell aggregations (Walker and Percival 2000). Temperature and pH effects on oxidizing biocides have been well documented (refer to White 1999), whereas the most important parameter responsible for determining biocidal availability for killing is the organic content of water, which is a site- and season-specific dynamic parameter for which no specific value could be assigned. In this context the influence of organic matter on the efficacy of biocides is of utmost practical importance. The presence of even small quantities of organic matter reduces the efficiency of oxidizing biocides to varying degrees. The types of action that may occur are as follows:

- The biocide may react chemically with the organic material, giving rise to a complex that is in many instances non-biocidal, or it may form an insoluble compound with the organic matter, thus rendering it inactive
- Particulate and colloidal matter in suspension may absorb biocides so that it is subsequently, if not totally, removed from solution
- Naturally occurring fats, phospholipids etc. may dissolve or absorb biocides preferentially, rendering them inactive
- Organic and suspended particulate matter may form a coating on the surface that may render the fluid in the immediate vicinity rather more viscous, and so tend to prevent the ready access or penetration of biocides to the cell before any biocidal activity can occur

Antifouling efficacy on mixed population biofilms in low nutrient environments revealed a relationship between the nature of organic matter and disinfection efficiency. Chlorine was effective in removing natural biofilms with low organic carbon content, whereas it was ineffective with biofilms grown using amino acids and carbohydrates as the nutrient source (Butterfield et al. 2002). Organic load requires additional dosing of biocides to compensate for the demand in the system and to make available the biocide for reaction with biofilms. The price will be an increased concentration of chlorination by-products. Compared to chlorine, monochloramine-was found to be stable and is used in many recirculating and drinking water systems and is effective against biofilms (Murthy et al. 2008). Biofilm bacteria challenged with monochloramine retained significant respiratory activity even though they could not be cultured (Huang et al. 1995).

Application of biocides to industrial cooling water systems is done either on a continuous or on an intermittent basis. It is important when applying biocides to a cooling water circuit that the concentration developed within the system exceeds the minimal inhibitory concentration for the microbiological contaminants present and that it also has a sufficient contact time to exert its activity. Unless the system has a low retention time, there will be little difference between the inhibitory concentrations, whether dosing is continuous or intermittent. Conventionally, before the advent of surfactants, intermittent dosing along with an increase in velocity was practised in industrial cooling water systems where fouling caused by biofilms was found to be a problem to be overcome. This is dependent upon the generation of a relatively high concentration of microbiocide within the system at regular intervals of time and the use of high velocities intermittently to slough off biofilm layers.

Due to a wide diversity and varying population of microorganisms that can be present in any cooling system, it is impossible to establish definitive dosage figures that will have universal application. In general, however, high dosages are necessary in the case of severe microbial fouling. In effect, dosages are frequently applied in a two-phased manner. The initial dosage is usually high and aims at disrupting and dispersing any biomass present in the system, in addition to reducing the microorganisms to an acceptable level. Once the load is within the threshold limit then a lower concentration of biocides will inhibit further growth. In this context, cooling systems operate on a continuous low dose biocidal treatment with an intermittent shock dosing.

4.2.3 Efficacy of Biocides in Drinking Water Systems

Experience from drinking water systems can be adopted at least partially to biofouling control of heat exchanger circuits. However, drinking water disinfection has a different goal (i.e. the control of hygienically relevant microorganisms) while antifouling measures in heat exchanger systems do not have to meet such high hygienic standards but rather focus on limitation of microbial growth. Therefore, the term "disinfection" has a strictly hygienic connotation in drinking water, while in heat exchanger systems it refers in a more loose sense to partially inactivating the overall microbial biofilm population, while cells in suspension usually do not represent the dominant problem.

In drinking water distribution systems, growth of biofilms generally exceeds the growth of their planktonic counterparts (Camper 1996). Biofilms in drinking water systems are thin and patchy (Characklis 1988; Wingender and Flemming 2004). Control of biofilms in potable water systems is straightforward and usually achieved by establishing stable water through control of biologically degradable organic carbon (BDOC). This keeps the naturally occurring microbial population in drinking water in an oligotrophic situation. Furthermore, the drinking water industry is continually seeking novel disinfection strategies to control biofouling in distribution systems where nutrient limitation cannot be secured.

Conventionally, chlorine and chloramines are used as disinfectants in potable water distribution systems (US Environmental Protection Agency (US EPA) 1992). The efficacy of different biocides on test organisms is listed in Table 6. The problem in drinking water distribution systems is similar to cooling circuits with respect to the development of multi-species biofilms. Studies by Williams et al. (2005) have shown that biofilm communities in distribution systems are capable of changing in response to disinfection practices. Comparing two different treatments using monochloramine and chlorine it was found that after 2 weeks, increased dosing was required to maintain monochloramine levels in the system. In monochloramine-treated systems *Mycobacterium* and *Dechloromonas* were dominant whereas in chlorine-treated systems proteobacteria were dominant. Hence, it is advisable to use a combination of biocides or to alternate between biocides in distribution systems in order to prevent microorganisms from developing resistance.

Table 6 Biocides used for disinfecting planktonic and sessile cells in drinking water systems

Biocide	Test system and organism	Concentration (mg L^{-1})	Effect
Planktonic cells			
Cl_2	*E. coli* *Legionella pneumophila* δ- and β-Proteobacteria	0.2 4 Monochloramine	Bacterial survival even after 2 weeks of continuous exposure (Williams et al. 2003)
Ozone	*P. fluorescence* Laboratory cultures	0.1 and 0.3	Effective at 10–3 min (Viera et al. 1999)
Biofilms			
Cl_2 and $NHCl_2$	*K. pneumoniae* *P. aeruginosa* Steel surfaces Natural biofilms Pipe surfaces	2	Respiratory activity observed deep in biofilm with CTC stain (Huang et al. 1995)
Chloramine T	*L. pneumophila*	0.1 – 0.3%	Reduction in planktonic cells only (Ozlem et al. 2007)
Oxsil 320 N	*P. aeruginosa*	3	Wood et al. (1996)
Potassium mono persulfate	*P. aeruginosa*	20	Eliminated total viable counts (Wood et al. 1996)
Oxsil 320N			A tenfold increase in concentration required to eliminate sessile cells (Surdeau et al. 2006)
Chlorine dioxide	Diverse microbes in a Chemostat	0.25 1.0	Percentage kill of 73.8% Percentage kill of 88.4% (Walker and Morales 1997)
		1.5	Percentage kill of 99.3%
	Heterotrophic Biofilms	0.25 low 0.5 high	Disinfection (Gagnon et al. 2005)
Chlorite ion	Heterotrophic Biofilms	0.1 low 0.25 high	Disinfection (Gagnon et al. 2005)
Ozone	Laboratory biofilms	0.15	Diminish sessile cell population by three orders of magnitude (Viera et al. 1999)

Another important observation is that discontinuous or intermittent addition of biocides increased the release of cells from the biofilm to bulk water. A tenfold increase in microbial cells in the water phase was observed in the absence of chlorine dosing (Codony et al. 2005). Intermittent dosing of biocides resulted in planktonic cells developing resistance, corresponding to the number of times layoff periods occurred. Results indicated that intermittent biocidal dosings may accelerate the development of microbial communities with reduced susceptibility to disinfection in drinking water systems (Codony et al. 2005).

Maintenance of a chlorine residual level does not inactivate all bacteria in a water distribution system (Momba et al. 1998). Biofilm formation was observed at residuals of 16.5 mg L^{-1} hydrogen peroxide, 1 mg L^{-1} monochloramine and 0.2 mg

L^{-1} free chlorine (Momba et al. 1998). Studies with chlorine have shown that 3–5 mg L^{-1} (Nagy et al. 1982) and 10 mg L^{-1} (Exner et al. 1987) of free chlorine eliminates biofilms in pure water systems.

Chlorine dioxide is another option for disinfection in distribution systems. Chlorite ion, a by-product generated in systems dosed with chlorine dioxide, was found to be less effective at concentrations between 0.20 and 0.34 mg L^{-1} in eliminating heterotrophic bacteria (Gagnon et al. 2005). Field trials at the East Bay Municipal Utility District (EBMUD) in California comparing the efficiency of UV/ClO_2, ClO_2, UV/Cl_2 and Cl_2 for biofilm control showed that UV/ClO_2 was most effective against suspended and sessile heterotrophic bacteria. ClO_2 was more effective than Cl_2 against suspended and sessile bacteria, and that UV treatment alone was not as efficient as ClO_2 and Cl_2 treatments (Rand et al. 2007). On the other hand, ozone has been a very effective agent for disinfecting potable water systems. The formation of by-products like iodate and bromate has been observed with ozonated waters. A low drinking water standard of 10 mg L^{-1} has been set for drinking water, and hence disinfection strategies should be designed to operate at these ranges (Gunten 2003). It is generally believed that increasing the concentration of a disinfectant should control regrowth but many instances exist where the opposite is seen (LeChevallier et al. 1987; Martin et al. 1982; Reilly and Kippen 1984; Oliveri et al. 1985).

4.2.4 Efficacy of Biocides and Resistance of Biofilm Organisms

It is well known that biofilm organisms display a resistance to biocides. For their inactivation, sometimes more than two orders of magnitude higher concentrations are required than for planktonic cells (for review see Schulte et al. 2005). The reasons for this phenomenon are under research and not fully elucidated. Among the mechanisms discussed in terms of increased resistance are:

- Influence of abiotic factors such as limited access of biocides to biofilms in crevices or in dead legs of water systems, and attachment to particles
- Diffusion–reaction limitation, due to the reaction of oxidizing biocides with EPS components (main inactivation factor for chlorine)
- Slow growth rate, which protects dormant organisms from biocides interfering with physiological processes
- Biofilm-specific phenotypes that express, e.g., copious amounts of EPS in response to biocides or enzymes such as catalase that inactivate hydrogen peroxide
- Persister cells, which is the term for the small number of organisms in a population that survive even the most extreme concentrations by mechanisms still unknown

Ranking of halogen biocides against biofilms of *Pseudomonas fluorescens* (a contaminant in cooling water circuits), *Pseudomonas aeruginosa* (a contaminant in potable water distribution systems) and *Klebsiella pneumoniae* (a contaminant in potable water distribution and hygiene systems) showed stronger resistance of biofilms than of planktonic cells (Tachikawa et al. 2005). Results of this study showed that efficacy of different biocides varied with respect to the microorganism. In the case of *P. fluorescens* biofilms exposed to various biocides, survival increased as follows:

$$NH_4Br > NH_2Cl > HOCl > STARBEX^{®} > Br_2Cl$$

with *K. pneumoniae* biofilms, percentage survival increased as follows:

$$Br_2Cl > HOCl > NH_2Cl > STARBEX^{®} > NH_4Br$$ (Tachikawa et al. 2005).

STARBEX is a stable liquid bromine-based antimicrobial compound marketed by NALCO (Naperville, IL). It is imperative from the results to ascertain the dominant microorganisms present in an industrial system before a biocidal regime can be put into place. Further, bacterial species having a high inherent susceptibility to water-treatment biocides become dominant in systems in the presence of biocides. This has been attributed to the formation of resistant cells. The effect was demonstrated by Brözel et al. (1995) with *P. aeruginosa*, *Pseudomonas stutzeri* and *Bacillus cereus* sub-cultured repeatedly in the presence of sub-inhibitory concentrations of biocides, and thus adapted to grow in the presence of increasing concentrations. Hence, in industrial water systems it is advisable to alternate between biocides to maintain biofilms within the threshold levels.

Ozone was found to be effective at concentrations between 0.1 and 0.3 ppm at eliminating planktonic cells of *Pseudomonas fluorescens* (a contaminant in industrial systems) (10^7–10^8 cells mL^{-1}) within a contact period of 10–30 min, whereas ozone at a concentration of 0.15 ppm was only able to diminish cells by two to three orders of magnitude (Viera et al. 1999). Biofilms have also been reported to develop resistance to quaternamonium compounds like benzalkonium chloride as a result of an increase in hydrophilicity of the bacterial cell surface by the production of exopolysacchrides in *P. aeruginosa* CIP A22. However, this change in hydrophobicity was intermediate as the cells returned to normalcy after washing (Campanac et al. 2002). This study shows that bacteria have similar mechanisms of resistance for oxidizing and non-oxidizing compounds, i.e. development of EPS. Quaternary ammonia compounds dosed along with a domestic detergent did not induce microbial resistance in long-term exposures (McBain et al. 2004).

Due to the enhanced resistance exhibited by biofilms towards biocides, novel approaches like dosing a combination of biocides are currently under investigation. A laboratory study by Son et al. 2005 using a mixture of biocides showed that combinations of Cl_2/O_3, Cl_2/ClO_2 and Cl_2/ClO_2 showed enhanced efficiency (52%) compared to a single biocide (Cl_2) in killing *Bacillus subtilis* spores. In comparison, a combination of Cl_2/H_2O_2 was not found to be as effective. This approach of a combination of biocides could be tried out in heat exchangers (targeted biocide addition) where improvement in threshold levels of biofilm would amount to significant savings. Another study supporting the concept of application of dual biocides was by Rand et al. (2007), who tested a combination of UV/ClO_2, UV/Cl_2, ClO_2 and Cl_2 and showed that the combination of UV/ClO_2 was the most effective against suspended (3.93 log reduction) and attached (2.05 log reduction) heterotrophic bacteria. In contrast, UV light alone was not effective in disinfecting suspended or sessile bacteria compared to both ClO_2 and Cl_2. Pretreatment with UV aided in increased disinfection efficiencies with both the biocides ClO_2 and Cl_2.

The approach of using a combination of biocideshas also been tested in pure water systems. Comparison of the disinfection efficiency of chlorine and chlorine dioxide against microbial cells revealed chlorine dioxide to be effective over a wide range of pH (Junli et al. 1997a). Further, disinfection efficiency of ClO_2 on algae (*Ulothrix* Cl_2 94.2%, ClO_2 100%; *Chlamydomonas* Cl_2 92.9%, ClO_2 75%; *Microphorimidum* Cl_2 81.3%, ClO_2 100%) was found to be the same or slightly better than liquid chlorine. Enhanced disinfection was observed with ClO_2 against virusesand zooplankton (Junli et al. 1997b). Chlorine dioxide inactivation of *Bacillus subtilis* spores in natural waters and spiked ultrapure waters were far more effective than chlorine (Barbeau et al. 2005). Intermittent application of chlorine dioxidewas found to be ineffective in disinfecting bacteria in dental unit water lines (Smith et al. 2001). Comparison of efficacies of non-oxidizing biocides, e.g. Macrotrol MT200, Microtreat AQZ2010 and Microbiocide 2594, assayed against 23 groups of bacteria showed susceptibility of Gram-positive (MIC <4 mg L^{-1}) and Gram-negative bacteria (MIC <16 mg L^{-1}) that were in ranges far lower than those for alkylated naphthoquinonederivate molecules (MIC 1–64 mg L^{-1}) (Chelossi 2005).

Another example of a multiple biocide strategy is using oxidizing biocides like hydrogen peroxide and potassium monopersulfate and a surface-active agent (copper and cobalt phthalocyanine) incorporated in the surface matrix, which reduced the quantity of the biocide (potassium monopersulfate) required (Wood et al. 1996). This successful approach, demonstrated for surfaces of medical importance, could be tried in industrial systems where a multiple strategy of using biocide and a low surface energy antifouling coating in tandem would reduce the amount of biocide to be dosed.

4.3 Cleaning of Surfaces: Role of Surfactants or Surface-Active Agents

Biocides have been used for killing both planktonic as well as sessile cells (Chen and Stewart 2000) but killing alone is not enough, as explained earlier. In addition to the killing action, oxidizing biocides like chlorine, ozone, hydrogen peroxide and peracetic acid are known to weaken the biofilm matrix (Flemming 2002). The basic concept is to apply shear forces to a weakened biofilm matrix for removal. This can be achieved by the use of mechanical forcessuch as an increase in water flow velocity or flushing with air or steam. Basically, biofilms are kept together by weak physico-chemical interactions (see above). Understanding the requirement, industry has adopted the use of surfactants or surface-active agents (the majority of which are biodegradable and less toxic) addressing van der Waals interactions, as well as complex-forming substances such as citric acid in order to overcome electrostatic interactions (Flemming et al. 1999).

Quaternary ammonium compounds(QACs) are amphoteric surfactants that are widely used for the control of bacterial growth in clinical and industrial environments (Brannon 1997). These are known to have broad-spectrum antimicrobial and surfactant properties, which have made QACs such as benzalkonium chloride the favoured agents (Shimizu et al. 2002). QACs are known to act on the cell membrane and rupture cells (Simoes et al. 2005a). Cetyltrimethylammonium bromide(CTAB), a cationic QAC is known to act on the lipid component of the membrane causing cell lysis as secondary effect (Gilbert et al. 2002). These are usually applied to open or closed recirculating systems and are non-toxic for short-term applications, against non-target organisms. QACs are dosed in small closed recirculating cooling systems where the water is inaccessible for potable and domestic purposes, as the effects of these compounds on the biota are yet to be worked out completely. These biocides are required at milligram levels and are dosed periodically, once a day or once in 8 h and are effective for short-term periods like 12 or 48 h after dosing.

Ortho-phthalaldehyde(OPA), an aromatic compound with two aldehyde groups (McDonnell and Russell 1999) having excellent microbiocidal and sporicidal activity (McDonnell and Russell 1999; Rutala and Weber 2001) has received clearance by the FDA (US Food and Drug Administration in 1999) and is currently being tested with different biofilm models (Simoes et al. 2003, 2007, 2008). Some commercial products based on quaternamonium compounds are also available (NALCO, Naperville, IL; GE-Betz Dearborn, Decatur, IL) for treating cooling water systems. Several detergents are available for disinfecting medical equipment for hygienic purposes, whereas their use in potable water distribution systems is non-existent. Investigations on the mode of action of these surface-active agents on biofilm components, their interaction with water systems and their degradation and by-product formation all need to be carried out before these can be recommended in real-time systems. The cleaning efficacyof QACs, however, is very limited.

Surfactants to a certain extent are also known to inactivate microbial cells, apart from removing them from the surface. The efficacy of CTAB (cetyltrimethyl ammonium bromide), a cationic surfactant, on *Pseudomonas fluorescens* biofilms (Simoes et al. 2005a; 2006a) grown under laminar and turbulent conditions revealed biofilms generated under laminar conditions to be more susceptible to CTAB than biofilms generated under turbulent conditions. Total inactivation of cells was not achieved for either flow condition. In comparison, an anionic surfactant sodium dodecyl sulfate(SDS) was effective in inactivation at higher concentrations, but neither CTAB nor SDS promoted detachment of biofilms from surfaces (Simoes et al. 2006b). These results indicate that surfactants alone are not sufficient to remove biofilms. Furthermore, post-surfactant treatments resulted in biofilms recovering respiritory activity to levels found in untreated controls. Subsequent studies demonstrated resistance of *P. fluorescens* cells attached to glass surfaces on treatment with CTAB and the aldehyde OPA (Simoes et al. 2008). The low cell detachment observed with CTAB treatments has been attributed to a change in the bacterial cell surface charge (it acquired a positive charge) and increased electrostatic interaction of the microbe to the surface (Azeredo et al. 2003). In comparison, the combined exposure to CTAB application and increased shear stresses promoted increased biofilm removal, demonstrating physical and chemical forces to be effective in removing biofilms (Simoes et al. 2005a). Alternatively, the response of biofilms to combined exposure to oxidizing biocides and surface-active agents needs to be evaluated to improve our understanding of their efficacies. Screening for biofilm detachment using other surface-active agents needs to be carried out and their mechanism of action with respect to their molecular and antimicrobial properties needs to be studied.

5 Monitoring the Effectiveness of Biocidal Dosings

Monitoring is of particular importance when water treatment is the primary approach to prevention of biofouling in industrial systems (Bruijs et al. 2001; Flemming 2002). Microbial growth can be prevented by a good biocidal (Maukonen et al. 2003; Simoes et al. 2005b; Meyer 2006) with a biofilm monitoring programme in place (Flemming 2003). Control or prevention of microbial attachment may form the basis of a successful treatment programme (Meyer 2003). Various monitoring techniques are available of which the following would be of practical use in industrial systems: (1) in-situ analysis where fouling deposits are collected and analyzed, (2) online monitoring devices and (3) side-stream monitoring devices.

In-situ analysis is a labour-intensive job requiring special laboratory skills for estimation of various physical, chemical and biological parameters. On the other hand, online monitoring techniques are found to offer an indication of surface deterioration to plant operators to review their dosing strategy (e.g., Flemming 2003; Jahnknecht and Melo 2003). Characklis proposed as early as 1990 an online biofouling monitoring system from which the data collected is relayed to a central processor system. This would allow for early warning, effective countermeasures and efficacy determination.

The requirements for online monitoring are very demanding: it should give the information online, in real time, non-destructively, automatically and possibly remotely sensed. In general, only physical methods can meet these requirements. The problem is that they usually detect a deposit but not its nature. Therefore, they will respond to abiotic fouling as well as to biofouling. This requires experience and advanced application research, which is not often performed.

Different types of online monitoring systems are available and it is up to the operator to choose between them. They have been systematically considered by Flemming (2003). These involve in-place monitors like test substrates (Yohe et al. 1986; Donlan et al. 1994); retractable bioprobes (Jones et al. 1993); an optical fouling monitor (Wetegrove et al. 1997; Tamachkiarow and Flemming 2003); the BioGeorge electrochemical biofilm activity monitoring system (Bruijs et al. 2001); and the Bridger Scientific Fouling Monitor described by Bloch and DiFranco (1995). Flemming et al. (1998) have described the design features and functioning of some of the online industrial fouling monitoring devices (fibre optical sensor FOS; differential turbidity measurement device DTM; Fourier transformation infrared spectroscopy flow cell).

An important aspect brought out by Donlan (2000) on the use of online fouling monitors in operational industrial units is that these devices may throw light on the extent of deterioration occurring in the system with respect to the current levels of biocidal dosing but may not mimic exactly the system condition (pipe or electrically conductive surface) where biofilms have been accumulating for years. Each of these online monitoring methods has its own strengths and weaknesses and the type of monitor should be carefully chosen for a particular application. As pointed out by Donlan (2000), online sensors and detection devices are indicative of surface deterioration rather than the nature of fouling (biological, inorganic fouling), which is essential in determining the biocidal action. These devices measure total fouling,which includes clay/silt, corrosion and scale deposits, and biofilms. Addressing this aspect the electrochemical sensor BioGeorge has been developed, which measures the change in electrochemical reactions produced by biofilms on stainless steel electrodes (Bruijs et al. 2001).

In spite of the large number of online devices available, the concept of online monitoring has not been widely adopted by the industry, partly because there is no real consensus on accepted biofilm monitoring techniques and the paucity of information regarding the concentration of biocides required to control biofilms in industrial systems (as opposed to laboratory data) (Donlan 2000). To resolve the concern an "expert system approach" has been proposed by Donlan (2000) that involves studies comparing biofilm levels using different techniques. In other words, it is the threshold levels of interference for a particular technical system that is the scope of an industrial operator and not the online monitoring equipment. Hence, the expert system approach should involve a study of threshold levels of interference on a site- and season-specific basis and the results extrapolated to the online monitoring device for its effective usage.

Compared to online monitoring devices, side-stream monitoring devices are more practical and offer data of real-time value to operators. Several types of side-stream monitoring devices are available: Robbins device (McCoy and Costerton 1982), annular reactors (Chexal et al. 1997) and parallel plate flow-through systems (Pedersen et al. 1982). Measuring devices (pressure gauges) for D*p* would also offer an indication of the effectiveness of the control measure in practice. However, this method is more appropriate for macrofouling organisms. The use of different methods to evaluate biocide efficacy can lead to different conclusions about the effects caused by the biocide (Simoes et al. 2005b). Simple flow-through systems housing the material of interest and connected to the main system would be the best method of understanding fouling development.

A regular monitoring programmeshould be a part of an antifouling programme. Since fouling follows an asymptotic pattern in industrial systems, this curve should be established for a given system to arrive at the sampling strategy. The next step is the sampling strategy, where three types of coupons needs to be introduced. For the time being, the most common method is to expose a short-term coupon (exposed for a period of 15–20 days in a system, retrieved and quantified). Later, a long-term coupon is exposed for a period of 30–40 days, retrieved and quantified. The time intervals cited are arbitrary and need to be standardized for a given geographical location based on the asymptotic fouling curve. Short-term exposure refers to the log phase of the curve (15–20 days) and long-term exposure refers to the plateau phase where deposition levels off (30–40 days). The third is a permanent coupon (for visual observation, which is to be observed by the naked eye to note seasonal changes). This is a less expensive and effective method compared to a simulated side-stream monitoring device where these sampling procedures can be overcome.

In power stations, when a more precise control over the process parameters is required, side-stream monitors incorporating both D*p* and D*t* measuring devices to determine the thermal resistance of fouling deposits are a more precise and accurate method of evaluating the effectiveness of the biocide. Data from such monitors could be logged and available online through a computer for operators to fine-tune their biocidal programmes. However, monitoring remains a highly neglected field in improvement of antifouling measurements and early warning systems, as well in minimizing the environmental burden of biocides. Still, preventive overdosing of biocides is very common, causing considerable damage to the environment due to interference with biological treatment of waste water and to excessive formation of by-products.

6 Concluding Remarks

Every industrial cooling water system is unique with respect to biological, chemical and process parameters. As pointed out in this chapter, cooling water treatment programmes have to meet a compromise between cost, cleanliness and environmental requirements, wherein the threshold factor is of importance. Biofilms are ubiquitous in industrial systems and have been demonstrated to be mechanically stable. Elimination of biofilms in industrial systems is not necessary. However, it is vital to learn how to live with biofilms and how to prevent their excessive development. For this point, the threshold of interference due to biofilms becomes important and has to be ascertained in order to evolve suitable control measures. This level is up to the subjective tolerance of the operator and is only operationally, not scientifically, based.

Increasing the biocide dose to combat biofilms is neither a sufficient nor a completely acceptable option as biofilms in industrial cooling circuits have been shown to develop resistance to biocides in the long run. Alternating between biocides would help solve the problem to a certain extent; however, it is not viable in huge industrial circuits where capital investments are involved. Although several mechanisms of resistance have been put forward, biofilm resistance observed in industrial systems is mainly due to failure of the most popular biocide, i.e. chlorine, to penetrate into deep biofilm layers before being consumed by EPS components and interaction with the process fluid. Quite often, cells deep in the biofilm are unaffected and multiply to reach levels expected in untreated systems. To tackle this problem, more persistant biocides are used, like monochloramine or bromine chloride, or a stronger oxidant like chlorine dioxide for power plants and ozone for potable water distribution systems, along with a surface-active agent to remove the biomass from surfaces.

Increased biocidal doses would initiate other problems like corrosion and by-product accumulation. Instead, fine-tuning of the biocide dose and regime based on continuous monitoring or surveillance should be adopted to keep biofilms under the threshold level. In addition, it must be kept in mind that killing is not cleaning and that it is imperative to use surfactants in a fouling control programme.

Biomass offers copious nutrients for increased colonization and regrowth of bacteria. Hence, cleaning of surfaces is an important aspect of an antifouling programme. As a consequence, heat exchanger systems should be designed to be cleaning-friendly, with surfaces easily accessible (e.g. for pigging) and with low adhesion forces of biofilms. Cleaning is more important than killing the organisms and leaving them in place. Nutrients are potential biomass but are not addressed by biocides – some biocides make nutrients even more bioavailable (e.g., by chlorination of humic substances). Treating of water for removal of nutrients is a non-viable option for power plants, whereas this can be used as a limiting factor for biofilm prevention in desalination membranes. Suitable devices for removal of organic load need to be developed for industrial applications. The concept of living with biofilms is a reality to be accepted, and it can be achieved by understanding the laws of biofilm development.

Currently, more is known about the action of oxidizing biocides like chlorine, chlorine dioxide and ozone than about the organic and synthetic biocides that are now flooding the market. Even though these organic biocides are toxic at low concentrations, long-term environmental effects on the receiving water bodies need to be assessed. This leaves us with chlorine dioxide and ozone as the potential biocides to replace chlorine because of increasing legislations on the upper discharge limits of chlorine. Comparatively little literature is available on the type, action and efficiencies of surfactants, the main reason being insufficient success. Under these conditions, chlorine dioxide promises an interesting alternative due to it high oxidizing nature and low by-product formation.

From earlier studies it is clear that biocides alone are not sufficient to control fouling. For efficient industrial operations, an integrated antifouling programme involving a practical and reliable monitoring programme, biocide dosing, biodispersant dosing, online cleaning and, eventually, off-line cleaning has to be put into practice. Online mechanical cleaning methods assist biocides in combating fouling. Offline cleaning methods should be included in the design of industrial systems. The frequency of offline cleaning is again dependent on the required threshold levels of interference. Side-stream monitoring devices simulating D*p* and D*t* with online data recording are a convenient method of fine-tuning biocide dosing with respect to spikes in biofilm formation. Comparatively, the use of coupons for periodic monitoring would offer a better understanding of the diversity and density of organisms at surfaces.

Basically, the most elegant way to prevent biofouling is always nutrient limitation. Considering biofouling as a "biofilm reactor in the wrong place", it can be put in the "right" place by using a biological filter ahead of the system to be protected. The biofilm develops here, "in the right place", where it does not disturb the process and can be handled easily (Flemming 2002). Of course, this cannot be achieved in all situations but certainly much more often than it is done now. It requires nothing but a little shift of perspective.

References

APHA (1995) Standard methods for the examination of water and wastewater 19th edn. American Public Health Association/American Water Works Association/Water Environment Federation, APHA Washington DC, 9:34–9.41

Azeredo J, Pacheco AP, Lopes I, Oliveira R, Vieira MJ (2003) Monitoring cell detachment by surfactants in a parallel plate flow chamber. In: Proceedings of the 2002 international specialized conference on biofilm monitoring. Water Sci Technol 47:77–82

Barbeau B, Desjardin R, Mysore C, Prevost M (2005) Impacts of water quality on chlorine and chlorine dioxide efficacy in natural waters. Water Res 39:2024–2033

Bloch KP, DiFranco P (1995) Preventing MIC through experimental on-line fouling monitoring. National Association of Corrosion Engineers annual conference; paper no 257

Blank LW (1984) Control of algal biofouling at High Marnham; 1981–83. CERL note no TPRD/L/2649/N84, Central Electricity Research Laboratories, Leatherhead

Blakke R, Olsson PQ (1986) Biofilm thickness measurements by light-microscopy. J Microbiol Methods 5:93–98

Bott TR, Pinheiro MMVPS (1977) Biological fouling – velocity and temperature effects. Can J Chem Eng 55:473

Bott TR, Gunatillaka M (1983) Nutrient composition and biofilm thickness In: Bryers RW, Cole SS (eds.) Fouling of heat exchanger surfaces. United Engineering Trustees New York, pp. 727–734

Bott TR, Melo LF (1992) Particle–bacteria interactions in biofilms In: Melo LF, Bott TR, Fletcher M, Capdeville B (eds.) Biofilms – science and technology. NATO ASI Series Kluwer Academic Netherlands pp. 199–206

Bott TR (1994) The control of biofilms in industrial cooling water systems In: Wimpenny W, Nichols D, Sticker, Lappin-Scott H (eds.) Bacterial biofilms and their control in medicine and industry Bioline, Cardiff

Bott TR (1995) Fouling of heat exchangers Elsevier, Amsterdam p. 524

Bott TR (1999) Biofilms in process and industrial waters: the biofilm ecology of microbial biofouling, biocide resistance and corrosion. In: Keevil CW, Godfree A, Holt D, Dow C (eds.) Biofilms in the aquatic environment. Royal Society of Chemistry, London pp. 80–92

Brannon DK (1997) Cosmetic microbiology: a practical handbook. CRC, Boca Raton

Bradford MM (1976) A rapid and sensitive method for the quantification of microgram quantities of protein utilizing the principle of protein dye binding. Anal Biochem 72:248–254

Bruijs MCM, Venhuis LP, Jenner HA, Licina GJ, Daniels D (2001) Biocide optimization using an on-line biofilm monitor. Power Plant Chem 3(7):400–405

Brözel VS, Pietersen B, Cloete TE (1995) Resistance of bacterial cultures to non-oxidizing water treatment bactericides by adaptation. Water Sci Technol 31(5–6):169–175

Busscher HJ, Handley PS, Bos R, van der Mei HC (1999) Physico-chemistry of microbial adhesion from an overall approach to the limits. In: Baszkin A, Norde W (eds.) Physical chemistry of biological interfaces Marcel Dekker, New York pp. 431–458

Busscher HJ, Bos HC, van der Mei H (1995) Initial microbial adhesion is a determinant for the strength of biofilm adhesion. FEMS Microbiol Let 128:229–234

Busscher HJ, van der Mei H (1997) Physico-chemical interactions in microbial adhesion and relevance for biofilm formation. Adv Dent Res 11:24–32

Butterfield PW, Camper AK, Ellis BD, Jones WL (2002) Chlorination of model drinking water biofilm: implications for growth and organic carbon removal. Water Res 36:4391–4405

Campanac C, Pineau L, Payard A, Baziar-Mouysset G, Roques C (2002) Interactions between biocide cationic agents and bacterial biofilms. Antimicrob Agents Chemother 46:1469–1474

Camper AK (1993) Coliform regrowth and biofilm accumulation in drinking water systems: a review In: Geesey GG, Lewandowski Z, Flemming HC (eds.) Biofouling/biocorrosion in industrial systems Lewis Chelsea, MI

Camper AK (1996) Factors limiting microbial growth in distribution systems: laboratory and pilot-scale experiments. AWWA Research Foundation, Denver

Cassels JM, Yahya MT, Gerba CP, Rose JB (1995) Efficacy of a combined system of copper and silver and free chlorine for inactivation of *Naegleria fowleri* amoebas in water. Water Sci Technol 31:119–122

Chambers CW, Protor CM, Kabler PW (1962) Bactericidal effect of low concentrations of silver. J Am Water Works Assoc 54:208–216

Chalut J, D'Arise K, Bodkin PM, Stodolka C (1995) Identification of cooling water biofilm using a novel ATP monitoring technique and their control with the use of biodispersants. In: Corrosion/88. NACE International Houston, paper no 211

Chandy JP, Angles ML (2001) Determination of nutrients limiting biofilm formation and the subsequent impact of disinfectant decay. Water Res 35:2677–2682

Characklis WG (1988) Bacterial regrowth in distribution systems. AWWA Research Foundation, Denver, p. 332

Characklis WG (1990): Microbial fouling. In: W.G. Characklis, K.C. Marshall (eds.) Biofilms. Wiley, New York, 523–584

Characklis WG, Marshall K (1990) Biofilms: a basis for an interdisciplinary approach. Wiley New York, pp. 3–15

Chen X, Stewart PS (2000) Biofilm removal caused by chemical treatments. Water Res 34:4229–4233

Chelossi E, Faimali M (2005) Comparative assessment of antimicrobial efficacy of new potential biocides for treatment of cooling and ballast waters Sci Total Environ

Chexal B, Horowitz J, Munson D, Spalaris C, Angell P, Gendron T, Ruscak M (1997) Biofilm monitoring in power plant waters for use in prediction and control of MIC EPRI. Presented at tenth service water reliability improvement seminar, Denver, CO

Chick H (1908) An investigation of the laws of disinfection. J Hygiene 8:92–158

Christensen BE, Characklis WG (1990) Physical and chemical properties of biofilms. In: Characklis WG, Marshall KC (eds.) Biofilms. Wiley, New York, pp. 93–130

Cloete TE, Jacobs L, Brozel VS (1997) The chemical control of biofouling in industrial water systems. Biodegradation 9:23–37

Cloete TE, Westaard D, van Vuuren SJ (2003) Dynamic response of biofilm to pipe surface and fluid velocity. Water Sci Technol 47(5):57–59

Codony F, Morato J, Mas J (2005) Role of discontinuous chlorination on microbial production by drinking water biofilms. Water Res 39:1896–1906

Daley RJ, Hobbie JE (1975) Direct counts of aquatic bacteria by a modified Epifluorescence technique. Limnology and Oceanography 20:875–883

Donlan RM (2000) Biofilm control in industrial water systems: Approaching an old problem in new ways In: Evans LV (ed.) Biofilms: recent advances in their study and control Harwood Academic Netherlands, pp. 333–360

Donlan RM, Pipes WO, Yohe TL (1994) Biofilm formation on cast iron substrata in water distribution systems. Water Res 28:1497–1503

Donlan RM, Forster T, Murga R, Brown E, Lucas C, Carpenter J, Fields B (2005) Legionella pneumophila associated with the protozoan *Hartmannella vermiformis* in a model multi-species biofilm has reduced susceptibility to disinfectants. Biofouling 21(1):1–7

Dubois M, Gilles KA, Hamilton JK, Rebers A, Smith F (1956) Colorimetric method for determination of sugars and related substances. Anal Chem 28:350–356

Ewans DW, Griffiths JS, Koopmans R (1992) Options for controlling zebra mussels. Ontario Hydro Res Rev 7:1–25

Exner M, Tuschewitski GJ, Scharnagel J (1987) Influence of biofilms by chemical disinfection and mechanical cleaning. Zentrablatt Bakteriologie Mikrobiologie Hygeine B. 183:549–563

Fields BS, Benson RF, Besser RE (2002) Legionella and Legionnaire disease: 25 years of investigation. Clin Microbiol Rev 15:506–526
Flemming H-C (2002) Biofouling in water systems-cases causes and countermeasures. Appl Microbiol Biotechnol 59(6):629–640
Flemming H-C (2003) Role and levels of real time monitoring for successful anti-fouling strategies. Water Sci Technol 47 (5):1–8
Flemming H-C, Griebe T (2000) Control of biofilms in industrial waters and processes In: Walker J, Surman S, Jass J (eds.) Industrial biofouling. Wiley, New York, pp. 125–141
Flemming H-C, Leis A (2002) Sorption properties of biofilms. In: Bitton G (ed.) Encyclopedia of environmental microbiology, vol 5. Wiley, New York, pp. 2958–2967
Flemming H-C, Schaule G (1996) Biofouling. In: Heitz E, Sand W, Flemming HC (eds.) Microbially influenced corrosion of materials-scientific and technological aspects. Springer, Heidelberg Berlin New York, pp. 39–54
Flemming H-C, Wingender J (2002): Extracellular polymeric substances: structure, ecological functions, technical relevance. In: Bitton G (ed.) Encyclopedia of environmental microbiology, vol 3. Wiley, New York, pp. 1223–1231
Flemming H-C, Griebe T, Schaule G (1996) Antifouling strategies in technical systems: a short review. Water Sci Technol 34(5–6):517–524
Flemming H-C, Tamachkiarowa A, Klahre J, Schmitt J (1998) Monitoring of fouling and biofouling in technical systems. Water Sci Technol 38(8–9):291–298
Flemming H-C, Wingender J, Moritz R, Mayer C (1999): The forces that keep biofilms together. In: Keevil W, Godfree AF, Holt DM, Dow CS (eds.) Biofilms in aquatic systems. Royal Society of Chemistry, Cambridge, pp. 1–12
Fliermans CB, Bettinger GE, Fynsk AW (1982) Treatment of cooling systems containing high levels of Legionella pneumophila. Water Res 16(6):903–909
Gagnon GA, Stawson RM (1999) An efficient biofilm removal method for bacterial cells exposed to drinking water. J Microbiol Methods 34:203–214
Gagnon GA, Rand JL, O'Leary KC, Rygel AC, Chauret C, Andrews RC (2005) Disinfectant efficacy of chlorite and chlorine dioxide in drinking water biofilms. Water Res 39:1809–1817
Geesey GG, White GC (1990) Determination of bacterial growth and activity at solid–liquid interfaces. Ann Rev Microbiol 44:579–602
Gilbert P, Das JR, Jones MB, Allison D (2001) Assessment of resistance towards biocides following the attachment of micro-organisms to, and growth on, surfaces. J Appl Microbiol 91(2):248–254
Gilbert P, Allison DG, McBain AJ (2002) Biofilms in vitro and in vivo: do singular mechanisms influx cross-resistance? J Appl Microbiol 92:98S–110S
Goysich MJ, McCoy WF (1989) A quantitative method for determining the efficacy of algicides in industrial cooling towers. J Ind Microbiol 4:429–434
Griebe T, Flemming HC (1998) Biocide-free antifouling strategy to protect RO membranes from biofouling. Desalination 118:153–156
Gunten Urs von (2003) Ozonation of drinking water: Part II. Disinfection and by-product formation in presence of bromide, iodide or chlorine. Water Res 37:1469–1487
Harris A (1999) Problems associated with biofilms in cooling tower systems. In: Keevil CV, Godfree A, Holt D, Dow C (eds.) Biofilms in the aquatic environment. Springer, Berlin Heidelberg New York, pp. 139–144
Huang CT, Yu FP, McFeters GA, Stewart PS (1995) Non-uniform spatial patterns of respiratory activity with biofilms during disinfection. Appl Environ Microbiol 61:2252–2256
Jahnknecht P, Melo L (2003) Online biofilm monitoring. Rev Environ Sci Biotechnol. 2:269–283
Jenner HA, Whitehouse JW, Taylor CJL, Khalanski M (1998) Cooling water management in European power stations: biology and control of fouling. Hydroecologie Appliquee 10(1–2)225:

Jones DS, O'Rourke PC, Caine CS (1993) Detection and control of microbiologically influenced corrosion in a Rocky Mountain oil production system-a case history National Association of Corrosion Engineers annual conference, paper no 311

Junli H, Li W, Nanqi R, Fang MA, Juli (1997a) Disinfection effect of chlorine dioxide on bacteria in water. Water Res 31(3):607–613

Junli H, Li W, Nenqi R, Li LX, Fun SR, Guanle Y (1997b) Disinfection effect of chlorine dioxide on viruses, algae and animal planktons in water. Water Res 31(3):455–460

Keevil CW, Mackerness CW, Colbourne JS (1990) Biocide treatment of biofilms. Int Biodeter Biodeg 26:169–179

Keevil CW, Walker JT, Maule A, James BW (1999) Persistence and physiology of Escherichia coli O157:H7 in the environment. In: Duffy G, Garvey P, Coia J, Wasteson W, McDowell D(eds.) Verocytotoxigenic *E. coli* in Europe: survival and growth. Teagasc, Dublin, pp. 42–52

Kim BR, Anderson JE, Mueller SA, Gaines WA, Kendall AM (2002a) Literature review-efficacy of various disinfectants against *Legionella* in water systems. Water Res 36:4433–4444

Kim J, Cho M, Chung WK, Kang SJ, Yoon JY (2002b) Biogrowth control in cooling tower with EEKO-BALL. Fall conference of Korean Society of Water and Wastewater, abstract A-47

Korstgens V, Flemming HC, Wingender J, Borchard W (2001) Uniaxial compression measurement device for investigation of the mechanical stability of biofilms. J Microb Method 46:9–17

Kusnetsov J, Livanainem E, Elomaa N, Zacheus O, Martikainen PJ (2001) Copper and silver ions more effective against *Legionella* than against mycobacteria in a hospital warm water system. Water Res 35:4217–4225

LeChevallier MW, Babcock TM, Lee RG (1987) Examination and characterization of distribution system biofilms. Appl Environ Microbiol 53:2714–2724

LeChevallier MW, Cawthon CD, Lee RG (1988) Inactivation of biofilm bacteria. Appl Environ Microbiol 54:2492–2499

Lehtola JM, Laxander M, Miettinen TI, Hirvonen A, Vartiainen T, Martikainen PJ (2006) The effects of changing water flow velocity on the formation of biofilms and water quality in pilot distribution system consisting of copper or polyethylene pipes. Water Res 40:2151–2160

Lowe MJ (1988) The effect of inorganic particulate materials on the development of biological films. PhD thesis, University of Brimingham

Ludensky M (2003) Control and monitoring of biofilms in industrial applications. Int Biodeter Biodeg 51:255–263.

Lutz P, Merle G (1983) Discontinuous mass chlorination of natural draft cooling towers. Water Sci Tech15:197–213

Martin RS, Gates WH, Tobin RS, Sumarah R, White P, Forestall P (1982) Factors affecting coliform bacterial growth in distribution systems. J Am Water Works Assoc 74:34–37

Maukonen J, Matto J, Wirtanen G, Raaska L, Matila-Sandholm T, Saarela M (2003) Methodologies for the characterization of microbes in industrial environments: a review. J Ind Microbiol Biotech 30:327–356

McBain AJ, Ledder RG, Moore LE, Catrenich CE, Gilbert P (2004) Effect of quaternary ammonium based formulations on bacterial community dynamics and antimicrobial susceptibility. Appl Environ Microbiol 3449–3456

McCoy WF, Costerton JW (1982) Fouling biofilm development in tubular flow systems. Dev Ind Microbiol 23:551–558

McDonnell G, Russell AD (1999) Antiseptics and disinfectants: activity action and resistance. Clin Microbial Rev 12:147–179

Merle G, Montanat M (1980) Essai d'utilisation du dioxyde de chlore dans la boucle TERA installee a Montereau: resultats des tests d'efficacite EDF DER report HE/31–80.045

Melo LF, Bott TR (1997) Biofouling in water systems. Exp Therm Fluid Sci 14:375–381

Meyer B (2003) Approaches to the prevention, removal and killing of biofilms. Int Biodeter Biodeg 51:249–253

Meyer B (2006) Does microbial resistance to biocides create a hazard to food hygiene? Int J Food Microbiol 112:275–279

Mojica K, Elsey D, Cooney MJ (2007) Quantitative analysis of biofilm EPS uronic acid content. J Microb Methods 71(1):61–65
Momba MNB, Cloete TE, Venter SN, Kfir R (1998) Evaluation of the impact of disinfection processes on the formation of biofilms in potable surface water distribution systems. Water Sci Technol 38(8–9):283–289
Murphy HM, Payne SJ, Gagnon GA (2008) Sequential UV- and chlorine-based disinfection to mitigate *Escherichia coli* in drinking water biofilms. Water Res 42(8–9):2083–2092 doi:10.1016/j.watres.2007.12.020
Murthy PS, Venkatesan R, Nair KVK Inbakandan D, Syed Jehan S, Magesh Peter D, Ravindran M (2005) Evaluation of sodium hypochlorite for fouling control in plate heat exchanger for seawater application. Int Biodeter Biodeg 55:161–170
Nagy LA, Kelly AJ, Thun MA, Olson BH (1982) Biofilm composition formation and control in the Los Angeles aqueduct system. In: Proceedings of the American Water Works Association water quality technology conference. American Water Works Association, Denver, pp. 141–160
Nebot E, Casanueva JF, Casanueva T, Sales D (2007) Model for fouling deposition on power plant steam condensers cooled with seawater: Effect of water velocity and tube material. Int J Heat Mass Transfer 50:3351–3358
Oliveri VP, Bakalian AE, Bossung KW, Lowther ED (1985) Recurrent coliforms in water distribution systems and the presence of free residual chlorine. In: Jolley RL, Bull RJ, Davis WP, Katz S, Roberts MH, Jacobs VA (eds.) Water chlorination chemistry environmental impact and health effects Lewis Boca Raton
Ozlem N, Sanli-Yurudu, Ayten Kimiran-Erdem, Aysin C (2007) Studies on the efficacy of chloramine T trihydrate (*N*-chloro-*p*-toluene sulfonamide) against planktonic and sessile populations of different *Legionella pneumophila* strains. Int J Hyg Environ-Health 210:147–153
Pedahzur R, Shoval HI, Ulitzur S (1997) Silver and hydrogen peroxide as potential drinking water disinfectants their bactericidal effects and possible modes of action. Water Sci Technol 35:87–93
Pedersen K (1982) Factors regulating microbial biofilm development in a system with slowly flowing seawater. Appl Environ Microbiol 44:1196–1204
Percival SL (2000) Detection of biofilms in industrial waters and processes. In: Walker J, Surman S, Jass J (eds.) Industrial biofouling detection prevention and control. Wiley, Toronto, pp. 103–124
Petrucci G, Rosellini X (2005) Chlorine dioxide in seawater for fouling control and ost disinfection in potable waterworks. Desalination 182:283–291
Peyton BM (1996) Effects of shear stress and substrate loading rate on *Pseudomonas aeruginosa* biofilm thickness and density. Water Res 30(1):29–36
Pinheiro MM, Melo LF, Bott TR, Pinheiro JD, Leitao L (1988) Surface phenomena and hydrodynamic effects on deposition of *Pseudomonas fluorescens*. Can J Chem Eng 66:63–67
Prince EL, Muir AVG, Thomas WM, Stollard RJ, Sampson M, Lewis JA (2002) An evaluation of the efficacy of Aqualox for microbiological control of industrial cooling tower systems. J Hosp Infect 52:243–249
Pujo M, Bott TR (1991) Effects of fluid velocities and Reynolds numbers on biofilm development in water systems. In: Keffer JF, Shah RK, Ganie EN (eds.) Experimental heat transfer, fluid mechanics and thermodynamics. Elsevier, New York pp. 1358–1362
Rand JL, Hofmann R, Alam MZB, Chauret C, Cantwell R, Andrews RC, Gagnon GA (2007) A field study evaluation for mitigating biofouling with chlorine dioxide or chlorine integrated with UV disinfection. Water Res 41:1939–1948
Reilly KJ, Kippen JS (1984) Relationship of bacterial counts with turbidity and free chlorine in two distribution systems. J Am Water Works Assoc 75:309–314
Rickard HA, McBain AJ, Stead AT, Gilbert P (2004) Shear rate moderates community diversity in freshwater biofilms. Appl Environ Microbiol 70:7426–7435
Rosa DES, Sconza F, Volterra L (1998) Biofilm amount estimation by fluorescein diacetate. Water Res 32(9):2621–2626

Rupp CJ, Fux CA, Stoodley P (2005) Viscoelasticity of *Staphylococcus aureus* biofilms in response to fluid shear allows resistance to detachment and facilitates rolling migration. Appl Environ Microbiol 71(5):2175–2178

RutalaWA, Weber DJ (2001) New disinfection and sterilization methods. Emerging Infect Dis 7(2):348–353

Sanden G, Fields BS, Barbaree JM, Feeley JC (1989) Viability of *Legionella pneumophila* in chlorine free waters at elevated temperatures. Curr Microbiol 18:61–65

Sauer K, Camper AK (2001) Characterization of phenotypic changes in *Pseudomonas putida* in response to surface associated growth. J Bacteriol 183:6579–6589

Schaule G, Flemming HC, Ridgway HF (1993) Use of 5-cyano-2,3-ditolyl tetrazolium chloride (CTC) for quantifying planktonic and sessile respiring bacteria in drinking water. Appl Environ Microbiol 59:3850–3857

Schulte S, Wingender J, Flemming HC (2005) Efficacy of biocides against biofilms. In: Paulus W (ed.) Directory of microbicides for the protection of materials and processes. Kluwer Academic, Doordrecht, pp. 90–120

Shimizu MK, Okuzumi A, Yoneyama T, Kunisada M, Araake H, Ogawa, Kimura S (2002) In vitro antiseptic susceptibility of clinical isolates from nosocomial infections. Dermatology 204(Suppl 1):21–27

Simoes M, Carvalho H, Pereira MO, Vieira MJ (2003) Studies on the behaviour of *Pseudomonas fluorescens* biofilms after ortho-phthalaldehyde treatment. Biofouling 19:151–157

Simoes M, Pereira MO, Vieira MJ (2005a) Action of a cationic surfactant on the activity and removal of bacterial biofilms formed under different flow regimes. Water Res 39:478–486

Simoes M, Pereira MO, Vieira MJ (2005b) Validation of respirometry as a short term method to assess the toxic effect of a biocide. Biofouling 47:217–223

Simoes M, Pereira MO, Machado I, Simoes LC, Vieira MJ (2006a) Comparitive antibacterial potential of selected aldehyde based biocides and surfactants against planktonic *Pseudomonas fluorescens*. J Ind Microbiol Biotech 33:741–749

Simoes M, Simoes LC, Machado I, Pereira MO, Vieira MJ (2006b) Control of flow generated biofilms with surfactants - Evidence of resistance and recovery. Inst Chem Eng 84(c4):338–345

Simoes LC, Simoes M, Oliveira R, Vieira M (2007) Potential of the adhesion of bacteria isolated from drinking water to materials. J Basic Microbiol 47:174–183

Simoes M, Simoes LC, Cleto S, Pereira MO, Vieira MJ (2008) The effects of a biocide and a surfactant on the detachment of *Pseudomonas fluorescens* from glass surfaces. Int J Food Microbiol 121:335–341

Smith AJ, Bagg J, Hood J (2001) Use of chlorine dioxide to disinfect dental unit waterlines. J Hosp Infect 49:285–288

Son H, Cho M, Kim J, Oh B, Chung H, Yoon J (2005) Enhanced disinfection efficiency of mechanically mixed oxidants with free chlorine. Water Res 39:721–727

States S, Kuchta J, Young W, Conley L, Ge J, Costello M, Dowling J, Wadowsky R (1998) Controlling *Legionella* using copper-silver ionization. J Am Water Works Assoc 90:122–129

Stoodley P, Boyle J, Cunningham AB, Dodds I, Lappin-Scott HM, Lewandowski Z (1999) Biofilm structure and influence on biofouling under laminar and turbulent flows In: Keevil CW, Godfree A, Holt D, Dow C (eds.) Biofilms in the aquatic environment. Royal Society of Chemistry Cambridge pp. 13–24

Surdeau N, Laurent-Maquin X, Bouthors S, Gelle MP (2006) Sensitivity of bacterial biofilms and planktonic cells to a new antimicrobial agent, Oxsil 320 N. J Hosp Infect 62:487–493

Suci PA, Vrany JD, Mittelman MW (1998) Investigation of interactions between antimicrobial agents and bacterial biofilms attenuated total reflection Fourier transform infrared spectroscopy. Biomaterials 19:327–339

Surman SB, Walker JT, Goddard DT, Morton LHG, Keevil CW, Weaver W, Skinner A, Hanson K, Caldwell D, Kurtze J (1996) Comparison of microscope techniques for the examination of biofilms. J Microbiol Method 2(5):57–70

Tamachkiarow A, Flemming H-C (2003) On-line monitoring of biofilm formation in a brewery water pipeline system with a fibre optical device (FOS). Water Sci Technol 47(5):19–24
Tachikawa M, Tezuka M, Morita M, Isogai K, Okada S (2005) Evaluation of some halogen biocides using a microbial biofilm systems. Water Res 39:4126–4132
Tsai YP (2006) Interaction of chlorine concentration and shear stress on chlorine consumption, biofilm growth rate and particle number. Bioresource Technol 97:1912–1919
US Environmental Protection Agency (US EPA) (1992) Control of biofilm growth in drinking water distribution systems. EPA/625/R-92/001. USEPA, Washington, DC
Verran J, Hissett T (1999) The effect of substratum surface defects upon retention of and biofilm formation by micro-organisms from potable water. In: Keevil CW, Godfree A, Holt D, Dow C (eds.) Biofilms in the aquatic environment. Royal Society of Chemistry Cambridge, pp. 25–33
Vieira MJ, Oliveira R, Melo LF, Pinheiro MM, van der Mei HC (1992) Biocolloids and biosurfaces. J Dispersion Sci Tech 13(4):437–445
Viera MR, Guiamet PS, de Mele MFL, Videla HA (1999) Use of dissolved ozone for controlling planktonic and sessile bacteria in industrial cooling systems. Int Biodeter Biodeg 44:201–207
Walker JT, Mackerness CW, Rogers J, Keevil CW (1995) Biofilm – a haven for waterborne pathogens. In: Lappin-Scott HM, Costerton JW (eds.) Microbial biofilms. Cambridge University Press, London, pp. 196–204
Walker JT, Morales M (1997) Evaluation of chlorine dioxide (ClO2) for the control of biofilms. Water Sci Technol 35 (11–12):319–323
Walker JT, Percival SL (2000) Control of biofouling in drinking water systems. In: Walker J, Surman S, Jass J (eds.) Industrial biofouling. Wiley, New York, pp. 55–76
Watson HE (1908) A note on the variation of the rate of disinfection with change in the concentration of the disinfectant. J Hygiene 8:536
Williams MM, Braun Howland EB (2003) Growth of *Escherichia coli* in model distribution system biofilm exposed to hypochlorous acid or monochloramine. Appl Environ Microbiol 5463–5271
Williams MM, Jorge W, Domingo S, Meckes MC (2005) Population diversity in model potable water biofilms receiving chlorine or chloramines residual. Biofouling 219 (5/6):279–288
Wetegrove RL, Banks RH, Hermiller MR (1997) Optical monitor for improved fouling control in cooling systems. J Cooling Tower Inst 18:52–59
White GC (1999) Handbook of chlorination and alternative disinfectants, 4th edn. Wiley, New York, p. 1568
Wood P, Jones M, Bhakoo M, Gilbert P (1996) A novel strategy for control of microbial biofilms through generation of biocide at the biofilm surface interface. Appl Environ Microbiol 2598–2602
Wingender J, Flemming HC (2004): Contamination potential of drinking water distribution network biofilms. Water Sci Technol 49: 277–285
Yohe TL, Donlan RM, Kyriss KK (1986) Sampling device for determining conditions on the interior surface of a water main. US patent no 4,631,961

Biofilm Control: Conventional and Alternative Approaches

H.-C. Flemming (✉) and H. Ridgway

Abstract "Biofouling" is referred to as the unwanted deposition and growth of biofilms. This phenomenon can occur in an extremely wide range of opportunities ranging from colonization of medical devices, during production of ultrapure drinking and process water, and fouling of ship hulls, pipelines and reservoirs. Although biofouling occurs in such different areas, it has a common cause, which is the biofilm. Biofilms are the most successful form of life on earth and tolerate high concentrations of biocidal substances. Conventional anti-fouling approaches usually rely on the efficacy of biocides, aiming for inhibition of biofilm growth. It is important to keep in mind that killing of biofilm organisms usually does not solve biofouling problems as mostly the biomass is the problem and must be removed. Therefore, cleaning is at least equally important. However, for a sustainable anti-fouling strategy, an advanced approach is suggested, which includes the analysis of the fouling situation, a selection of suitable components of the "anti-fouling menu" and an effective and representative monitoring of biofilm development. One important part of this menu is nutrient limitation, which could be implemented on a much broader scale than is practiced today. Other items on the menu include methods to monitor unwanted biofilm development and assessment of the efficacy of anti-fouling measures. Also, natural anti-fouling strategies are worth exploring and learning from – and nature never relies on only one defence line but on integrated approaches.

1 What is Biofouling?

The term "biofouling" is referred to as the undesired development of microbial layers on surfaces. This operationally defined term has been adapted from heat exchanger technology where "fouling" is defined generally as the undesired deposition of material on surfaces, including:

H.-C. Flemming
IWW Centre for Water, Moritzstrasse 26, 45476, Muelheim Germany
e-mail: hanscurtflemming@compuserve.com

Springer Series on Biofilms, doi: 10.1007/7142_2008_20

- *Scaling, mineral fouling*: deposition of inorganic material precipitating on a surface
- *Organic fouling:* deposition of organic substances (e.g. oil, proteins, humic substances)
- *Particle fouling:* deposition of, e.g., silica, clay, humic substances and other particles
- *Biofouling*: adhesion of microorganisms to surfaces and biofilm development

In the first three kinds of fouling, the increase of a fouling layer arises from the transport and abiotic accumulation on the surface of material from the water phase. What is deposited on the surface originates quantitatively from the water. In these cases, fouling can be controlled by eliminating the foulants from the liquid phase. However, this is different in the case of biofouling: microorganisms are "pseudo-particles", which can multiply. Even if 99.99% of all bacteria are eliminated by pre-treatment (e.g. microfiltration or biocide application), a few surviving cells will enter the system, adhere to surfaces, and multiply at the expense of biodegradable substances dissolved in the bulk aqueous phase. Thus, microorganisms convert dissolved organic material into biomass locally, through metabolic transformations. These metabolic processes, i.e. biodegradation and surface growth, form the basis of biofilm reactors (e.g. membrane bioreactors) that have been introduced in the past decade. Biofouling can be considered as a "biofilm reactor in the wrong place and time". Substances suitable as nutrients, which would not act as foulants per se, will support fouling indirectly. Although most current anti-fouling measures target the microorganisms directly (e.g. chlorine disinfection of a potable water system), the role of nutrients as a potential source of biomass is frequently overlooked.

Moreover, biocides tend not to decrease the nutrient level that ultimately supports the biofilm. On the contrary, nutrients released into solution by the oxidative breakdown of normally recalcitrant organics can support rapid post-biocide (LeChevallier 1991). As it is virtually impossible to keep a common industrial system completely sterile, microorganisms on surfaces will always be present, "waiting" for traces of nutrients. Thus, all biodegradable substances must be considered as potential biomass.

Usually, the different kinds of fouling mentioned above occur together. The proportion of biofouling can be considerable. An example is the development of dental plaque, i.e. mineral depositions on teeth which is favoured by biofilms. In algal biofilms, precipitation of calcium carbonate is increased, mainly due to the rise in pH resulting from photosynthesis (Callow et al. 1988). However, other mechanisms may also play a role such as changing of the water activity by EPS molecules. Generally, biofouling has to be considered as a biofilm problem. In order to understand the effects and dynamics of biofouling and to design appropriate countermeasures, it is important to understand the natural processes of biofilm formation and development.

From a microbiological point of view, there is no "typical" fouling organism. If excessive biomass or non-specific contamination of the water phase is the problem, it will be the most abundant organism in a given site that will be the main fouling organism. If metabolic products cause the problem, such as low-chain fatty acids, hydrogen sulphide or inorganic acids, the organisms producing these substances will cause the fouling. Again, "fouling" is an operational expression, which is defined by the specific physicochemical and biological characteristics of a system.

Nearly all microorganisms are capable of forming biofilms as this is a universal way of microbial life. Practical observations revealed that particularly slimy strains of environmental bacteria may prevail in water system biofilms (Wingender and Flemming, personal observation). Usually, the composition of fouling biofilms is dominated by the autochthonic flora, which can differ profoundly with different fouling sites and conditions, including those systems whose microbial flora has been perturbed by the application of biocides.

Biofouling in the sense of the given definition can occur in extremely diverse situations ranging from space stations (Koenig et al. 1997) to profane explanations for religious miracles like that of Bolsena, which is attributed to the growth of *Serratia marcescens* on sacramental bread and polenta. Communion cups have been identified as potential infection risks due to biofilms on the chalices (Fiedler et al. 1998). In medicine, implant devices such as catheters are prone to biofouling. Dental waterlines can be seriously contaminated by pathogens (Barbeau et al. 1998). In general, it is acknowledged now that biofilms are a common cause for infections (Costerton et al. 1999). Cases, causes and countermeasures have been reviewed (e.g., Flemming 2002) and more are presented in this book.

In technical systems, a less considered problem is the fact that biofilms can provide a habitat for pathogenic microorganisms. Biofouling, therefore, may be not only a technical problem but can also imply the exposure of working personnel to such pathogens released from biofilms. The contamination risk can occur from skin contact and from inhalation of aerosols. *Klebsiella, Mycobacterium, Legionella, Escherichia coli* and coliform organisms have been found in water system biofilms (LeChevallier et al. 1990) from where they can detach and will be found in the water phase. In the presence of corrosion products, pathogens seem to be particularly protected. This is the conclusion of the study of Emde et al. (1992), which found a much higher variety of species in corrosion product deposits, called "tubercles", compared to the free water phase, even after extended periods of chlorination. The fate of viruses in biofilms is still in question. Reasoner (1988) reports very occasional incidence of pathogens in drinking water biofilms. This is confirmed by a large study on drinking water distribution system biofilms carried out currently in Germany. First results indicate that some pathogens seem to be even eliminated by the autochthonic biofilms (Wingender and Flemming, personal observation). In distribution systems, due to the surface to volume ratio, more than 95% of the entire biomass is located at the walls and less than 5% in the water phase (Flemming 1998). These biofilms contribute considerably to the overall purification process because they degrade diluted organic matter. There is no correlation between the cell concentration in the water and in the biofilm, although most of the cells found in the water phase originate from biofilms. However, ongoing large field research reveals that biofilms developing on certified materials seem not to represent a threat to drinking water and in general do not harbour potentially pathogenic organisms (Kilb et al. 2003). The situation is different if materials are involved that support microbial growth. An example is shown in Fig. 1a: massive biofilm formation on synthetic elastomers in drinking water pipelines. Figure 1b shows a scanning electron micrograph of the same biofilm. The size of the cells indicates very good growth conditions, in contrast to starving microcolonies, which are usually found on

a

1μm

Mag = 1.00 K X 10μm EHT = 15.00 kV WD = 10 mm Signal A = SE2 Photo No. = 2615 Date :13 Feb 2001 Time :12:11

b

Fig. 1 **a** Massive biofilm development on an elastomer coating of a valve in a drinking water system (Kilb et al. 2003). **b** Scanning electron micrograph of a section of **a**. *Right:* magnification of a section of the micrograph on the *left* (courtesy of G. Schaule)

materials that do not leach nutrients. This biofilm harboured coliform bacteria, which were detected in downstream drinking water samples.

2 Countermeasures

In biofouling cases, it is reasonable to follow a three step protocol:

1. Identification of the cause and localization of the problem
2. Sanitation (cleaning is as important as killing the microorganisms)
3. Prevention

This has been described in detail earlier (Flemming 2). Usually, if a problem arises in a process, the diagnosis "biofouling" will be attributed if other causes do not explain the phenomena. In order to design the most effective countermeasures, it is important, to verify this diagnosis. This has to be performed by sampling of the surfaces, which requires a set of more sophisticated techniques (Schaule et al. 2000) than sampling of the water phase, although the latter is unfortunately performed exclusively in most cases. The most common countermeasure against unwanted microbial growth is the use of biocides (Flemming and Schaule 1996). This line of thinking expands a medical paradigm to technical systems: the colonization by bacteria is considered as a kind of "illness" that has to be cured by some means of disinfectant, antibiotic or other biocide. However, it is well known that biofilm organisms display a much higher tolerance to biocidal agents than their freely suspended counterparts (LeChevallier et al. 1991). Various mechanisms are discussed that may protect biofilm organisms (McBain 2001). The most plausible explanation is based on a diffusion-limitation of the biocide by the EPS matrix. However, recent measurements have revealed that this cannot be the case. Small molecules experience practically no diffusion limitation in a biofilm matrix. Only if they react with EPS components (as is the case with oxidizing biocides such as chlorine or ozone) is consumption of the biocide and, thus, a concentration gradient observed, caused by reaction with EPS components (Schulte et al. 2005). Tolerance against hydrogen peroxide is frequently accompanied by an enhanced catalase activity. In general, enhanced biocide tolerance must be taken into account in anti-fouling applications (Morton et al. 1998).

2.1 An Integrated Anti-fouling Strategy

A more complex and hopefully more effective approach to combating biofilms may be stimulated by an increasingly restrictive legislation towards biocides, particularly in the EU, although the relevant literature cannot be exhaustively reviewed here (for further details see Flemming and Greenhalgh 2008). It is important not to rely only on one "wonder weapon" but to analyse all fouling factors and to develop an integrated approach, based on detailed knowledge of biofilm development. The basic idea is "to live with biofilms", an approach that may well inspire creativity in new directions (Flemming 2002).

Biofouling is an operational definition, referring to that amount of biofilm development that interferes with technical, aesthetic or economical requirements. Research on reverse osmosis (RO) membrane biofouling revealed that biofilms commence development within the first moments of operation, thereby contributing to the demise of the separation process without any knowledge or forewarning that such processes are at work (Griebe and Flemming 1998). Only after observing a certain reduced membrane permeability is the "level of interference" passed and biofouling is said to have occurred. This motif can be transferred to other water systems; they practically all carry biofilms, but not all of them suffer from biofouling. Figure 2 shows schematically the development of biofilms in a system.

What are the options for keeping biofilm development in a system below the individual level of interference? Basically, the extent of biofilm growth is grossly ruled by the availability of nutrients and the shear forces. Thus, nutrients must be considered as potential biomass. This is an important issue as, usually, biocidal approaches do not take this aspect into account and do not limit nutrients; to the contrary, some biocides increase the nutrient content by oxidizing recalcitrant organics and rendering them more bioavailable (LeChevallier 1991). Nutrient limitation has been demonstrated successfully as a countermeasure to biofouling (Griebe and Flemming 1998). By using biological sand filters prior to RO membranes it was possible to suppress the extent of biofilm growth below the threshold of interference, although the membrane was not completely free of a biofilm (Table 1).

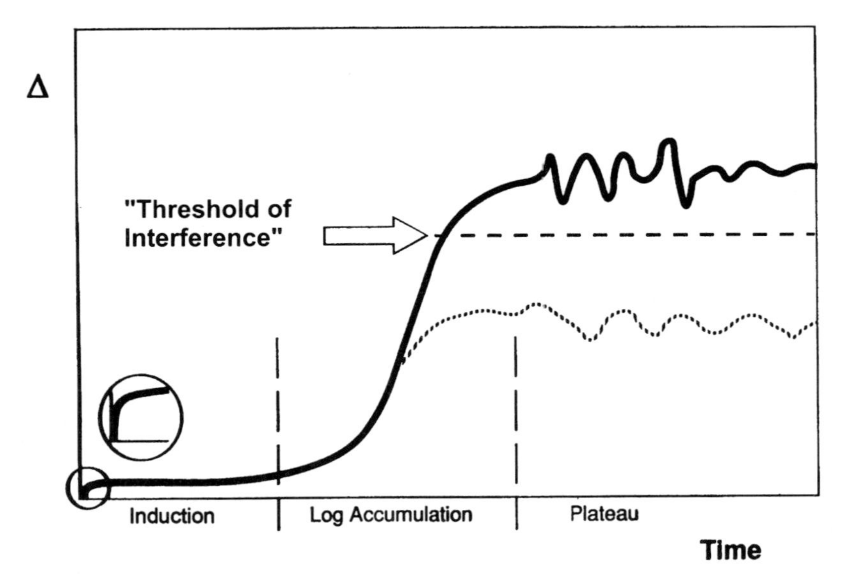

Fig. 2 Schematic of biofilm development. *Dotted line* arbitrary threshold of interference (after Flemming et al. 1994)

Table 1 Deposit data from membranes before and after biological filter

Parameter	Unit	Before filter	After filter
Cell number	$\#/cm^2$	1.0 × 108	5.5 × 106
CFU	$\#/cm^2$	1.0 × 107	1.2 × 106
Protein	$\mu g\ cm^{-2}$	78	4
Carbohydrates	$\mu g\ cm^{-2}$	26	3
Uronic acids	$\mu g\ cm^{-2}$	11	2
Humic substances	$\mu g\ cm^{-2}$	41	12
Biofilm thickness	µm	27	3
Flux decline	%	35	<2

Obviously, this approach cannot be applied in all cases. However, there still remain plenty of opportunities where it provides a suitable and realistic alternative to adding biocides for prevention of biofouling. This approach would certainly reduce the burden of wastewater with environmentally problematic substances and certainly deserves more attention.

2.1.1 Surface Design and Primary Adhesion

Clearly, rough surfaces are more prone to microbial colonization than smooth surfaces. This has been confirmed with stainless steel surfaces (Faille et al. 2000). However, even on the smoothest surface, bacteria can attach. This is the result of unsuccessful approaches to prevent biofouling in heat exchangers by electropolishing. In order to understand what happens when a bacterial cell comes into contact with a surface, it is helpful to take the entire situation in account. As shown in Fig. 3 for the example of a Gram-negative organism, cells are surrounded by extracellular material. Also, surfaces immersed in water become within seconds covered with a so-called conditioning film consisting of macromolecules such as humic substances, polysaccharides and proteins, which are present in trace amounts in water. This has long since been known (Loeb and Neihof 1975) but not taken into account. The cells do not need to be viable for adhesion, the already present EPS are sufficient for adhesion (Flemming and Schaule 1988)

Many approaches have been followed in order to prevent microbial adhesion. Until now, only three of them have been successful:

1. *Tributyl tin anti-fouling compounds*. However, these are so toxic to marine organisms that they have been widely banned from use.
2. *Natural anti-fouling compounds*. Such compounds have been isolated mainly from marine plants that are not colonized by bacteria (Terlezzi 2000). Steinberg et al. (1997) have isolated signalling molecules from an Australian seaweed exhibiting anti-colonizing activity. More marine anti-fouling products have been investigated by Armstrong et al. (2000) and Tirrschof (2000). The problem with all these compounds is that most of them are only scarcely available, they are difficult to apply on a constant basis on a surface, and they will select for organisms

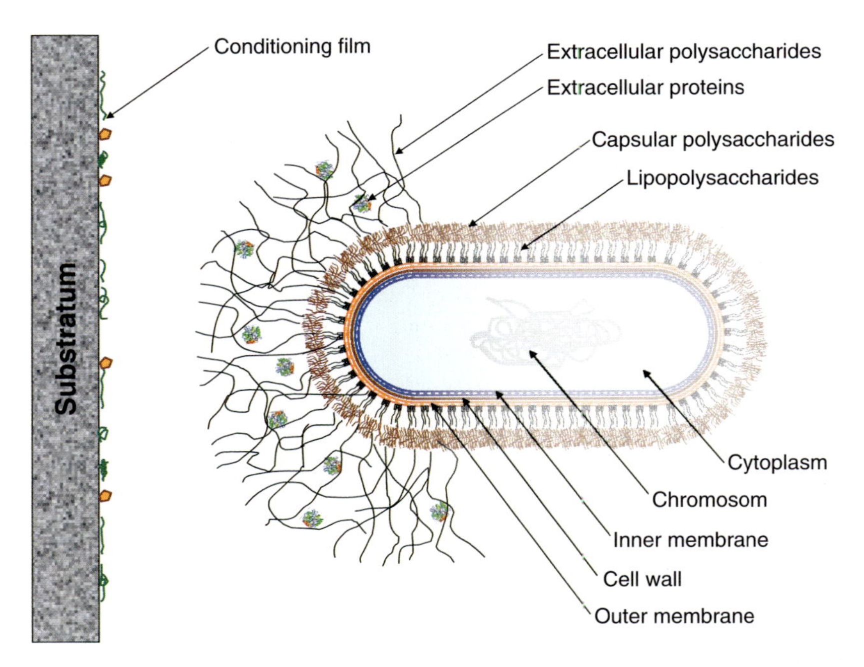

Fig. 3 Situation of a Gram-negative bacterium encountering a non-biological surface in water (after Flemming 2003)

that can overcome the effect. Apart from that, they will have to undergo the EU biocide guideline procedure, which is assessed to cost about 5–10 million € per substance. Further aspects of natural anti-fouling compounds are discussed in other chapters in this volume.

3. *Surfaces with lotus effect* (Neinhuis and Barthlott 1997). This effect relates to the "purity of the sacred lotus" (which is shared by less sacred cabbage leaves as well) based on the particular structure of the wax layers on the leaf surface. A highly hydrophobic pattern of needles will prevent water from moistening the surface due to the physicochemical interactions of three phases: solid, liquid and gaseous. By nature, this effect is not possible with immersed surfaces. Also, as soon as surface-active substances cover the hydrophobic pattern, surface tension decreases and water is no longer repelled. Thus, the lotus effect can be taken advantage of only on solid–air interfaces and only if no surfactants are used. However, other approaches, such as the influence of surface texture microstructure, have been pursued (Bers and Wahl 2004). They found that the surface structure of *Ophiura texturata* significantly repelled primary adhesion. However, the translation of this effect into technical application has not yet succeeded.

As a consequence, we will have to live with the fact that most surfaces can and will be colonized by microorganisms, which will cause fouling given the right conditions. Nevertheless, surfaces susceptible to biofouling may be "re-engineered" to discourage fouling. For example, Louie et al. (2006) recently demonstrated that thin-film composite

RO membranes could be coated with a polyether–polyamide copolymer (PEBAX 1657), which penetrated deeply into the membrane surface resulting in a smoother hydrophilic surface. Compared to uncoated controls, the coated RO membranes displayed a significant reduction in fouling by an oil/surfactant/water emulsion in trials lasting more than 100 days.

A more novel approach to designing low-fouling surfaces that is still in its early stages of development involves the application of molecular simulations to observe and measure *in silico* the dynamics of surface fouling by macromolecular substances. An example of this approach is illustrated in Fig. 4 in which a hydrated oligomer of bacterial alginate is shown undergoing rapid adsorption to the "surface" of an aromatic cross-linked polyamide RO membrane. The system potential energy is shown to decline substantially in this molecular dynamics simulation (inset), suggesting this type of adsorption interaction is energetically favourable. The aim of such modelling exercises is to introduce chemical modifications into the (polyamide) surface that will inhibit or impede such rapid macromolecular fouling.

The alginate oligomer is positioned initially (t = 0) above the membrane surface fragment (left vertical panel). The t = 0 positions are viewed from three spatial perspectives: side view (top), oblique view (middle), and top-down view (lower).

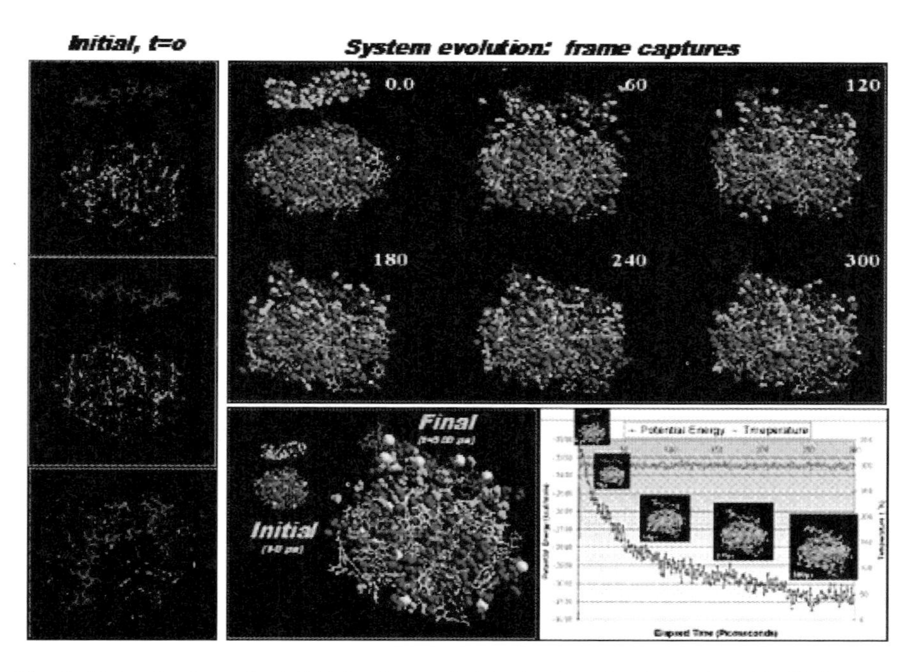

Fig. 4 Molecular dynamics (MD) simulation of alginate adsorption to a polyamide (PA) reverse osmosis membrane "surface" (H. Ridgway, AquaMem Scientific Consultants and Stanford University, unpublished data). The MD simulation shows rapid adsorption of a hydrated oligomer of bacterial alginate (*red*) to the PA surface (*brown*). Water molecules associated with the alginate are *green*. Membrane-associated water is *blue*. For more details see text. Frame capture times are given in picoseconds

Water and sodium counter ions have been hidden in the t = 0 images to better observe the alginate and membrane atoms. The right-hand upper panel indicates that alginate adsorption occurred relatively rapidly over a period of about 300 ps. As indicated by a decline in the system potential energy (inset graph), alginate adsorption was thermodynamically favourable. Alginate and membrane bonds are represented as sticks; water atoms are given as blue or green depending on whether they were initially associated with membrane or alginate, respectively.

2.1.2 Biofilm Management

In actual practice in a variety of systems, biofilm development can be successfully controlled through the application of a combination of cost-effective strategies,

Such a multi-factorial approach can be described as "biofilm management" and focuses on the limitation of factors that support biofilm growth above the "level of interference" (see Fig. 2). A successful example is the use of sand filters in order to remove biodegradable matter from cooling water in order to protect membrane units from biofouling (Griebe and Flemming 1998). Nutrient limitation is meanwhile an accepted approach to minimize fouling. A thorough fouling factor analysis is necessary, which must include in the first place the assessment of the nutrient situation. It has been explained earlier that nutrients have to be considered as potential biomass. High shear forces will limit excessive biofilm development, although they will not prevent it. Under high shear stress, there will be a selection for organisms that produce mechanically stable biofilms. Limiting the access of microorganisms will also be helpful; however, it must be taken into account that cells are particles that can multiply.

Cleaning is an important issue in biofilm management. For cleaning, cohesion of the biofilm and adhesion to surfaces have to be overcome, which are both aspects of the mechanical stability of biofilms. Koerstgens et al. (2001) have developed a film rheometer that allows for the quantification of biofilm stability with the apparent elasticity module ε as a relevant parameter. This research revealed that the EPS matrix is kept together by weak physicochemical interactions, which result in a fluctuating network of adhesion points. In compression experiments it was shown that until a yield point σ is reached, biofilms behave as gels with constant partner groups responsible for the adhesion. After exceeding σ, the gel breaks down, the partners of the adhesion points change and the biofilm behaves as a highly viscous fluid. This is why biofilms are slippery. In a model system with *Pseudomonas aeruginosa*, it was shown that Ca^{2+} ions increase the stability of the network by bridging alginate molecules, which are the main component of *P. aeruginosa* EPS. Mg^{2+} did not show such an effect, but Fe^{2+}, Fe^{3+} and Cu^{2+} did. Most commercial cleaners and biodispersants, however, proved ineffective in this testing system. An effective weakening of the EPS matrix can be achieved by enzymes (Johansen et al. 1997). However, this is not a fast effect and, in practice, it has proven transient and ineffective in many cases (e.g. Klahre et al. 1998). This is not surprising as EPS, like other structural biopolymers, are not readily biodegradable. Also, continuous use of enzymes will select for organisms producing EPS that are not susceptible to these enzymes.

An important aspect in cleaning is the use of surfaces to which biofilms do not attach strongly. Such materials have been developed and tested for anti-fouling on ship hulls and fishing nets, with silcones as a promising class of compound (Terlezzi et al. 2000; Estarlich et al. 2000; Holm et al. 2000). Anti-fouling polymer coatings were mentioned above (Louie et al. 2006). Modelling surface–foulant interactions (as described above) should help elucidate how anti-fouling coatings and anti-fouling surface treatments prevent primary macromolecular adsorption.

Electric fields have been used both for prevention of microbial adhesion and for inhibition of biofilm growth (Matsunaga et al. 1998, Kerr et al. 1999; Schaule et al., 2008). Practical observation, however, has shown that all kinds of electrodes immersed into water can be colonized and fouled by biofilms. Another approach to slow down biofouling processes and to facilitate cleaning is the use of coatings that can change their surface properties reversibly, induced by external stimuli such as light, temperature or pH value (Flemming, current research). Very interesting is the observation that surfaces with pulsed polarization show significantly lower biofilm growth over time (Schaule et al. 2008).

2.1.3 Biofilm Monitoring

It is of great importance to monitor biofilm development in order to optimize the time-course and effectiveness of countermeasures. This is not possible by sampling of the bulk water phase. Such samples give no information about the site, the extent and the composition of a biofilm and they generally underestimate by orders of magnitude the true microbial (surface) burden of a system. Although biofilms contaminate the water phase, they do so not on a constant basis but very irregularly. Biofilm cells may erode, but sloughing events may happen as well, leading to intermittent high cell numbers in the bulk water phase. Thus, biofilm monitoring must be performed using representative surfaces.

Conventional methods rely on sampling of defined surface areas or on exposure of test surfaces ("coupons") with subsequent analysis in the laboratory. A classical example is the "Robbins device" (Ruseska et al. 1982), which consists of plugs smoothly inserted into pipe walls, experiencing the same shear stress as the wall itself. After given periods of time, they are removed and analysed in the laboratory for all biofilm-relevant parameters. The disadvantage of such systems is the time-lag between analysis and result. Jacobs et al. (1996) described a simple spectrophotometric monitoring method using a nucleotide fluorescent stain (DAPI) and automatic measurement.

Other methods have been invented that report biofilm growth on-line, in real time and non-destructively. They all are based on physical methods. One example is the fibre optical device (FOS), which is based on a light fibre integrated in the test surface, measuring the scattered light of material deposited on the tip. The principle of the sensor is schematically depicted in Fig. 5a, a typical graph is shown in Fig. 5b (Tamachkiarow and Flemming 2003). Detection of autofluorescence of

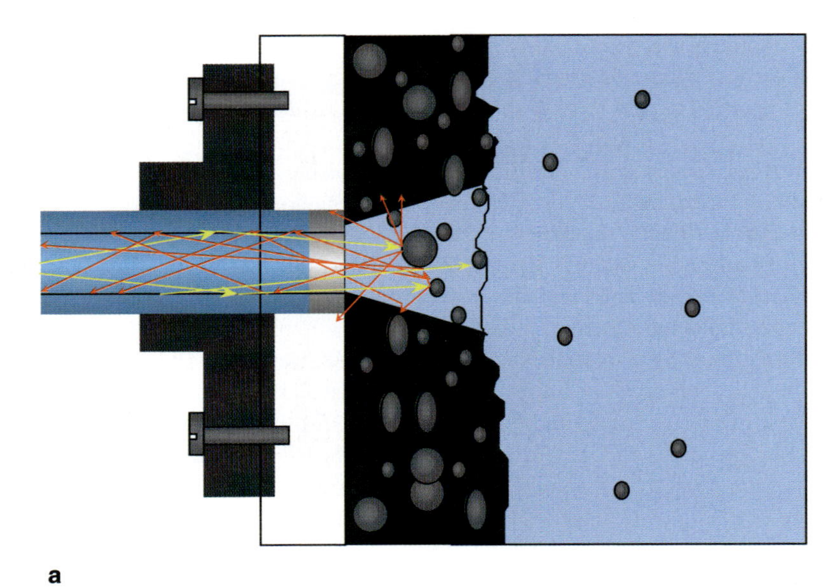

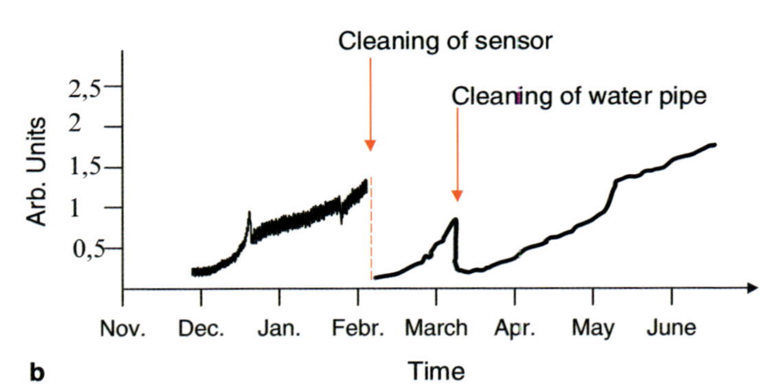

Fig. 5 **a** Schematic depiction of a fibre optical device (FOS). The tip of the fibre is integrated into the water-exposed surface. Light is coupled in by the sending fibre. Material deposited on the tip will scatter light, which is collected in the reading fibre. **b** Typical graph of intensity of backscattered light as provided by the FOS (after Tamachkiarow and Flemming 2003)

biomolecules by spectroscopy allows differentiation of biological material in the deposit from abiotic material.

Another method uses two turbidity measurement devices, one of which is constantly cleaned. The difference of the signals is proportional to the biomass developing on the non-cleaned window (Klahre et al. 2000). Nivens et al. (1995) have given an excellent overview on continuous non-destructive biofilm monitoring techniques, including FTIR spectroscopy, microscopic, electrochemical and piezoelectric techniques, which have also been systematically described by Flemming (2003).

3 Conclusions

An "integrated anti-fouling strategy"will not aim to kill all organisms in a system but keep them below a threshold of interference. The strategy has to be based on:

1. Multi-factorial analysis of the fouling situation
2. Installation of early warning systems
3. Limiting nutrient availability where ever possible (raw water, materials, additives etc.)
4. Prioritizing cleaning over killing
5. Effective and representative monitoring of cleaning measures

Any step towards a better understanding of biofilm growth and properties will add to the "menu" and expand the possibilities for a flexible, effective and environmentally suitable response to biofouling.

References

Armstrong E, Boyd KG, Pisacane A, Peppiatt CJ, Burgess JG (2000) Marine microbial natural products in antifouling coatings. Biofouling 16:215–224

Barbeau J, Gauthier C, Payment P (1998) Biofilms, infectious agents, and dental unit waterlines: a review. Can J Microbiol 44:1019–1028

Bers AV, Wahl M (2004) The influence of natural surface microtopographies on fouling. Biofouling 20:43–51

Callow ME, Pitchers RA, Santos R (1988) Non-biocidal anti-fouling coatings. In: Houghton DR, Smith RN, Eggins HOW (eds) Biodeterioration, 7, Elsevier Applied Sciences, New York, pp. 43–48

Costerton JW, Stewart PS, Greenberg EP (1999) Bacterial biofilms: a common cause of persistent infections. Science 284:1318–1322

Emde KME, Smith DW, Facey R (1992) Initial investigation of microbially influenced corrosion (MIC) in a low temperature water distribution system. Water Res 26:169–175

Estarlich FF, Lewey SA, Nevell TG, Thorpe AA, Tsibouklis J, Upton AC (2000) The surface properties of some silicone and fluorosilicone coating materials immersed in seawater. Biofouling 16:263–275

Faille C, Membre JM, Tissier JP, Bellon-Fontaine MN, Carpentier B, Laroche MA, Benezech T (2000) Influence of physicochemical properties on the hygienic status of stainless steel with various finishes. Biofouling 15:261–274

Fiedler K, Lindner M, Edel B, Wallbrecht F (1998) Infection from the communion cup: an underestimated risk? Zbl Hyg Umweltmed 201:167–188

Flemming HC, Greenhalgh M (2008) Concept and consequences of the EU biocide guideline. Springer Ser Biofilms. doi: 10.1007/7142_2008_12

Flemming HC, Schaule G (1996) Measures against biofouling. In: Heitz E, Sand W, Flemming H-C (eds.) Microbially influenced corrosion of materials – scientific and technological aspects. Springer, Heidelberg Berlin New York, pp. 121–139

Flemming HC, Tamachkiarowa A, Klahre J, Schmitt J (1998) Monitoring of fouling and biofouling in technical systems. Water Sci Technol 38:291–298

Flemming HC, Schaule G (1988) Biofouling on membranes – a microbiological approach. Desalination 70:95–119

Flemming H-C, Schaule G, McDonogh R, Ridgway HF (1994) Mechanism and extent of membrane biofouling. In:Geesey GG, Lewandowski Z, Flemming H-C (eds.)Biofouling and biocorrosion in industrial water systems.Lewis, Chelsea, MI, pp. 63–89

Flemming HC (2002) Biofouling in water systems – cases, causes, countermeasures. Appl Environ Biotechnol 59:629–640

Flemming H-C (2003) Role and levels of real time monitoring for successful anti-fouling strategies. Water Sci Technol 47(5):1–8

Griebe T, Flemming HC (1998) Biocide-free antifouling strategy to protect RO membranes from biofouling. Desalination 118:153–156

Holm ER, Nedved BT, Phillips N, Deangelis KL, Hadfield MG Smith CM (2000) Temporal and spatial variation in the fouling of silicone coatings in Pearl Harbour, Hawaii. Biofouling 15:95–107

Jacobs L, De Bruyn EE, Cloete TE (1996) Spectrophotometric monitoring of biofouling. Water Sci Technol 34(5–6):533–540

Johansen C, Falholr P, Gram L (1997) Enzymatic removal and disinfection of bacterial biofilms. Appl Environ Microbiol 63:3724–3728

Kerr A, Hodgkiess T, Cowling MJ, Smith MJ, Beveridge CM (1999) Effect of galvanically induced surface potentials on marine fouling. Lett Appl Microbiol 29:56–60

Kilb B, Lange B, Schaule G, Wingender J, Flemming HC (2003) Contamination of drinking water by coliforms from biofilms grown on rubber-coated valves. Int J Hyg Environ Health 206(6):563–573

Klahre J, Flemming HC (2000) Monitoring of biofouling in papermill water systems. Water Res 34:3657–3665

Klahre J, Lustenberger M, Flemming HC (1998) Microbial problems in paper production. Part III: monitoring. Das Papier 52:590–596

Koenig DW (1997) Microbiology of the space shuttle water system. Water Sci Technol 35(11–12):59–64

Koerstgens V, Wingender J, Flemming HC, Borchard W (2001) Influence of calcium-ion concentration on the mechanical properties of a model biofilm of *Pseudomonas aeruginosa*. Water Sci Technol 43(6):49–57

LeChevallier MW (1990) Coliform regrowth in drinking water: a review. J Am Water Works Assoc 92 (11):74–86

LeChevallier MW (1991) Biocides and the current status of biofouling control in water systems. In: Flemming HC, Geesey GG (eds) Biofouling and biocorrosion in industrial water systems. Springer, Heidelberg Berlin New York, pp. 113–132

Loeb FI, Neihof RA (1975) Marine conditioning films. In: Baier RE (ed) Applied chemistry at protein interfaces. Am Chem Soc, Washington, pp. 319–335

Louie JS, Pinnau I, Ciobanu I, Ishida KP, Ng A, Reinhard M (2006): Effects of polyether–polyamide block copolymer coating on performance and fouling of reverse osmosis membranes. J Membr Sci 280:762–770

Matsunaga T, Nakayama T, Wake H, Takahashi M, Okochi M, Nakamura N (1998) Prevention of marine biofoling using a condictive paint electrode. Biotechnol Bioeng 59:374–378

McBain AJ (2001) Do biofilms present a nidus for the evolution of antibacterial resistance?. In: Gilbert P, Allison D, Brading M, Verran J, Walker J (eds.) Biofilm community interactions: chance or necessity? BioLine Press, Cardiff, pp. 341–351

Morton LHG, Greenway DLA, Gaylarde CC, Surman SB (1998) Consideration of some implications of the resistance of biofilms to biocides. Int Biodeterior Biodegradation 41:247–259

Neinhuis C, Barthlott W (1997) Characterization and distribution of water-repellent, self-cleaning plant surfaces. Ann Bot 79:667–677

Nivens DE, Palmer RJ, White DC (1995) Continuous nondestructive monitoring of microbial biofilms: a review of analytical techniques. J Ind Microbiol 15:263–276

Reasoner DJ (1988) Drinking water microbiology research in the United States: an overview of the past decade. Water Sci Technol 20:101–107

Ruseska I, Robbins J, Lashen ES, Costerton JW (1982) Biocide testing against corrosion-causing oilfield bacteria helps control plugging. Oil Gas J 253–264
Schaule G, Griebe T, Flemming HC (2000) Steps of biofilm sampling and characterization in biofouling cases. In: Flemming HC, Griebe T, Szewzyk U (eds) Biofilms, investigative methods and applications. Technomics, Lancaster, PA, pp. 1–21
Schaule G, Rumpf A, Weidlich C, Mangold C, Flemming H-C (2008) Influence of polarization of electric conductive polymers on bacterial adhesion. Poster at IWA conference on biofilms, Wat Sci Techn, in press
Schulte S, Wingender J, Flemming H-C (2005) Efficacy of biocides against biofilms. In: Paulus W (ed.) Directory of microbicides for the protection of materials and processes. Kluwer, Doordrecht, pp. 90–120
Steinberg PD, de Nys R, Kjelleberg S (1997) Chemical defenses of seaweeds against microbial colonization. Biodegradation 8:211–220
Tamachkiarow A, Flemming HC (2003) On-line monitoring of biofilm formation in a brewery water pipeline system with a fibre optical device (FOS). Water Sci Technol 47(5):19–24
Terlezzi A, Conte E, Zupo V, Mazzella L (2000) Biological succession on silicone fouling-release surfaces: long term exposure tests in the harbour of Ischia, Italy. Biofouling 15:327–342
Tirrschof D (2000) Natural product antifoulants: one perspective on the challenges related to coatings development. Biofouling 15:119–127

An Example: Biofouling Protection for Marine Environmental Sensors by Local Chlorination

L. Delauney (✉) and C. Compère

Abstract These days, many marine autonomous environment monitoring networks are set up in the world. Such systems take advantage of existing superstructures such as offshore platforms, lightships, piers, breakwaters or are placed on specially designed buoys or deep sea fix stations. The major goal of these equipments is to provide in real time reliable measurements without costly frequent maintenance. These autonomous monitoring systems are affected by a well-known phenomenon in seawater condition, called biofouling. Consequently, such systems without efficient biofouling protection are hopeless. This protection must be applied to the sensors and to the underwater communication equipments based on acoustic technologies. This paper presents the results obtained in laboratory and at sea, with various instruments, protected by a localised chlorine generation system. Two other major protection techniques, wipers and copper shutters, are presented as well.

1 Introduction

Monitoring networks commonly use various sensors such as dissolved oxygen, turbidity, conductivity, pH or fluorescence units and, for specific matters, some underwater video systems such as cameras, video equipments and lights. For surface application the data gathered are generally transmitted in real time via satellite and for deep sea application data logger or wired networks are involved. In most cases the monitoring stations are autonomous, especially concerning the energy needs.

In addition to the numerous environmental monitoring stations used along continents, some specific measuring stations are deployed for other purposes in some specific areas where biofouling is very much present.

L. Delauney
Ifremer–*In Situ* Measurements and Electronics, B.P. 70, 29280, Plouzané, France
e-mail: laurent.delauney@ifremer.fr

Springer Series on Biofilms, doi: 10.1007/7142_2008_9

For example, systems for the monitoring of polluting wrecks (Marvaldi et al. 2006) are based on autonomous and real-time stations deployed in order to measure critical data nearby wrecks and to transmit them. These stations are equipped with conventional seawater physicochemical sensors and with acoustic transducers for underwater data communication (for an example, see Fig. 1). They are generally deployed from 15 m depth down to whatever is needed, and for long-term monitoring they are deployed for 1 month up to 6 months during which no maintenance is possible.

For deep sea research, down to 3,000 m, specialized autonomous stations measure physicochemical parameters and record pictures and movies. Some areas of interests are, for example, fumes of hydrothermal sites (Sarrazin et al. 2007). For these applications the autonomy must be provided up to 1 year. The compactness of these stations is crucial, since the equipment is deployed by a remotely operating vehicle.

These autonomous monitoring systems are affected by a well-known phenomenon in seawater condition, called biofouling. The major goal of these equipments is to provide in real time reliable measurements without costly frequent maintenance. In deep sea conditions this maintenance is nearly impossible to provide. For coastal applications it is quite well accepted now, that for economically viable in situ monitoring systems, the maintenance must not be performed more frequently than 2 month (Blain et al. 2004). Consequently, such systems without efficient biofouling protection are hopeless. This protection must be applied to the sensors and to the underwater communication equipments that are based on acoustic technologies.

Biofouling in seawater, during productive period, can occur very rapidly and lead to poor data quality in less than 2 weeks. As shown in Figs. 2 and 3, the biofouling species involved differ very much from one location to another (Le Haitre et al. in press).

Very often, this biofouling gives rise to a continuous shift in the measurements. Consequently, the measurements can be out of tolerance and the data become useless. Video systems such as cameras, video equipments and lights are also

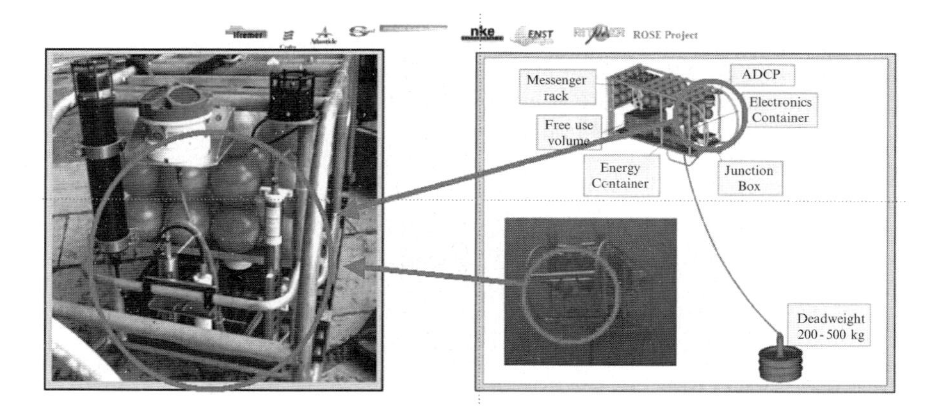

Fig. 1 Autonomous monitoring station for polluting wreck surveillance (Ifremer – ROSE project)

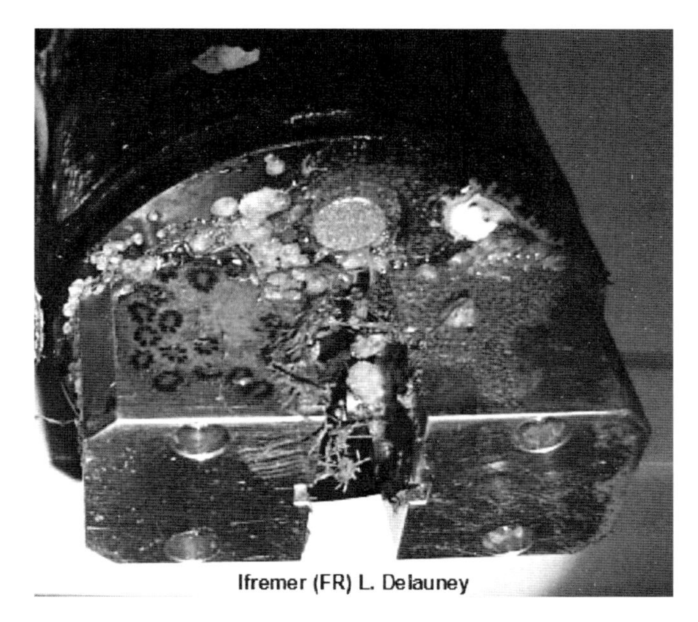

Fig. 2 Fluorometer after 30 days in Helgoland (Germany) during summer

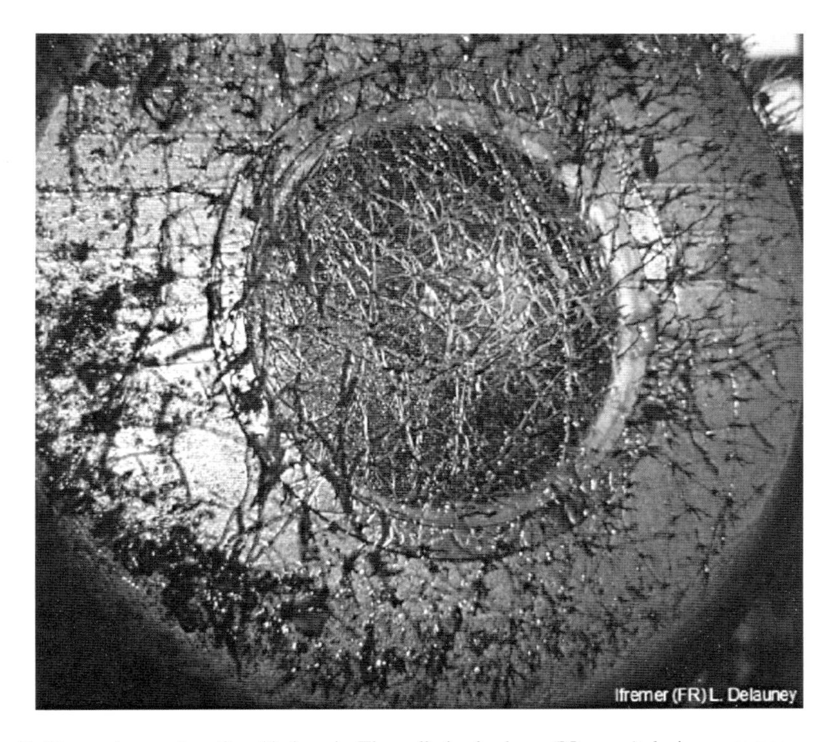

Fig. 3 Transmissometer after 40 days in Throndheim harbour (Norway) during summer

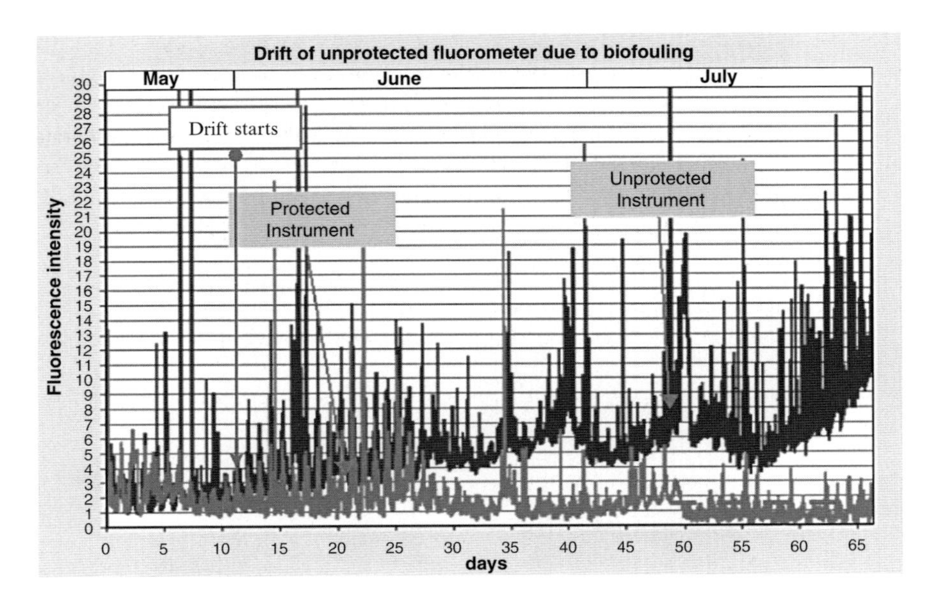

Fig. 4 Drift of an unprotected fluorometer due to biofouling development on the optics

compromised by biofouling. Pictures become blurred or noisy, and lights lose efficiency since their intensity decreases because of the scattering effect of biofilm and macro-fouling.

As shown in Fig. 4, after 7 days, owing to biofouling on the sensitive part of the sensor, a drift can be observed in the measurements produced by a fluorescence sensor (Delauney and Cowie 2002). This kind of optical sensor is very sensitive to biofouling since even a very thin biofilm on the optics interferes with the measurement process and gives rise to over-evaluated measurements.

2 Biofouling Protection Methods for Oceanographic Sensors

Biofouling protection for oceanographic sensors is a difficult task requiring specifications driven by three important characteristics:

1. It should not affect the measurement.
2. It should not consume too much energy in order to preserve the initial autonomy of the autonomous monitoring system.
3. It should be reliable even in aggressive conditions for technological systems (seawater corrosion, sediments, hydrostatic pressure, etc.).

Consequently, few techniques are actually used, and none of them are based on an antifouling paint because the area to be protected cannot be coated with any opaque substance.

Otherwise the measurements taken by the sensor can be completely compromised. We must know that the goal of a biofouling protection is to limit as best as possible the growth of biofouling on the sensitive part of the sensor. For every type of sensor, such as optical sensors (fluorometer, turbidimeter, transmissometer, dissolved oxygen), membrane sensors (pH, dissolved oxygen) or electrochemical sensors (conductivity), the interface between the media to measure and the sensitive area of the sensor must remain clear.

Currently, three biofouling protection systems for oceanographic sensors are in use for operational deployments:

4. Strictly mechanical devices – wipers
5. An "uncontrolled" biocide generation system based on a copper auto-corrosion mechanism
6. A "controlled" biocide generation system based on a localized seawater electro-chlorination system

These three techniques are commonly used on oceanographic sensors, each having both advantages and disadvantages.

2.1 Mechanical In Situ Wiper Systems

Biofouling protection system using wipers are based on a mechanical process that has to be adapted to the instrument from the early stages of design. Consequently, such systems can be found on the instruments if the sensor manufacturers have taken biofouling problem into account. Figure 5 shows such a biofouling protection system based on wipers on a commercial multiparameter probe (YSI 6600 EDS).

Fig. 5 Multiparameter probe with a mechanical biofouling protection based on wipers (photo: L. Delauney, Ifremer, France)

Fig. 6 Wiper biofouling protection after 150 days of operation (photo: L. Delauney, Ifremer, France)

The device consists of two distinct wipers that use three "scrapers". Two of them are made of sponge and directly wipe the sensors' optical interfaces for fluorescence and turbidity measurements and a brush with long bristles has been implemented to clean the non optical sensors such as pH and oxygen sensors that are based on membrane techniques.

This biofouling protection technique is effective as long as the scrapers are in good condition and as long as the geometry of the sensor head is suitable for this cleaning process. The problems with this technique are mainly the mechanical complexity of the system. The water tightness of the wipers' axles as well as the short-term robustness of the wiper motion device are major weaknesses. As said earlier, sensors with a non-flat measurement interface, as shown on top right of Fig. 6, cannot be protected with this technique. Currently, sensor manufacturers are searching for new, alternative biofouling protection techniques to simplify their instruments and consequently to improve their reliability.

2.2 Biofouling Protection by "Uncontrolled" Biocide Generation: Copper Release

Copper is known for its biocide properties (Manov et al. 2004). As copper corrodes in seawater, oxidized molecules are released into the water rather than remaining on the metal surface. Copper interferes with enzymes on cell membranes and prevents cell division.

Copper is toxic at high concentrations, and to achieve this, the principle is to catch in a "copper cell" a small volume of seawater on the sensor measurement interface. In this way, the sensor interface will be in contact with a solution having increasing concentration of Cu^{2+} ions as long as the cell is closed.

Many manufacturers use this protection technique. Some of them build the sensor head totally in copper and add a wiper system to scrap the optics.

A specific equipment can be found that allows to equip any sensor with a copper cell system more commonly named a "copper shutter." A motor drives the mechanism for shutters that open for measurements and close for biofouling protection over the optical windows. It keeps the sensor very close to the copper system releasing toxic copper, and the closed cell allows darkness, thereby reducing biofouling.

Such protection is not easy to implement on an existing sensor. The copper screen with the stepper motor needs to be placed on the sensor in a way that the copper screen includes a small volume of water over the sensor measurement interface. An example of such a system on a Seapoint fluorometer is shown in Fig. 7. To maximize the effectiveness of the protection, it was necessary to implement a copper cell and to coat the entire sensor head with copper.

Results obtained with such a system (Delauney and Compère 2006), when the implementation is made exactly as described earlier, can be satisfactory for long-term deployment. The optics remained clean during the 3 months of deployment in coastal area in Brest (FR) and during summer season.

Some results obtained with copper tubing and copper shutter on optical instruments are presented by Manov et al. (2004). They conclude that "copper-based antifouling systems have shown marked improvement in obtaining long-term dataset for acquisition of optical measurement."

However, this method can lead to the following problems:

- Copper corrosion produces copper oxide precipitates, which can interfere with the measurements.
- Copper corrosion produces bubbles on the copper-coated surfaces, which are trapped in the copper cell close to the measurement interface. This can interfere

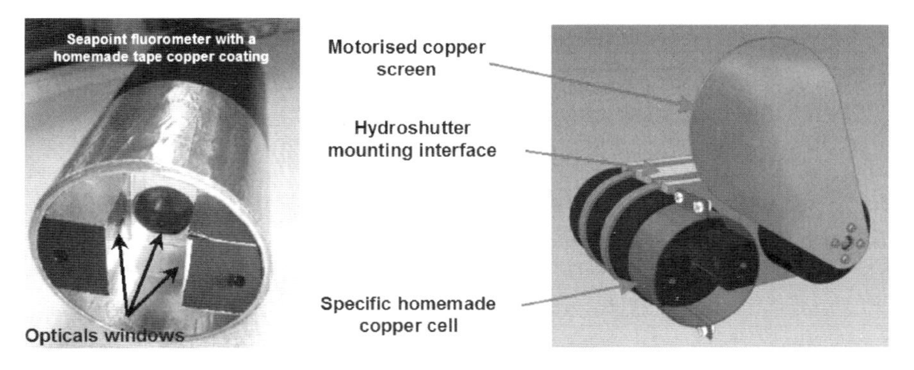

Fig. 7 Biofouling protection with a HOBI Labs copper shutter HydroShutter-HS (HOBI Labs, http://www.hobilabs.com) (photo and drawing: L. Delauney, Ifremer, France)

with the measurements, especially if the sensor is based on an optical technology. Bubbles are trapped easily since the system is based on a closed cell.

- When the copper screen is closed, it is of course impossible to take any measurements. The screen must be closed for sufficient time in order to get an effective protection, but then, it is impossible to take high-frequency measurements. And actually, with the wide band data transmission systems getting more and more common, and because of tide duration for which scientists need a good time resolution, copper shutter can be a limitation.
- The mechanical system involved is quite fragile. It is based on a stepper motor that cannot tolerate any mechanical obstacles; otherwise, the fragile gear box system will break. Consequently, the copper screen must be adjusted very precisely in order to fit sufficiently watertight to the copper cell. Any misplacement of the copper screen with the copper cell can lower the biofouling protection or interfere with the mechanism.

2.3 Biofouling Protection by "Controlled" Biocide Generation: Localized Seawater Electro-Chlorination System

This technique is the adaptation for biofouling protection of in situ oceanographic sensors, of a largely used technique to protect seawater cooling system for industry (Satpathy 2006). For our application, only the sensor transducing interface area will be protected, which explains the term "localized." Biocide generation is obtained by seawater electrolysis. With this technique, we can achieve a powerful biocide generation, hypochlorous acid, which can be concentrated as best as possible, on the sensor transducing interface area.

This technique has many advantages:

- Biocide generation is controlled. Consequently, the biocide quantity can be adjusted and on/off periods can be arranged as needed. On/Off periods are useful in arranging biocide-free periods so as to obtain the measurements in good environmental conditions. Moreover, the control of the biocide generation intensity is very important in order to adapt the biocide generation in function of the biofouling colonization.
- The energy needed for such systems is fully compatible with autonomous coastal monitoring systems and deep sea autonomous monitoring stations.
- The system is very robust and reliable since no mechanical parts are in motion.
- The system is easily adaptable to existing sensors even for usage at high depth.
- The system can be integrated to the sensors by manufacturers.

As shown in Fig. 8, the system is made of an electrode placed around the sensor transducing interface area, in this case the optic. This electrode is connected to an electro-chlorination unit. This unit can be a separate electronic container as shown in Fig. 8 or can be integrated inside the instrument.

Fig. 8 Biofouling protection of a fluorometer by localized seawater electro-chlorination (©Ifremer) – Protection system under Ifremer licence – NKE, Hennebont (56), France (photo: Ifremer, France)

This biofouling protection technique has been successfully used for many in situ coastal monitoring systems (Delauney et al. 2002) even immersed at low depth, 2 or 3 m, where biofouling development is intense (cf. Fig. 4), as well as for medium-depth (15–100 m) stations (Marvaldi et al. 2006) or even for high-depth stations down to 2,000 m where biofouling can appear close to hydrothermal vents (Sarrazin et al. 2007).

3 Localized Electro-Chlorination Biofouling Protection Results

The local chlorination technique was applied on various instrumental technologies, optic (turbidity, fluorescence, oxygen), electrodes (conductivity) and glass membrane (pH). For every test, in the laboratory or at sea, two sensors were placed simultaneously, one unprotected and one protected by the local chlorination device. The measurements were internally recorded or when possible recorded by a laptop for real time data analysis. When possible, some water was sampled and reference analysis was done in order to follow the eventual drift of the sensors in real time.

3.1 Determination of Possible Interference of Electro-Chlorination with the Measurement

Before implementing the system on the instruments, it is necessary to check possible interfering effects on the measurements. Electrodes in the vicinity of a sensor may perturb measurement. Consequently, a "Laboratory check" and a specific calibration is necessary. In the same way, biocide molecules can interfere with membranes or induce local water property modifications, and this effect must be studied even if it can be overcome by scheduled chlorination.

All instruments have been tested in the laboratory with standard solutions or standard analytical methods in order to calibrate the signal of a protected instrument vs. an unprotected one (Delauney and Compère 2006). Depending on the parameter measured, the following standard methods were used:

- Oxygen: Winkler titration (Aminot and Kerouel 2004)
- Fluorescence: Ifremer Fluorescein protocol (Delauney and Le Guen 2003)
- Conductivity: natural seawater sampling and Reference Guildline salinometer analysis (Aminot and Kerouel 2004; Fofonoff and Millard 1983)

The laboratory check for interference of chlorine with the measurement consists in comparing the responses of two instruments, one of them equipped with the local chlorination device. Two steps are involved. The first one determines the adverse effect of the local chlorination hardware. This can possibly be overcome by a specific calibration. The second determines the adverse effect of the chlorine generation, which can possibly be overcome by a scheduling of the chlorination generation.

3.2 *Biofouling Protection Field Test on Conductivity Sensor*

Local chlorination protection device was tested on the conductivity instrument at St. Anne du Portzic, Brest, in France. Two instruments were placed on site, one protected and one unprotected. The local chlorination scheduler is adjusted to last for 3 months with no maintenance.

Figure 9 shows the measurement obtained during the field test in St. Anne du Portzic, Brest, in France. The dark top curve shows measurement from the protected instrument. The light top curve that then drops shows measurement from the unprotected instrument. The bottom curve shows the difference between the two signals. The drift started after 80 days; it remained linear up to the 110th day and then became exponential until the end (133 days).

The reference measurements obtained from water withdrawn and subjected to Guildline salinometer conductivity analysis (large dots in Fig. 9) show a slight shift of the protected instrument (0.5 PSU). This drift is probably due to a stop in the chlorination process after the 100th day due to a lack of energy (failure of battery).

Figure 10 shows the unprotected conductivity sensor (left) and the protected one (right) after 133 days of deployment. Visually, we can perceive the effectiveness of the biofouling protection. It is even surprising how the local chlorination system placed inside the white probe housing has protected the outside.

The local chlorination biofouling protection for the conductivity sensor is efficient as was clearly shown during St. Anne du Portzic Brest test for a continuous period of 133 days. The drift of the unprotected instrument started after 80 days, in August.

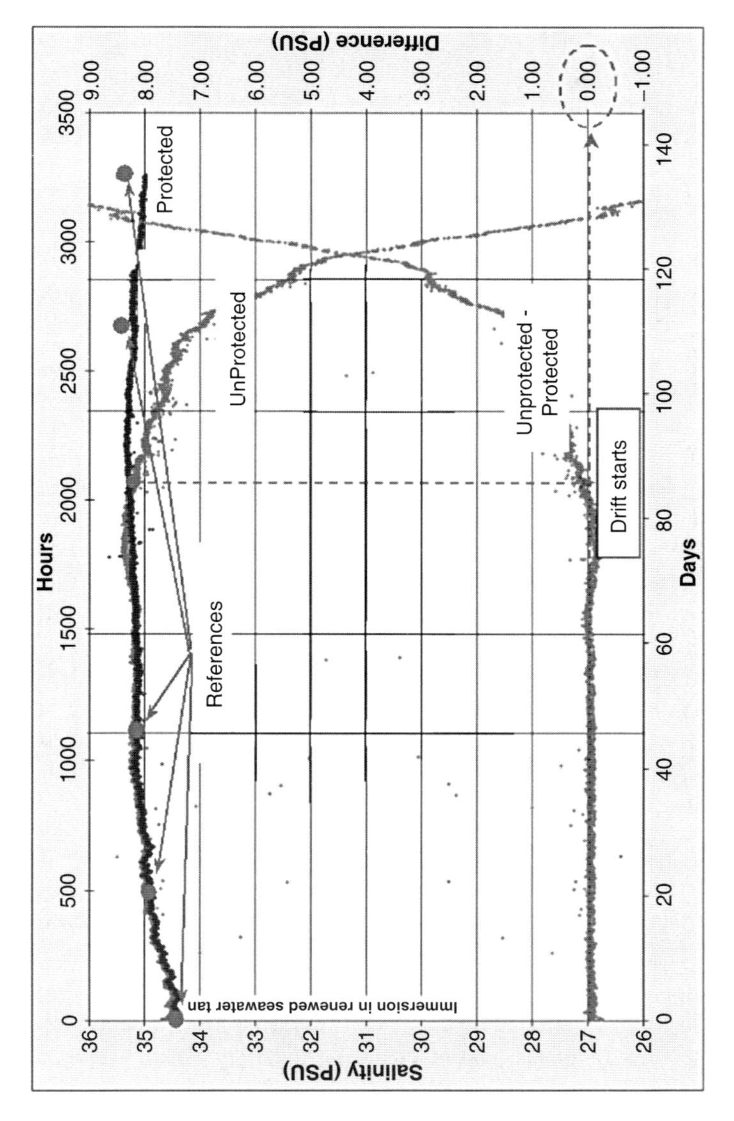

Fig. 9 Conductivity sensor in situ results (133 days duration), 3 June–16 October 2003, Brest

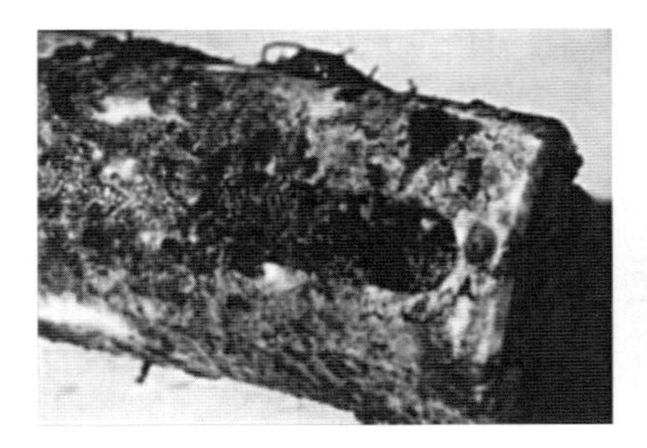
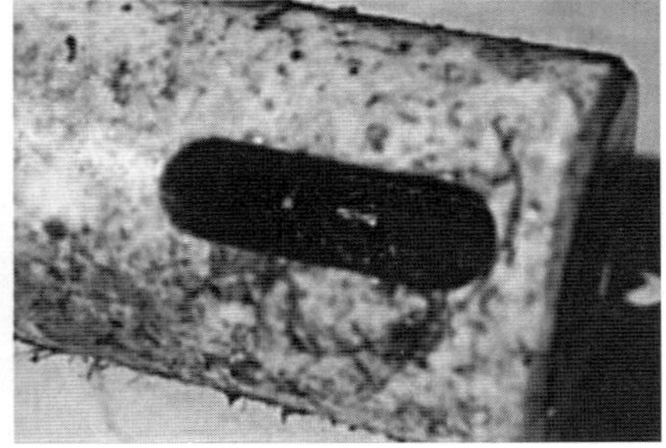

Fig. 10 Conductivity sensor: unprotected (*left*), protected (*right*) (photo: Ifremer, France)

3.3 Biofouling Protection Field Test on Optical Oxygen Sensor

The local chlorination protection device was tested on the oxygen instrument at St. Anne du Portzic, Brest, France. Two instruments were placed on site, one protected and one unprotected. As previously, the local chlorination scheduler was adjusted to last for 3 months with no maintenance.

Figure 11 shows the measurement obtained from day 110 to day 140 during the field test in St. Anne du Portzic, Brest.

The top dark curve shows measurement from the protected instrument. The top light curve that then drops shows measurement from the unprotected instrument. The bottom curve shows the difference between the two signals. The drift started after 127 days. The protected optode signal is very good up to day 160. The Winkler analysis, which was done until the 160th day, confirms this result.

The local chlorination biofouling protection for the oxygen optode sensor is efficient as was clearly shown during St. Anne du Portzic Brest test for a continuous period of 160 days. The drift started after 130 days, in April. The protected sensor showed a temporary failure from day 160 to day 170.

3.4 Biofouling Protection Field Test on Fluorescence Sensor

A local chlorination protection device was tested on fluorescence measurement instruments at Millport island, Scotland, for 100 days. Two instruments were placed on site, one protected and one unprotected. The local chlorination scheduler was adjusted to last for 3 months with no maintenance.

Figure 12 shows the measurement obtained during a field test at Millport Island. For this experiment the two fluorometers were immersed at 1.5 m depth on Millport island (Scotland) in 2004, with the collaboration of Dr P. Cowie (GMTC, UK) during the European BRIMOM Project. The dark curve shows measurement from the unprotected instrument. The light curve shows measurement from the protected

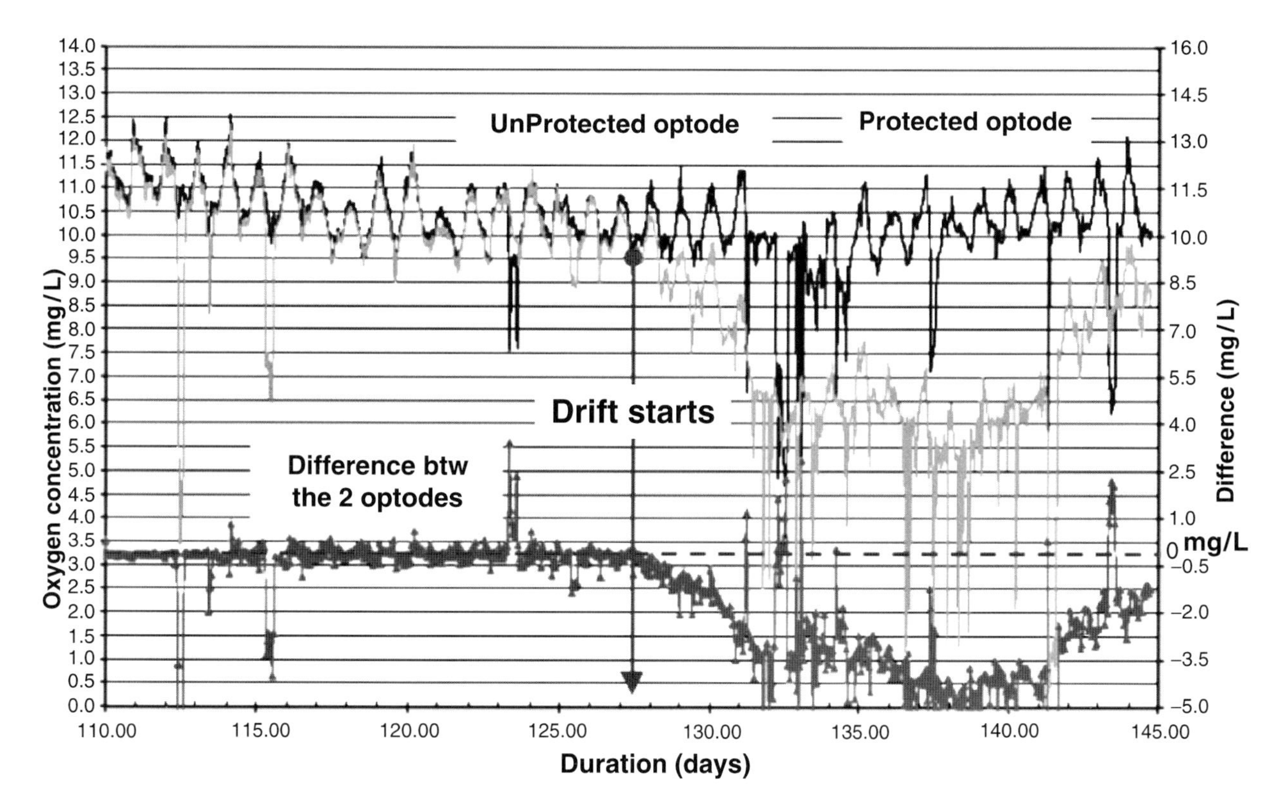

Fig. 11 Oxygen optode sensor in situ results (190 days duration), January–July 2004, Brest

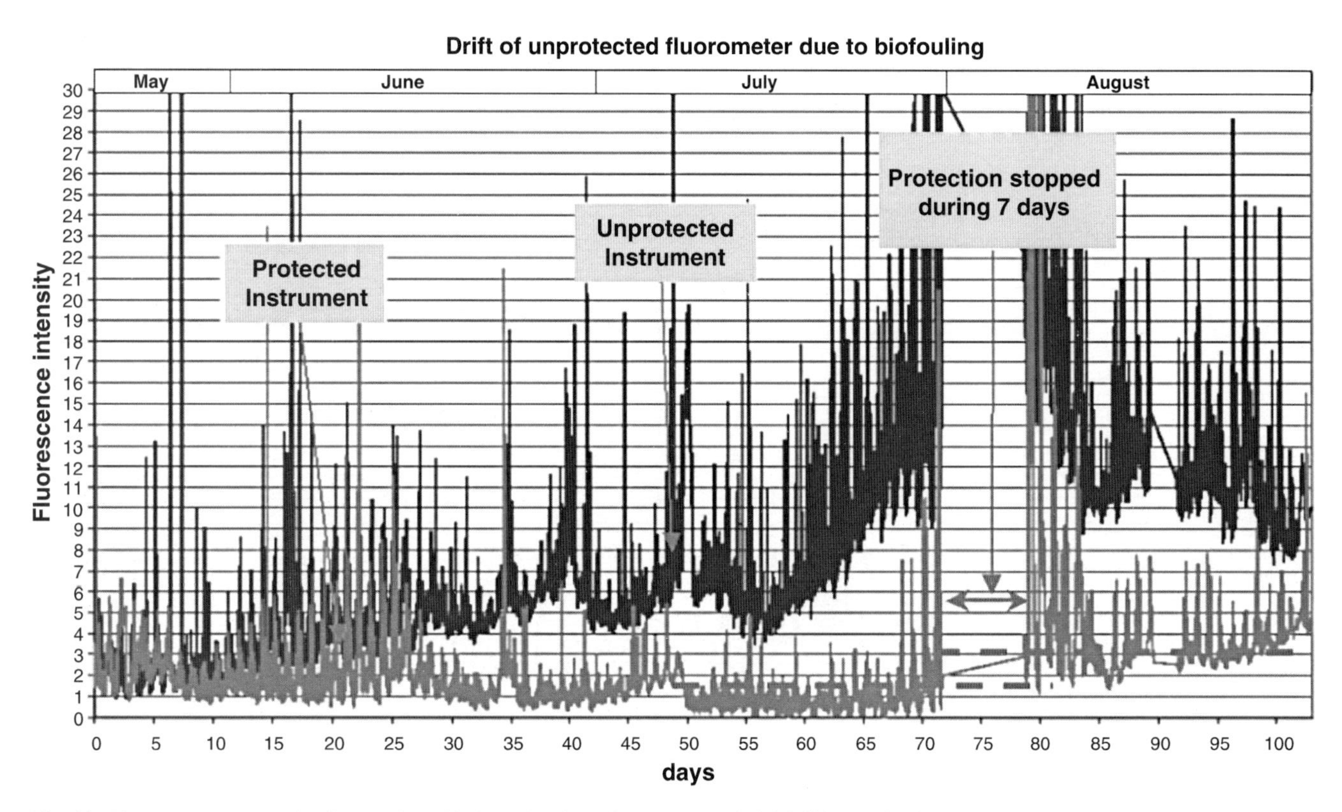

Fig. 12 Fluorescence sensor in situ results (100 days duration), May–August 2004, Millport Island

instrument. The unprotected fluorometer starts to drift after 7 days of immersion and keeps drifting during the 80 days of experiment. The protected fluorometer does not show any drift until day 70. It is very interesting to see that after day 70, the protected instrument was not protected for 7 days. Consequently, biofouling started and a non-negligible measurement drift was observed. This incident shows that if the protection system is not mechanical, but rather chemical, any beginning biofouling will induce a bias on the measurement which will be difficult to remove.

The local chlorination biofouling protection for the fluorometer sensor was effective as was clearly shown during the Millport Island test for a continuous period of 70 days. After 70 days, the small drift observed is mainly due to a chlorinator batteries failure for 7 days.

4 Conclusion

For the last 10 years, oceanographic sensor biofouling protection has improved quite a lot. Owing to the intense technological development of in situ autonomous monitoring systems, the biofouling problem for such systems has been a technological one which needed to be solved. The prohibition of tributyl tin as a biocide, which was used by some manufacturers to protect their sensors, has pushed researchers to find alternative methods to protect sensors from biofouling.

Wipers, scrapers, and other mechanical systems are interesting solutions but very often lead to mechanical failure, resulting in water leakage inside the instruments and, thus, destroying the entire equipment.

Copper shutter scan work quite well but are still complicated to implement, and the biocide generation is uncontrolled, which can lead to problems if the biocide formation interferes with the sensor measurements.

Localized electro-chlorination biofouling protection is actually a promising and an advanced solution for in situ oceanographic sensors, since many successful in situ results have been obtained and sensor manufacturers can integrate in their instruments a compact, simple, robust and low energy requiring solution.

This technique has been tested on many oceanographic instruments for coastal and deep sea monitoring. Very encouraging results have been obtained for the parameters commonly measured for marine monitoring. Every deployment has been a success for a duration of up to 160 days. Various types of biofouling, such as biofilm, algae, and barnacles, have been prevented on different types of instruments and different types of measurement technologies. The system can be adapted to many kinds of instruments quite easily. The energy requirement is compatible with autonomous monitoring.

Special care should be taken for some sensitive parameters such as oxygen or fluorescence. The chlorination period must be scheduled in order to leave free time intervals to take the measurements. The system is now used by Ifremer for autonomous coastal monitoring and allows a reasonable maintenance frequency of 3 months, with high-quality measurements obtained.

Acknowledgements Ifremer thanks Phil Cowie from UMBS, Millport, Scotland, for taking very good care of the Fluorometer experiment in Millport.

References

Aminot A, Kerouel R (2004) Hydrologie des écosystèmes marins, paramètres et analyses, mesure de la concentration en oxygène dissous, Chap. VII. pp 92–118. Publisher: INRA Edition. F-78026 Versailles France

Blain S, Guillou J, Tréguer P, Woerther P, Delauney L (2004) High frequency monitoring of the coastal marine environment using the MAREL buoy. J Environ Monit 6:569–575

BRIMOM (biofouling resistant infrastructure for measuring, observing and monitoring), project no. EVRI-CT-2002-40023

Delauney L, Compère C (2006) Biofouling protection for marine environmental optical sensors. International conference on recent advances in marine antifouling technology (RAMAT), Chennai (Madras), India. Proceedings of the International Conference on recent advances in marine antifouling technology. Allied Publishers Pvt. Ltd

Delauney L, Cowie P (2002) BRIMOM (biofouling resistant infrastructure for measuring, observing and monitoring) report (project no. EVR1-CT-2002-40023). http://cordis.europa.eu/data/PROJ_FP5/ACTIONeqDndSESSIONeq112362005919ndDOCeq221ndTBLeqEN_PROJ.htm

Delauney L, LeGuen Y (2003) Calibration procedure for in-situ marine fluorometer. In: International metrology conference 2003 proceedings, Toulon, France. CD Publisher: Collège Français de métrologie. 75015 Paris

Delauney L, Lepage V, Festy D (2002) BRIMOM (biofouling resistant infrastructure for measuring, observing and monitoring), project no. EVR1-CT-2002-40023. http://cordis.europa.eu/data/PROJ_FP5/ACTIONeqDndSESSIONeq112362005919ndDOCeq221ndTBLeqEN_PROJ.htm

Delauney L, Compère C, Lehaitre M (2005) Office for official publications of the european communities - Luxembourg. Proceeding of the fourth international conference on EuroGOOS. pp 397–403

Fofonoff NP, Millard RC (1983) Algorithms for computation of fundamental properties of seawater, UNESCO Technical papers in marine science 44, UNESCO Division of Marine Science, Paris, France, 53 pp

Lehaitre M, Delauney L, Compère C (in press) Oceanographic Methodology series, Real Time Coastal Observing Systems for Marine Ecosystem Dynamics and Harmful Algal Blooms: Theory, Instrumentation and Modelling. 2008, ISBN 978-92-3-104042-9. UNESCO Publishing

Manov DV, Chang GC, Dickey TD (2004) Methods for reducing biofouling on moored optical sensors. J Atmos Ocean Technol 21(6):958–968

Marvaldi J, Coail Y, Legrand J (2006) ROSE, CMM'06 Caractérisation du Milieu Marin, 16–19. SeatechWeek 2006 - Brest -France

Sarrazin J, Blandin J, Delauney L (2007) TEMPO: a new ecological module for studying deep-sea community dynamics at hydrothermal vents (IEEE catalogue no. 07 EX). Oceans 2007 IEEE conference proceedings, Aberdeen, Scotland, 18–21 June 2007

Satpathy KK (2006) Biofouling and its control in power plants cooling water system. International conference on recent advances in marine antifouling technology (RAMAT), Chennai (Madras), India. Proceedings of the international conference on recent advances in marine antifouling technology. Allied Publishers Pvt. Ltd

YSI 6600 EDS (Extended Deployment System) – Clean SweepTM – http://www.ysi.com

Surface Modification Approach to Control Biofouling

T. Vladkova

Abstract There are three principal approaches to control biofouling: (1) mechanical detachment of biofoulers if possible; (2) killing or inactivation of biofouling organisms using antibiotics, biocides, cleaning chemicals, etc. and (3) surface modification turning the substrate material into a low-fouling or non-sticking (non-adhesive) one. Such modification usually alters the surface chemical composition and morphology, surface topography and roughness, the hydrophilic/hydrophobic balance, as well as the surface energy and polarity.

In marine applications especially, current non-toxic biofouling control strategies are based mainly on the third approach, i.e., on the idea of creating low-fouling or non-adhesive material surfaces, an approach that includes development of strongly hydrophilic "water-like" bioinert materials. Strongly hydrophobic low-energy surfaces are preferable in industrial and marine biofouling control because of their relative stability in aqueous media and reduced interactions with living cells.

This chapter presents a brief overview of some possibilities for biofouling control by surface engineering. A number of related ideas will be discussed in this chapter, including: (1) the use of protein adsorption as a mediator of bioadhesion and biofouling, (2) physicochemical parameters influencing these phenomena, (3) theoretical aspects of cell/surface interactions, (4) some popular surface modification techniques, and (5) examples of successful biofouling control approaches.

1 Introduction

Biofouling may be defined as any non-desirable accumulation and growth of living matter on material surfaces (see Pasmore 2008). It is a worldwide problem affecting a multitude of industrial water-based processes, including pulp and paper manufacturing,

T. Vladkova
Department of Polymer Engineering, University for Chemical Technology and Metallurgy, 8 "Kliment Ohridsky" Blvd. 1756, Sofia, Bulgaria
e-mail: tgv@uctm.edu

Springer Series on Biofilms, doi: 10.1007/7142_2008_22

food processing and packaging, cooling towers, biomaterials production, membrane technologies, underwater constructions and sensors, ship hulls, fishing farms, heat exchangers, and water desalination systems. The cumulative cost of biofouling due to lost production and only partially successful remedial efforts may run into billions of dollars per year worldwide, which explains the outstanding interest in the development of effective and economical control measures.

There are three principal approaches to solving the biofouling problem: (1) mechanical detachment of biofouling organisms and/or adsorbed biomolecules (i.e., biofoulers and biofoulants, respectively), (2) inactivating or killing the biofilm using antibiotics, biocides, cleaning chemicals, etc., and (3) surface modification with the aim of turning the substrate material into a low- or non-sticking (i.e., non-adhesive) one. Such modification usually alters the surface chemical composition and morphology, surface topography and roughness, the hydrophilic/hydrophobic balance, and the surface energy and polarity. The most effective antifouling coatings available today contain toxic biocides and will therefore be banned by the year 2008 (Brady 2003).

Current non-toxic biofouling control is based mainly on the third approach, i.e., on creation of low-fouling or non-adhesive material surfaces, an approach first applied to the development of bioinert materials where strongly hydrophilic "water-like" surfaces appear more promising. Strongly hydrophobic low-energy surfaces are preferable in industrial and marine biofouling control settings because of their stability in aqueous media and reduced interactions with living cells (Abarzua and Jacubowski 1995).

Biofouling is perceived as a multistage process starting with a "conditioning film" (biofilm) formation in which the adhesion of microfoulers (i.e., bacteria, diatoms, algae, etc.) to the surface is an important step. The industrial and marine biofouling usually continues with settlement of soft and hard macrofoulers such as algae, barnacles, mussels, tubeworms, etc. In principle, it should be possible to prevent or at least reduce biofilm formation by creating material surfaces to which bacteria cannot initially attach. In practice, synthetic materials that are capable of preventing bacterial adsorption are still rather elusive, despite a significant volume of research (Callow and Fletcher 1994).

Surface modification approaches leading to the creation of low-energy, low-adhesive, non-sticking surfaces is accepted nowadays as the most promising ecology-friendly alternative to the use of toxic biocides.

2 Biofouling and Bioadhesion

Upon submersion in a non-sterile aqueous liquid, most surfaces become rapidly colonized by bacteria and other microorganisms. These attached cells and extracellular polymeric substances (EPS), along with extracellular material (e.g., extracellular biopolymers) constitute a biofilm (Costerton et al. 1987). Biofilm frequently forms even on antifouling surfaces containing biocides. According to Brusscher et al. (1995), initial (reversible) bacterial adhesion to a surface is the primary determinant for biofilm formation. The reversible interactions between a bacterium and a substrate depend on the physicochemical properties of the bacterial cells and substrate surface, as well as the medium. On the other hand, bacteria adhering onto a surface usually secrete a matrix (EPS), in order to cement themselves (irreversible adhesion) to the surface (Van deVivere and Kirchman 1993; see Smeltzer 2008).

The nature of the bioadhesive interactions in the biofouling process is one of the main questions of many current researchers because a fundamental understanding of their molecular mechanisms can guide the creation of material surfaces preventing or reducing biofouling. Several modes of adhesion bonding, such as chemical and electrostatic interactions, mechanical interlocking, diffusion, etc. are currently known (see Smeltzer 2008). The adhesion of microorganisms is very complicated because not just one motif is followed, but rather a combination of modes has been demonstrated. In addition, the types of adhesion mechanisms that are manifested by a biofilm are largely dependent upon the species composition and physiology of the biofilm. The specific surface structures of microorganisms such as pili, cell wall components, and EPS are also known to influence biofilm formation. Finally, bioadhesion and biofouling can be strongly dependent on surface hydrodynamic conditions (Casse and Swain 2006).

Ikada et al. (1984) theoretically predicted that the work of adhesion (i.e., the interfacial surface energy) in aqueous media, $W_{12,w}$, approaches zero when the water contact angle, Q, approaches zero or 90°, i.e., there are two possibilities for a mate-

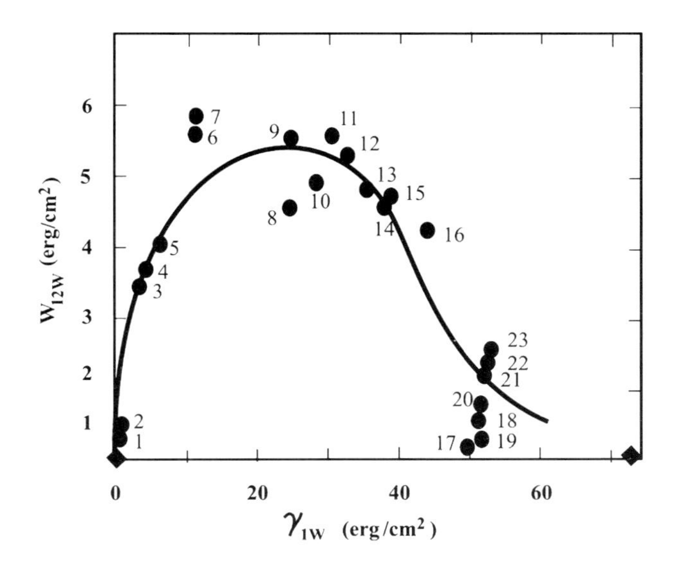

Fig. 1 Work of adsorption of bovine serum albumin (BSA) in water, $W_{12,w}$ against interfacial free energy between water and polymer, γ_{1w}: *1* canine vein; *2* canine artery; *3* poly(vinyl alcohol); *4* cellulose; *5* polyurethane S; *6* polyurethane T; *7* polyurethane E; *8* poly(vinyl chloride); *9* nylon 6,6; *10* poly(methylmethacrylate); *11* poly(ethyleneterephtalate) A; *12* poly(ethy-leneterephtalate B; *13* PVC; *14* nylon 11; *15* poly(styrene); *16* poly-(styrene); *17* poly(trifluoroethylene); *18* paraffin; *19* PTFE; *20* PP; *21* LDPE; *22* MDPE; and *23* HDPE (Ikada et al. 1984)

rial surface to have $W_{12,w}$ approaching zero. In other words, the two possibilities to be non-adhesive are (1) to create a super-hydrophilic, i.e., water-like surface (surface energy, g_{1w}~0) and (2) to create a super-hydrophobic surface (see Fig. 1)

This theoretical prediction, experimentally confirmed by BSA adsorption on various polymer surfaces in phosphate buffer, represented a point of departure in the development of strongly hydrophilic or strongly hydrophobic, low-adhesive, protein-repellent, bioinert materials and non-sticking fouling release surfaces.

Many attempts to explain bioadhesion have been made in terms of the thermodynamic theory of Derjaguin, Landau, Verwey and Overbeek (DLVO theory) that explains the stability of lyophobic colloids at solid–liquid interfaces (Derjaguin 1955). In addition to DLVO forces (i.e., Van der Waals and electrostatic), other types of interactions (such as hydrophobic interactions in polar media, ion bridging, and steric interactions) as well as some external cellular appendages are thought to play an important role. The net effect is a balance between all possible interactions (Oliviera 1997.)

Admittedly, such considerations treat the microorganisms as "living colloids," disregarding the specific roles of bacterial structures such as pili and cell wall components that play an important role in bioadhesion at the latter stages of biofilm formation (Morra and Cassinelli 1997). According to a number of authors, these structural features are of less significance in the initial stages of the attachment process than the intrinsic thermodynamic factorsinvolved (Van Loosdrecht et al. 1990; An and Friedman 1998; Morra and Cassinelli 1997). The study of Cunliffe et al. (1999) shows that over the time period of protein adsorption and initial cell attachment (1–24 h), this assumption is reasonable since the overall pattern of cell adhesion to different substrates was similar among the various cell types, although the absolute numbers varied considerably.

A review of bioadhesion-resisting surfaces(Kingshott and Griesser 1999) indicates that, in spite of the very large amount of studies on polyethylene glycol (PEG) coatings, there is still controversy about what exactly the properties and modes of action of an "ideal" PEG coating should be. While some studies have reported no irreversible protein adsorption, other very similar coatings appear less able to resist bioadhesion. Vladkova et al. (1999) found that not only the amount but also the conformation of proteins adsorbed on PEG layers is important for the cellular interactions.

The biofilm protects its habitants from predators, dehydration, biocides, and other environmental extremes while regulating population growth and diversity through primitive cell signals (see Smeltzer 2008). Dunne (2002) has demonstrated that from a physiological standpoint, surface-bound bacteria behave quite differently from their planktonic counterparts.

Recently, some researchers have attempted to replace the relatively crude macroscopic measurements used to describe bacterial adhesion to surfaces (e.g., cell hydrophobicity via contact angle measurements or water–hexadecane partitioning) with methods that directly measure cell-surface interaction forces (repulsive and attractive) using, for example, atomic force microscopy (AFM) to observe homogeneity and topography of bacterial surfaces as well as to directly measure adhesion forces (Velegol and Logan 2004; Li and Logan 2004; Xu and Logan 2005.)

Unfortunately, the mechanisms underlying the adhesion of bacteria to the material surface are still not properly predictable from these theories. Electrostatic and Lifshitz–van der Waals forces are usually considered responsible for the interactions at the biomaterial interface. Studying the role of chemical interactions in bacterial adhesion to polymer surfaces, Speranza et al. (2004) found an important role of the acid/base (Lewis) interaction. Pedry (2005) employed contact angles to relate interfacial free energies between the interacting surfaces. The thermodynamic approach treats the bacteria/substrate interaction as an equilibrium process. This investigation indicates that the more hydrophobic a substrate is, the less of an equilibrium process the interaction becomes. The relationship obtained for the free energy change occurring when bacteria attach to a surface considers only dispersion forces. But, other forces such as electrostatic forces probably also contribute to the interaction and the simple model should include other interactions in order to accurately model bacterial adhesion.

Lately, the concept of the "theta surface" (Baier 2006) has been developed as a contribution to the need predicted by Larry Hench for a robust "general field" theory that supports bioengineering solutions for biocompatibility and biofouling control. For such a theory, there will be no need to know the names or identities of specific biological substances that will be encountered, since all biological systems share the same fundamental chemistry and pattern of events.

A "theta surface" defines the characteristic expression of outermost atomic features least retentive of depositing proteins, and identified by the bioengineering criterion of having measured critical surface tension(CST) between 20 and 30 mN m^{-1}. Material applications requiring strong bioadhesion must avoid this range, whereas those requiring easy release of accumulating biomass should have "theta surface" qualities. More than 30 years of empirical observations of the surface behavior of various materials in biological settings, when correlated with contact angle-determined CST for the same materials, support the definition of the "theta surface". It derives from the concept of "theta solvents" for macromolecules, that is, suspending liquid phases that allow large, complicated molecules such as proteins to retain their thermodynamically most stable conformations, resisting "denaturation" in 3-dimensional suspensions. The theta surface is that atomic force expression controlled from solid surfaces, placed into aqueous biological media, that will least denature glycoproteinaceous macromolecules encountering those surfaces. It is the adsorbed configurations, and strengths of binding/retention of biomass to contacting materials under water, that determine resistance to shear-induced re-entrainment of that matter into the biological stream. So, maintenance at the interface of near-solution-state conformations of the first arriving macromolecules is the most effective approach to thrombus-resistant materials for long-term contact with flowing blood, and to fabrication of "easy-release" coatingsfor exposure to any other biological system (from sea water to dairy products, and from water purification units to sewage flow lines). Universal features of all such systems are the presence of water and glycoproteinaceous macromolecules or their refractory remnants (e.g., surface-active humic substances in the sea) as the dominant "conditioning film," forming water-displacement agents entropically favored as the new interfacial occupants.

Bioadhesion manifests at every stage of biofouling, including the settlement of macrofoulers such as barnacles, mussels, algae, etc. (see Flemming and Greenhalgh 2008; Nedved and Hadfield 2008; Harder 2008). Marine mussels are experts in bonding to a variety of solid surfaces in wet, saline and turbulent environments. The bonding is rapid, permanent, versatile, and protein based. In mussels, adhesive bonding takes the form of a byssus (a bundle of extracorporal threads) each connected to living tissues of the animal at one end and secured by an adhesive plaque at the other. Trying to perform reverse engineering of bioadhesion in marine mussels, Waite (1999) investigated the composition and formation of byssal plaques and threads with the hope of discovering technologically relevant innovations in chemistry and materials science. All proteins isolated from the byssus to date share the quality of containing the amino acid, 3,4-dihydroxyphenylalanine. This residue appears to have a dual functionality, with significant consequences for adsorption and cohesion. On the one hand, it forms a diverse array of weaker molecular interactions such as metal chelates, H-bonds, and pi-cations: these appear to dominate in surface behavior (adsorption). On the other hand, 3,4-dihydroxyphenylalanine and its redox couple, dopaquinone, can mediate formation of covalent cross-links among byssal proteins (cohesion). One of the challenges in making functional biomimetic versions of byssal adhesion is to understand how these two reactivities are balanced.

Flammang and Jangoux (2004) have studied bioadhesion models of marine invertebrates relating to biomechanical, morphological, biochemical, and molecular processes involved in the adhesion. They have found that the adhesion of Cuvenian tubules, which are specialized adhesive defense organs of some sea cucumber species, is instantaneous. The results of this investigation suggest that the underwater adhesive is in the form of a low molecular weight precursor protein in the secretory granules of the adhesive cells. Upon release, these proteins instantly polymerize, with no enzymatic curing required.

Holm et al. (2006) have found inter-specific variation in patterns of adhesion of marine fouling, studying the adhesion of six hard fouling organisms (barnacles, mollusks, and tubeworms) to 12 silicone fouling release surfaces. Removal stress (adhesion strength) varied among the fouling species and among the surfaces. None of the silicone materials generated a minimum in removal stress for all the organisms tested. These results suggest that fouling release materials do not rank (in terms of adhesion strength) identically for all fouling organisms, and thus development of a globally effective hull coating will continue to require testing against a diversity of encrusting species.

3 Physicochemical Parameters Influencing Bioadhesion and Biofouling

Despite the fact that there is no general theory of bioadhesion, a lot of factors influencing bioadhesion and biofouling are now known that could be used in biofouling control. Even though the ability to resist protein, glycoprotein, and polysaccharide adsorption remains imperative for a coating to prevent marine and industrial fouling, it has been suggested that surface free energy, mechanical properties, and wettability also play an important role in defining the extent to which a surface can resist biofouling and facilitate fouling release (Finlay et al. 2002; Brady and Singler 2000; Sigal et al. 1998).

3.1 Surface Energy and Related Parameters

Leading theories attempt to correlate the kind and intensity of biological responses to surface and interfacial energetics (Brusscher et al. 1995). Surface thermodynamic characteristics, such as hydrophobic/hydrophilic balanceand van der Waals and donor/acceptor forces, are determined by contact anglemeasurements. The surface free energy of the substratum is now accepted as one of the main factors influencing microbial adhesion. Adhesion to surfaces with different surface free energies has been studied by a large number of research groups. For a homogenous solid, the critical surface tensiong_c is the same as the surface free energy, i.e., surface tension assuming that there are no other forced elastic strains on the solid and no solvent adsorption (Good 1992).

Baier (1973) and Dexter (1979) were among the first researchers to correlate the adhesion of fouling organisms with the surface free energy of the substratum. Hamza et al. (1977) have showed that bacterial adhesion is less on hydrophobic surfaces with a low surface energyand that they are easier to clean because of weaker binding at the interface. McGuire and Swartzel (1987) found an optimum surface free energy, of 30–35 nM m^{-1}, at which milk protein adsorption is minimal. There are also researchers that have drawn the opposite conclusion that hydrophilic membranes have smaller biofouling tendency than hydrophobic ones (Pasmore et al. 2001).

Bacteria adhere to almost any surface, despite continuing arguments about the importance of the physicochemical properties of substratum surfaces, such as hydrophobicity and charge. Bos et al. (2000) demonstrated that bacteria do not have a strong preference for adhesion to hydrophobic or hydrophilic surfaces but that the substratum hydrophobicity is a major determinant of bacterial retention while it hardly influences bacterial adhesion.

Studying the attachment of bacteria, like Salmonella, etc. to different surfaces and the influence of their free energy, Sinde and Carballo (2000) found that the bacterial adherence could not be correlated with surface free energies or contact angles of bacteria.

In the case of soft fouling species (e.g., *Ulva* spores) using non-polar, self-assembled monolayers, it has been shown that adhesion is strongly influenced by critical surface tension (or "wettability") (Finlay et al. 2002).

A generalized relationship between surface tension (i.e., the free energy of a surface, which is commonly referred to as "surface energy") and the relative amount of bioadhesion has been established as shown in Fig. 2. This is commonly known as the "Baier curve" (Anderson et al. 2003). The key feature of this curve is that the minimum in the relative adhesion, at 22–24 nM m^{-1}, (mJ m^{-2}), does not occur at the lowest surface energy.

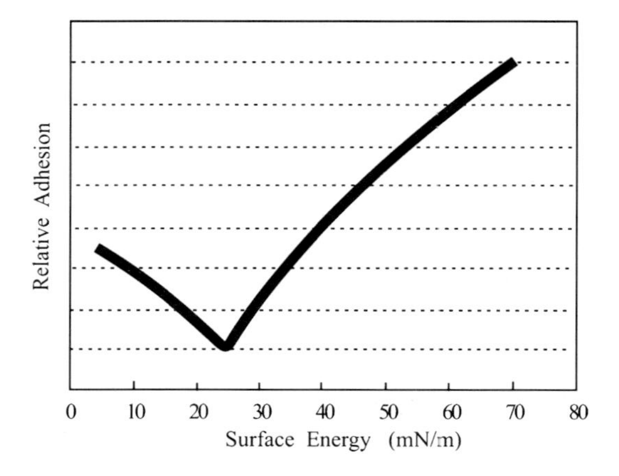

Fig. 2 The "Baier curve" (Anderson et al. 2003)

Recently, Zhao et al. (2004) investigated the effect of surface free energy on bacterial adhesion and reported the optimum surface free energy at which the bacterial adhesion force is minimal to be about 20–30 nM m^{-1}.

Schmidt et al. (2004) studyied adhesive and marine biofouling release properties of coatings containing surface-oriented perfluoroalkyl groups andhave found that the fouling release properties of the low-surface-energy surfaces cannot be evaluated by using only static or advancing contact angle. Contact angle hysteresis appears to be a direct indication of the liquid or adhesive penetration and to correlate with marine biofouling resistance.

Dahlström et al. (2004) have shown that the initial surface wettabilityis of importance in the settlement of macrofouling larvae, such as barnacles, bryozoans and hydroids in both field and laboratory assays. Studying the settlement on surfaces with different wettability, they concluded that the wettability might cause a biological inhibition by interacting with chemo-receptors when the larva is making surface contact, or that the inhibition might be of a physicochemical nature and, thus, surface contact is impeded by repulsive chemical forces.

Holland et al. (2004) have found that fouling release poly(dimethylsiloxane) (PDMS) coatings accumulate diatom slimes, which are not released even from vessels operating at high speeds (>30 knots). Fouling diatoms adhere strongly to a hydrophobic PDMS surface and this feature maybe contributes to their successful colonization of the fouling release coatings.

Studying the effect of substratum surface energy and chemistry on attachment of marine bacteria and algal spores, Ista et al. (2004) and Walker et al. (2005) have found that a number of macroorganisms, such as algae, exploit unicellular forms for attachment and colonization of surfaces. Surface coverage by both macro- and microorganisms depends initially on the ability of single cells to adsorb and adhere to the attachment substratum.

Meyer et al. (2006) have confirmed that silicone coatings with critical surface tension(CST) between 20 and 30 nM m^{-1} more easily release diverse types of biofouling than materials of higher or lower CST. Oils added to these coatings further selectively diminish the attachment strength of different marine fouling organisms, without significant modification of the initial CST. They have also demonstrated some contact angle anomalies indicating that surface-active eluates from silicone coatings inhibit the adhesive mechanisms of fouling organisms.

3.2 Elastic Modulus

In order to control adhesion of biological organisms to a substrate, some fundamental fracture mechanics have to be considered. For prediction of the force required to break an adhesive from a silicone elastomer substrate, basic fracture mechanics should be examined. Griffith (1921) formulated the following equation for the critical stress (s_c) required to propagating a crack in a plate for a uniaxial direction:

$$\sigma_c = \sqrt{EG_c / \pi a(1-\upsilon^2)} \quad (1)$$

where E, G_c, a, and υ are the elastic modulus, Griffith's critical fracture energy per area, half the crack length, and Poisson's ratio, respectively. Griffith then applied this equation to the stress over a set crack area (A = pa^2), known as the critical pull-off force (P_c): (2)

$$P_c = \sqrt{\pi E G_c a^3 / (1-\upsilon^2)} \quad (2)$$

A few years later, Kendall (1971, 1994), following Griffith's fracture analysis to model adhesion of elastomer substrates, derived the critical pull-off force for thin elastomer film and radius of the disc being smaller than the size of the elastomer film:

$$P_c = \pi a^2 (2G_c K / t)^{\frac{1}{2}} \quad (3)$$

where: G_c, a, t and K are the critical fracture energy, radius of the contact area, elastomer thin film thickness, and bulk modulus [$K = E/3(1–2)$]. These equations show that there is a proportional relationship between the critical pull-off forceand $(EG_c)^{1/2}$, which are material properties. In this case, fracture energy is directly related to the work of adhesion, which is then equal to the critical surface tension (g_c) of the elastomer (Silberzan et al. 1994; Kendall 1994). As a result, the adhesion correlates with $(Eg_c)^{1/2}$. The elastic modulus and surface energy are parameters that can be engineered with the material.

Thus, from a fracture mechanics study it has been shown that the elastic modulus is a key factor in bioadhesion and the ability of organisms to release from a surface (Brady and Singler 2000; Berglin et al. 2003).

Figure 3 demonstrates that the adhesion correlates better with $(Eg_c)^{1/2}$ than with either surface energy or elastic modulus on their own, despite some scatter in the data.

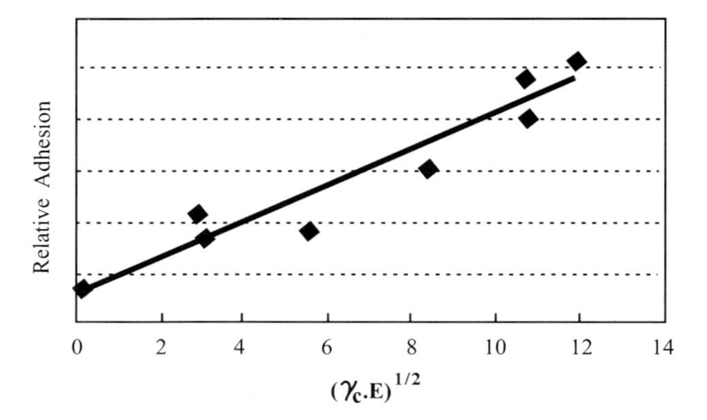

Fig. 3 Relative adhesion as a function of the square root of the product of critical surface free energy (γ_c) and the elastic modulus (*E*) (Brady and Singler 2000)

For this reason, siloxane elastomers are the major commercial candidates for environmentally benign fouling release coatings, as they possess both low modulus and low surface energy (Wynne et al. 2000). Commercial antifouling silicone elastomers such as RTV11 or Intersleek have modules in the 3–1.4 MPa range (Arce et al. 2003).

3.3 Thickness of Coating

Thickness is another characteristic of low surface energy coatings that plays an important role in bioadhesion (Anderson et al. 2003; Sun et al. 2004). It has been found that below ~100 mm dry film thickness, barnacles can "cut through" to the underlying coats and thus establish firm adhesion. Above this thickness there is no marked improvement in fouling release properties.

Determining the elastic modulus of silicon rubber coatings and films by depth-sensing indentation, Zhili et al. (2004) have observed hard substrate effects when the indentation displacement is less than 10% of the total coating thickness (of 1 mm).

Chaudhury et al. (2005) confirmed the effect of film thickness and modulus on the release of adhered spores and sporelings of the green alga *Ulva*.

3.4 Surface Chemistry

Silicone polymer systems that have generally shown the lowest adhesion of biofoulants are characterized by a specific chemical structure that determines their special behavior and, in particular, their durability and fouling release ability. Siloxane polymers have a backbone of repeating (–Si–O–) units with saturated organic moieties attached to the two non-backbone valences of the silicon. The Si–O bond is stronger than a C–C bond (451 kJ mol^{-1} compared to 348 kJ mol^{-1}), which helps to explain the long-term durability of these compounds under field conditions. With a length of 1.63 Å, Si–O is longer than the C–C bond in most organic polymers (Baney and Voigt 1977).

This property presents large bond rotation and thus large chain mobility and restructuring ability. As a result, the non-polar and polar groups can reorient on the surface to their most favorable position depending on the environment (Hillborg and Gedde 1999). PDMS) at its lowest energy state is in the all-*trans* conformation due to this arrangement having the most favorable van der Waals interactions where the methyl groups are separated by four bonds (Mark 1979). This ability for recovery to its lowest energy state is beneficial when trying to predict the material properties when exposed to diverse environments.

It has been found that incorporation of low molecular weight silicone polymers (oils) enhances the fouling release properties of PDMS polymers (Milne 1977; Stein, et al. 2003).

Krishnan et al. (2006) studied surfaces of novel block copolymers with amphiphilic side chains for their ability to influence the adhesion of marine organisms. The ability of the amphiphilic surface to undergo an environment-dependent transformation in surface chemistry when in contact with the EPS is a possible reason for its antifouling nature.

3.5 Slippage

The fracture mechanic equations assume that the applied force is in the normal direction to the elastomer surface. However, when the force is applied at an angle, as would be the case when a ship is moving, there is the additional component of interfacial slippage that has to be considered. According to Newby et al. (1995), the adhesion of a viscoelastic adhesive on silicone elastomers is controlled heavily by interfacial slippage rather than by thermodynamics. When peeling a viscoelastic adhesive, an extension deformation occurs behind and contraction occurs in front of the moving crack tip (Newby and Chaudhury 1997). If slippage is allowed to occur by the substrate, then the work required to move the crack tip is lowered, which results in a lower adhesion strength. The more mobile chains are on the substrate surface, the lower the friction observed. When force is applied in the normal direction, the release mode is peeling from nucleated voids within the contact area, but when force is directed at an angle, failure occurs by a fingering process (like webbed fingers) as described by Newby et al. (1995) that starts at the edges and moves inward (Kohl and Singler 1999). These finger adhesion release deformations were also shown to increase in length (i.e., amplitude) slightly as the modulus or the thickness of the elastomer coating increased, and significantly as the rigidity of the adhesive increased (Ghatak et al. 2000).

Friction/lubricity is a factor of biological adhesion. In order to engineer the elastomer matrix with lubricity, various non-functional silicone oil additives have been introduced into the bulk of the elastomer. Due to surface energetics, the lower surface energy silicone oils migrate on the surface creating a lubricious layer (Homma et al. 1999). Oils by their nature are lubricants but this is not the main reason for their efficiency. This is thought to be due to the surface tension and hydrophobicity changes that the oils affect during the curing process and after immersion. It has been shown that fouling release coatings do not rely on leaching of the oils for their fouling release properties. Both laboratory studies (Truby et al. 2000) and ships' trials have shown that performance was maintained for up to 10 years in service (Anderson et al. 2003).

3.6 Surface Roughness and Topography

In addition to surface free energy, elastic modulus, and surface chemistry, other factors, including surface roughness and topography, also significantly influence bacterial adhesion. Therefore the mechanism is very complex. Surface roughness influences the spreading of liquid cements secreted by organisms to increase adhesion on engineered topography. With the fracture mechanisms discussed previously, the surfaces were assumed to be completely smooth. However, even the smoothest substrate has molecular roughness on the surface. Depending on the viscosity of the liquid, the adhesive might not fill all the small crevices (Baier et al. 1968). When solidification of the adhesive occurs, there are stress concentrations that occur at the focal points of the roughness. The stress at these focal points is much higher than the applied force, so less applied force is needed to fracture the adhesion. If the voids are relatively close together then the fracture crack can propagate even more easily. The unfilled crevices can be on the molecular level or micron scale depending on the size of the organism.

The trends observed from many studies are that as the roughness increased, the advancing angle increased and the receding angle decreased. This means that when static conditions are examined, as the roughness increases, the contact angle increases and thus the critical surface tension calculated increases. However, this statement does not consider the size, shape, and the exact location of the droplet edge in reference to the rough features, but this roughness influences the spreading of liquid adhesives. Many reports on the cellular responses to topographical cues on both the nanometer and micrometer scales have appeared in the past few decades. However, it has been argued by a number of authors that these structural features are of less significance in the initial stages of the attachment process than the intrinsic thermodynamic factors involved (An and Friedman 1998; Morra and Cassinelli 1997), and a number of detailed studies have been carried out to support this assertion (Van Loosdrecht et al. 1990).

In the area of marine fouling, topography has been shown to alter settlement of bacteria(Scheuerman et al. 1998), barnacles(Berntsson et al. 2000; Berntsson 2001; Berntsson and Jonsson 2003) and algae(Callow et al. 2002; Hoipkemeier-Wilson et al. 2004) and to deter colonization of invertebrate shells(Scardino et al. 2003; Bers and Wahl 2004). The change in wettability of a surfacethat results from surface roughness, i.e., topography, is likely to be contributing factor to these responses.

Prior to adhesion, the swimming zoospore is able to select suitable surfaces on the basis of surface characteristics, such as topography, or on the basis of physicochemical properties, such as contact angle (Callow et al. 1997, 2000, see Nedved and Hadfield 2008).

Promising antifouling properties of microstructured surfaces have been reported by Bohringer (2003) and Bers and Wahl (2004). Hoipkemeier-Wilson et al. (2004) have studied the settlement and release of *Ulva* spores from microengineered topographies.

Influence of nanoscale topography (Griesser et al. 2002) on hydrophobicity (the contact angles and their hysteresis), including that of fluoro-based polymer thin films (Gerbig et al. 2005), is reported in the special literature. Brennan et al. (2005) have patented surface topography for non-toxic bioadhesion control.

Carman et al. (2006) have experimentally demonstrated the importance of wettability models in predicting cellular contact guidance for engineered topographies, but do not fully explain the process. Bioadhesion is a complex and specific process. The material modulus and surface elasticity of cell membranes are other factors to consider, in addition to the variety of adhesive proteins, glycoproteins, and polysaccharides that organisms secrete. The wettability models are limited by the assumption that the liquid droplet is much larger than the topographical features. This allows for line tension effects to be neglected. Measurements with smaller drop sizes are believed to enable the inclusion of line tension effects. Ultimately, the goal is to improve the predictive quality of an energy-driven model for bioadhesion.

4 Physicochemical Parameters Influencing the Cell/Material Surface Interaction

The ability of cells to adhere to each other or to the underlying substrate (called cell adhesion) is their main property. Biological adhesion is not fully explained by physical adhesion but is a much more general and complicated phenomenon determined by a number of interacting biological processes such as cell attachment and mobility, cell growth and differentiation, etc. (Bitton and Marshall 1980; Adams and Watts 1993).

Analyzing the special literature and studying experimentally the adhesion interaction of living cells with different model surfaces, Altankov (2003) has concluded that the initial interaction of cells with biomaterials is governed by the efficiency of the cell adhesion, the latter depending mainly on the surface properties of the substrate and the adsorbed proteins. Hydrophilic surfaces support cell adhesion and proliferation, cell growth, and the organization of the focal adhesion complex delivering the signal via integrin receptors. An optimum interaction with cells usually appears at moderate hydrophilicity (WCA ~ 60°). The chemical functional groups oppress it in the following manner:

$-NH_2 > -OH >$ epoxy $> -SO_3 > -COOH > -CF_3$

A relationship between the efficiency of the cell interaction and the total negative charge of the surface exists. This interaction is influenced not only by the chemically grafted functional groups but also by the adsorbed ions. The synthesis and organization of the fibronectin matrix by cells is better on surfaces that weakly bond fibronectin compared to other matrix proteins. The conformation of the adsorbed adhesive proteins also plays an important role in the adhesive interaction of strongly hydrophilic non-charged PEG surfaces (Vladkova et al. 1999).

Properties of the substrate, such as hydrophobicity (Schackenraad et al. 1992), hydrophilicity (Gölander 1986), steric hindrance (Kuhl et al. 1994), roughness (Kiaie et al. 1995), and the existence of a "conditioning layer" at the surface (Abarzua and Jacubowski 1995), are all thought to be important in the initial cell attachment process.

5 Protein Adsorptionas Mediator of Bioadhesion and Biofouling

Protein adsorption is the primary event in biofouling and in the interaction of foreign surfaces with tissue, blood, and cells (Corpe 1970). The biological cascade of industrial and marine biofouling as well as of all undesirable response reactions against biomaterials begins with deposition of proteins. Therefore, low protein adsorption is accepted now as the most important prerequisite for resistance against biofouling.

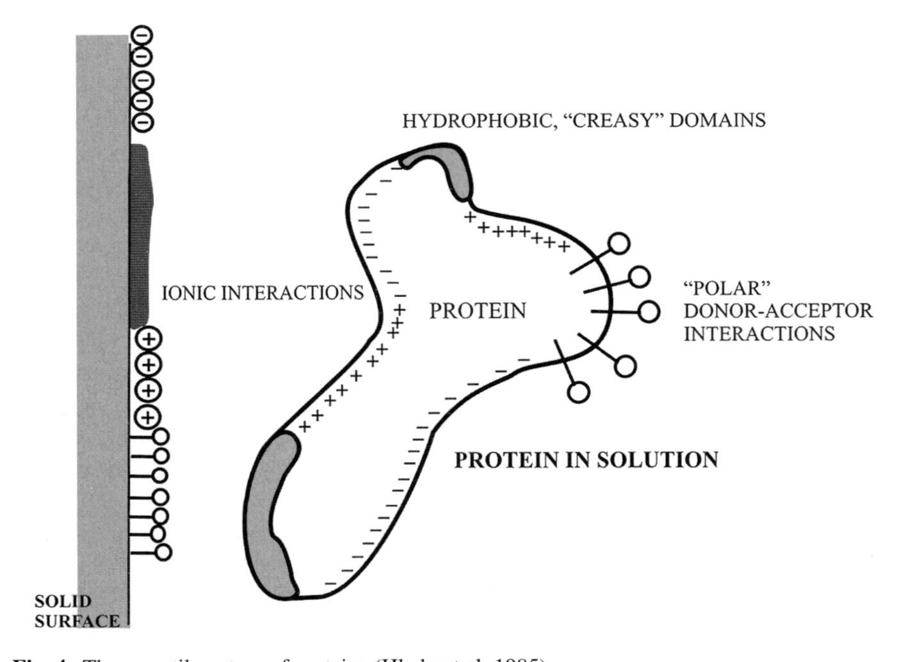

Fig. 4 The versatile nature of proteins (Hlady et al. 1985)

Because of their versatile nature (Fig. 4), different proteins can be adsorbed by various mechanisms when presented with a complementary surface, which makes the prevention of protein adsorption difficult.

Most investigations are devoted to the study of the adsorption of single well-defined proteins, adsorption from multicomponent systems, or from blood plasma and are aimed at identification of protein-repellent biomaterial surfaces (Gölander et al. 1986; Gölander 1986; Malmsten 1998; Pasche 2004; Atthoff 2006).

It is known that the protein adsorption and biocontact properties of polymers depend on surface chemical composition and topography, surface hydrophilic/hydrophobic balance and charge, the mobility of the surface functional groups, the thickness and density of the modifying layer and its adhesion to the substrate, etc. Hence, by changing some of these parameters we can control protein adsorption (Gölander 1986).

According to Loeb and Neihof (1975), and Baier (1980) the adsorption of organic molecules leads to formation of a "conditioning film" on a newly immersed surface, altering the physicochemical properties of this surface and providing a nutrient source for attachment of microbial flora. The primary mechanism in the attachment of marine organisms to surfaces involves secretion of protein or glycoprotein adhesives (Vreeland et al. 1998; Kamino et al. 2000; Stanley et al. 1999). Therefore, it is no surprise that significant attention has been directed toward development of efficient protein-resistant surfaces (Hester et al. 2002; Griesser et al. 2002; Ostuni et al. 2001, 2003; Bohringer 2003; Groll et al. 2004) for marine antifouling (Youngblood et al. 2003) as well as for biomedical applications (Gölander et al. 1984, 1986; Wagner et al. 2004; Vladkova 1995, 2001).

Identification of the type and amount of proteins adsorbing to the material surface could provide important information for the rational development of new materials that can resist biofouling. Adsorption of different organisms by adhesive proteins undergoing subsequent underwater curing is thought to be a mediator of bioadhesion and biofouling. Some recent investigations have focused on further study of the curing mechanisms of bioadhesive proteins as well as on the mechanical properties of bioadhesives such as spore adhesive glycoprotein of the green alga *Ulva*(Humphrey et al. 2005; Walker et al. 2005).

Using biomolecules and green alga as probes, comparative evaluations have been performed of the antifouling and fouling release properties of hyperbranched fluoropolymer (HBFP)–poly(ethylene glycol) (PEG) composite coatings and PDMS elastomers. The maximum resistance to protein, lipopolysaccharide, and *Ulva* zoospore adhesion, as well as the best zoospore- and sporling-release properties have been recorded for the HBFP–PEG coating containing 45%wt PEG. This material also exhibited better performance than did a standard PDMS coating (Gudipati et al. 2005).

It is expected that new analytical techniques and direct measurement of interfacial forces between proteins and surfaces will improve understanding of protein/surface interactions and open new possibilities for the guided design of surfaces intended to resist bioadhesion.

6 Protein Repellent Surfaces

6.1 Strongly Hydrophilic Surfaces

Many strongly hydrophilic and hydrophobic surfaces have been developed to decrease protein adsorption to biomaterials (Elbert and Hubbel 1996). A comparative protein adsorption study of different strongly hydrophilic surfaces, including positively charged (*N*-vinylpyrolidon), negatively charged (AA), and non-charged (PEG) have clearly demonstrated the advantages of non-charged strongly hydrophilic surfaces (Gölander et al.,1986).

PEGs, which currently represent the "gold standard" of biomaterials, are most often used in the creation of bioinert material surfaces. The bioinertness of PEG molecules is utilized also in the prevention of marine biofouling using water-resistant hybrid co-polymer networks containing PEG segments. Much research is devoted to study of the protein adsorption resistance mechanisms of different PEG-coated surfaces, for example surfaces with adsorbed PEG-graft copolymers (Pasche 2004).

The structural similarity of the $–CH_2CH_2O–$ unit to water and the strong hydrogen bonding to the O-atom have been used to rationalize its miscibility with water. The $–CH_2–$ groups are believed to be "caged" by a water network (Bailey and Koleske 1976), see Fig. 5. Hence, when a foreign moiety approaches a PEG-coated surface, that moiety behaves as if it was interacting with a hydrated surface and its adsorption is minimized.

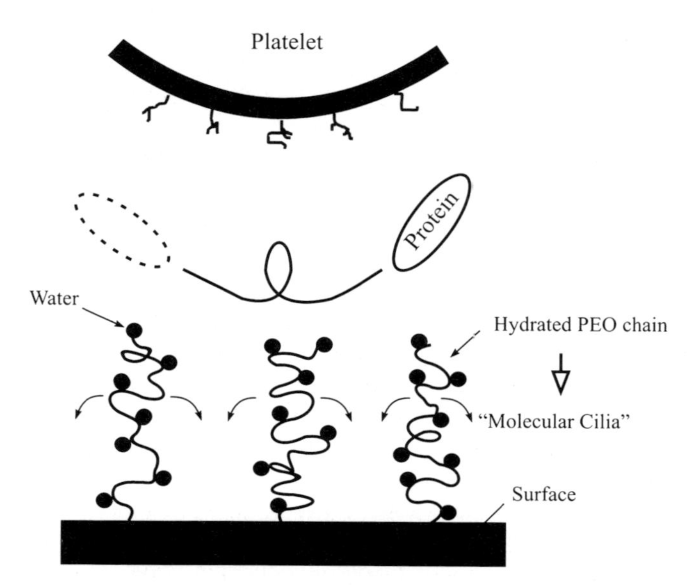

Fig. 5 "Molecular cilia" mechanism on PEG surface with hydrated poly(oxyethylene) chain (Mori and Nagaoka 1982)

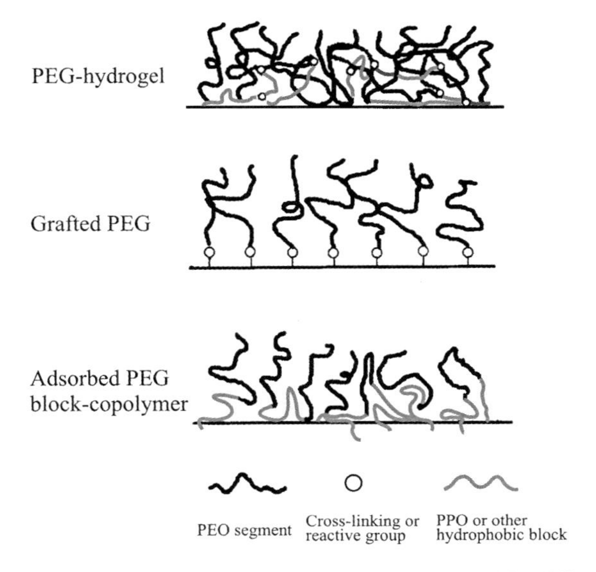

Fig. 6 Scheme showing the structural features of PEG layers obtained by different coating

A number of experimental techniques have been used to introduce PEG groups on different polymer surfaces, such as PE, PVC, PMMA, NR, PDMS, PS, etc., by wet chemistry or by plasma treatment (Vladkova 1995, 2001; Vladkova et al. 1999; Harris 1992). Wet chemistry methods include deposition of photopolymer hydrogel PEG coatings, including a two-step photopolymerization procedure to increase the surface density of PEG chains (Gölander et al. 1984), grafting, or adsorption of PEG chains on the substrate surface. Structural features of the PEG layers obtained in this way are presented schematically in Fig. 6. Concentrating PEG chains through creation of brush-type surface coatings using mono-functional PEG-acrylates and UV polymerization has been used to prepare super-hydrophilic (water contact angle $<10°$) surfaces with exclusively low (below 0.05 mg m^{-2}) protein adsorption (Gölander et al. 1984)

PEG-aldehyde (Gölander et al. 1987) and PEG-epoxide grafting or PEG-epoxide/PEI copolymer (Fig. 7) quasi-irreversible adsorption (Vladkova et al. 1999) at optimal reaction conditions also leads to the formation of surfaces with very low protein adsorption of below 0.05 mg m^{-2} (by ellipsometry). Figure 8 shows a simple sketch of PEG-aldehyde grafting by Schiff base reaction with surface NH_2 groups.

The examples of PEG-coated surfaces described herein are only a small part of those described in the special literature.

The bioinertness of PEG molecules is utilized in marine biofouling prevention using water-resistant hybrid co-polymer networks containing PEG segments, oriented toward the water in aqueous media (Gudipati et al. 2004, 2005). Surface-responsive materialsfor non-stick coatings are prepared by linking perfluoropolyether (hexafluoro-propylene oxide oligomer, PFPE), poly(dimethylsiloxane) (PDMS)

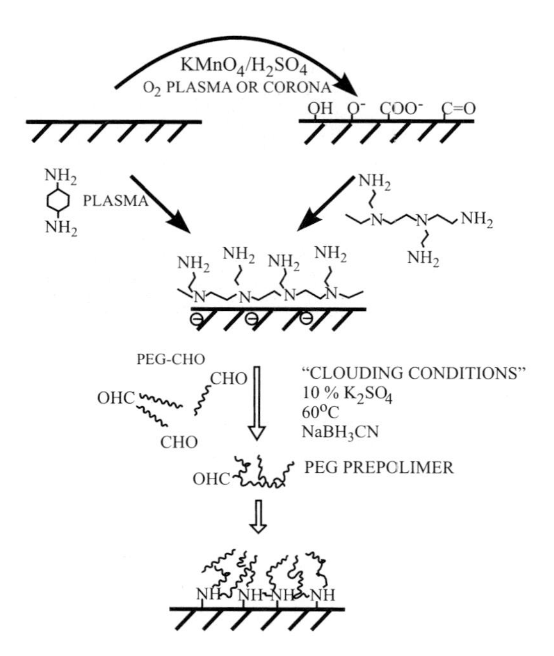

Fig. 7 Grafting of PEG by Schiff base reaction between PEG-CHO and surface-NH_2

and PEG segments. Non-stick properties exhibited in air are due to the presence of PDMS and PFPE at the air–solid interface. Upon exposure to water, this material becomes non-stick due to migration of PEG segments to the water–solid interface Russell (2002).

6.2 *Protein-Repellent Plasma Films*

Plasma treatment in vacuum or at normal pressure, in atmospheres of different gases, as well as ion- or electron beam, etc., are referred to as "dry" chemistry methods, and they represent another approach to surface modification aimed at creation of easily cleaned or non-fouling material surfaces (Ratner et al. 1990; Chan 1993; Sheu et al. 1995; Chan et al. 1996; Vladkova 2001). Comparative studies of plasma-deposited films indicated that both strongly hydrophobic silicon and strongly hydrophilic PEG surfaces result in very low protein adsorption, unusually weak complement system activation, and low cell and platelet adhesion (Kicheva et al. 1992; Vladkova 1995), which is in agreement with the prediction of Ikada et al. (1984). Similar "dry" chemistry also offers a possibility to turn the hydrophobic surfaces into hydrophilic surfaces and the opposite, to combine the stability of the hydrophobic materials in water with the advantages of the hydrophilic surfaces. For example, PDMS surface modification has been performed to alter the hydrophilic–hydrophobic balance on the surface and hence the interaction with living cells (Satriano et al. 2001, 2002; Vladkova et al.

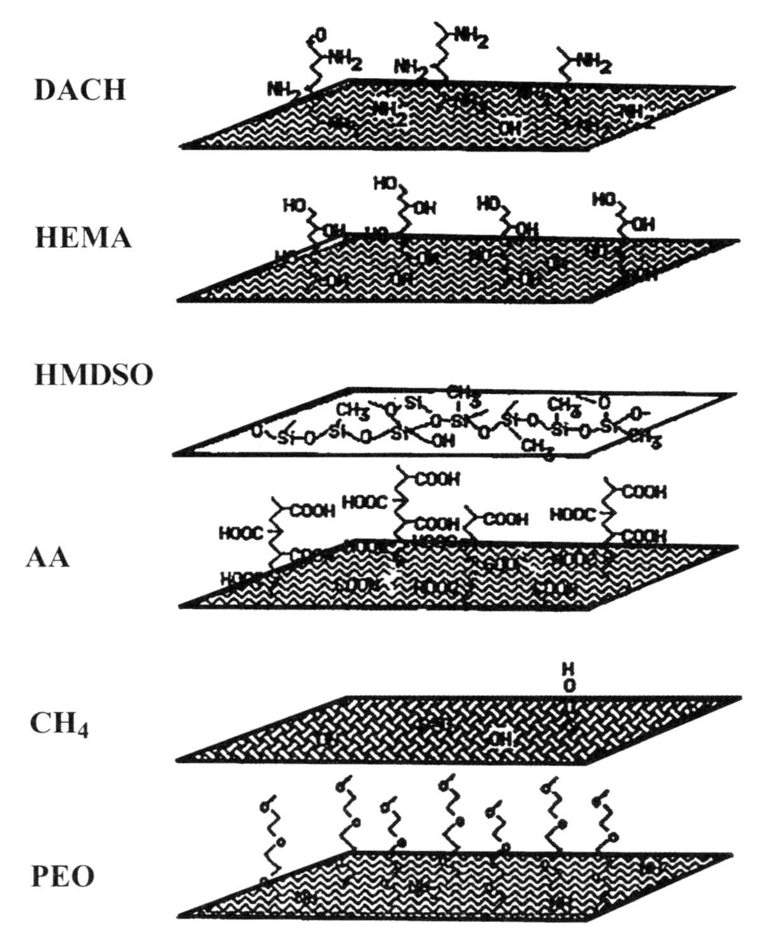

Fig. 8 Chemical composition of plasma-deposited polymer films: diaminocyclohexane (*DACH*), hydroxiethylmetacrylate (*HEMA*), hexamethyldisiloxane (*HMDS*), poly(ethylene oxide) (*PEO*)

2005). Radio frequency plasma discharge is considered as an important technique in the creation of protein-repellent surfaces.

7 Low Surface Energy Coatings to Control Biofouling

Considerable attention in recent decades has been focused on the concept for the creation of non-biocidal, non-toxic coating systems that prevent the attachment of fouling organisms. The objective for these minimally adhesive "fouling release" coatings was to create surfaces reducing the adhesion strength of attaching organisms and hence causing their detachment under their own weight as they grow or their dislodgement by water movement when a ship moves through the water (Linder 1992). The initial interest in the development of such types of coatings was focused on the fluoropolymers, but later it moved to the siloxane elastomers and their copolymers because of the combination of lower elastic modulus with low surface energy.

The most minimally adhesive polymer surfaces known currently are prepared from siloxanes (Swain and Schultz 1996; Pike et al. 1996; Kohl and Singler 1999; Uilk et al. 2002), fluoropolymers (Schmidt et al. 1994; Wang et al. 1997; Brady et al. 1999; Bunyard et al. 1999; Gan et al. 2003; Gudipati et al. 2004), and fluorosiloxanes (Johston et al. 1999; Mera et al. 1999; Uilk et al. 2002; Grunlan et al. 2004). Their non-adhesive nature is attributed to low surface energy g, low storage modulus *G*, and low glass-transition temperature T_g (Owen 1990; Newby et al. 1995; Brady 1999, 2000; Wynne et al. 2000). Low g values reduce polar and hydrogen-bonding interactions with the marine organism's adhesive, thereby decreasing the joint strength. *G* is also significant because the rupture of an adhesive bond involves viscoelastic flow at the coating surface (Kinloch and Young 1983).

Hybrid xerogel films have also been studied as novel coatings for antifouling and fouling release (Tang et al. 2005). They were found to inhibit settlement of zoospores of the marine fouling alga *Ulva*, hyperbranched fluoropolymer poly(ethylene oxide)

Hyperbranched fluoropolymer–poly(ethylene glycol) (HBFP–PEG) composite coatings have been identified as material exhibiting better antifouling and fouling release performance than standard PDMS coatings (Gudipati et al. 2005).

Minimally adhesive polymer surfaces (MAPSs) from star oligosiloxanes, star oligofluorosiloxanes, and α,ω-bis(3-aminopropyl) PDMS have been prepared by Grunlan et al. (2006). It was found that varying of the molecular weight of the star oligosiloxanes and star oligofluorosiloxanes, as well as altering the ratio to α,ω-bis(3-aminopropyl) PDMS, may enhance their fouling release behavior. Minimally adhesive, fouling release applications of surface-enriched perfluoropolyether (PFPE) graft terpolymer-based coatings are also in the research focus.

Silicone fouling release coatings, facilitating only weak adhesion of macrofouling organisms such as barnacles, tubeworms, and macroalgae (Kavanagh et al. 2003; Stein et al. 2003; Sun et al. 2004) currently represent the only viable commercial alternative to biocide antifouling coatings.

Milne (1977a,b) was among the first researchers who pointed out the antifouling properties of silicone polymers and also observed that the low molecular weight silicone polymers (oils) greatly enhanced their fouling release properties. These early observations constitute the basis of most silicone fouling release systems commercially available today. Intersleek (International Coatings) was the first commercial fouling release coating and its manufacturers recently celebrated the covering of their 100th ship. Hempasil (Hempel Company), EP 2000 and SN-1 HP (E Paint Company), Eco-speed (Chugoku Marine Paints), Phase Coat URF, Si-Coat 560 and 561, etc. are other practically applicable low-surface-energy coatings. However, none of them meet all of the desired performance characteristics.

The use of silicone fouling release coatingsis restricted to larger faster moving vessels. Many lack the toughness to withstand the rigorous physical demands of the marine environment, do not sufficiently self-clean or, due to polymer restructuring or other degradation pathways, lose many of the desirable surface properties with time and exposure to the marine environment. Therefore, many research groups are looking for new possibilities for solving these problems. For example, the largest research project, which includes 31 organizations across Europe (AMBIO 2006), aims to develop new types of nanostructured fouling-prevention polymeric surface coatings that mimic natural non-fouling surfaces (e.g., dolphin skin, lotus leaf effect).

More recently, Vladkova et al. (2006) succeeded in developing hard fouling-preventing silicone coatings based on industrially produced room temperature vulcanized (RTV) PDMS. Figure 9 demonstrates the biofouling of such sample coatings after 1-year exposure in the Indian Ocean (at the Fishing Harbour, Chennai) where the water salinity, temperature, and concentration of fouling organisms are very high and thus the biofouling is very heavy. No macrofouling, only slime formed mainly by brown diatoms, is observed on these samples exposed in static conditions. Engineered coatings such as these that disallow development of macrofouling under static conditions are suitable for macrofouling prevention of the hulls of slowly moving ships and static underwater constructions.

Fig. 9 Fouling of silicone samples after 1-year exposure in the Indian Ocean, Chennai Fishing Harbour

8 Conclusions

A general theory of bioadhesion remains illusive and the molecular details underlying the adhesion mechanisms of fouling organisms remain unclear.

Many physicochemical factors influencing bioadhesion and biofouling are unknown or remain nebulous, such as surface free energy and related parameters, water contact angle and contact angle hysteresis, elastic modulus, surface chemistry, surface roughness and topography, and biological response, etc. Creation of a "theta-surface" represents a new and fundamental antifouling concept, but it is still at the initial step of development.

Surface patterningseems to be a very promising anti-biofouling strategy but from a practical standpoint significant questions remain, such as ablation by water, abrasion by sand particles, etc., under actual field applications. Recently developed polymeric materials that resist such aggressive environmental impacts may eventually make surface-patterning approaches pragmatic, but currently they are too expensive to be commercially viable alternatives.

The creation of exceedingly smooth surfacesmay be a more realistic approach to decreased biofouling. Manipulating the surface topography to enhance smoothness and engineering local surface hydrodynamics should contribute to this approach.

The known fouling release coatings decrease the adhesion strength of the fouling organisms but no one surface is known to prevent the formation of biofilm.

In depth study of the adsorption of adhesive proteins secreted by fouling organisms may be the key to identification of a "universal" surface that would release all biological fouling systems, since the biological cascade of biofouling begins with deposition of such proteins. Surface modification to create material surfaces with suitable composition and morphology would be necessary to control biofouling.

References

Abarzua S, Jacubowski S (1995) Biotechnological investigation for the prevention of biofouling 1. Biological and biochemical principles for the prevention of biofouling. *Mar Ecol Prog Ser* 123:301–312

Adams J, Watts F (1993) Regulation of development and differentiation by the extracellular matrix. *J Invest Dermatol* 117:1183–1198

Altankov G (2003) Interaction of cells with biomaterial surfaces. DSc thesis, BAS, Institute of Biophysics, Sofia

Altankov G, Groth T (1994) Reorganization of substratum bound fibronectin on hydrophilic and hydrophobic materials is related to biocompatibility. *J Mater Sci: Mater Med* 5:732–737

AMBIO (2006) Advanced nanostructured surfaces for the control of biofouling. http://www.ambio.bham.ac.uk/. Last accessed 14 July 2008

An YH, Friedman RJ (1998) Concise review of mechanisms of bacterial adhesion to biomaterial surfaces. *J Biomed Mater Res* 43:338–348

Anderson C, Atlar M, Callow M, Candries M, Milne A, Townsin RL (2003) The development of fouling-release coatings for seagoing vessels. J Marine Design B4:11–23

Arce FT, Avci R, Beech IB, Cooksey KE, Wigglesworth-Cooksey B (2003) A comparative study of RTV11 and intersleek elastomers. *J Chem Phys* 119:1671–1682
Atthoff B (2006) Tailoring of biomaterials using ionic interactions. Synthesis, characterization and application, PhD thesis, Uppsala University, Sweden
Baier RE (1973) Influence of the initial surface condition of materials on bioadhesion. In: Acker RF, Brown BE, DePalma JR, Iverson WP (eds.) Proceedings third international congress on marine corrosion and fouling. Northwestern University Press, Evanston, IL, pp. 633–639
Baier RE (1980) Substrata influences on the adhesion of microorganisms and their resultant new surface properties. In: Bitton G, Marshal K (eds.) Adsorption of microorganisms to surfaces. Wiley, New York, pp. 59–104
Baier RE (2006) Surface behavior of biomaterials: the theta surface for biocompatibility. *J Mater Sci: Mater Med* 17:1057–1062
Baier RE, Shafrin EG, Zisman WA (1968) Adhesion: mechanisms that assist or impede it. *Science* 162:1360–1368
Bailey FE, Koleske JV (1976) Poly(ethylene oxide). Academic, New York
Baney RH, Voight CE, Mentele JW (1977) In: Harris FW, Seymour RB (eds.) Structure-solubility relationships in polymers. Plenum, New York, pp. 225–232.
Berglin M, Lönn N, Gatenholm P (2003) Coating modulus and barnacle bioadhesion. *Biofouling* 195:63–69
Berntsson KM (2001) Larval behaviour of the barnacle Balanus improvisus with implications for recruitment and biofouling control. PhD thesis, Dept. Marine Ecology, Göteborg University
Berntsson KM, Jonsson PR (2003) Temporal and spatial patterns in recruitment and succession of a temperate marine fouling assemblage: a comparison of static panels and boat hulls during the boating season. *Biofouling* 19:187–195
Berntsson KM, Jonsson PR, Lejhall M, Gatenholm P (2000) Analysis of behavioural reaction of micro textured surfaces and implications for recruitment by the barnacle Balanus improvisus. *J Exp Mar Biol Ecol* 251:59–83
Bers V, Wahl M (2004) The influence of natural surface micro topographies on fouling. *Biofouling* 20(1):43–51
Bitton G, Marshall KC (eds.) (1980) Adsorption of microorganisms to surfaces. Wiley, London
Bohringer KF (2003) Surface modification and modulation in microstructures: controlling protein adsorption, monolayer desorption and micro-self-assembly. *J Microtech Microeng* 13:S1–S10
Bos R, van der Mei HC, Gold J, Busscher HJ (2000) Retention of bacteria on a substratum surface with micro-patterned hydrophobicity. *Microbiol Lett* 189:311–315
Brady RF (1999) Properties which influence marine fouling resistance in polymers containing silicon and fluorine. *Prog Org Coat* 35:31–35
Brady RF (2000) Clean hulls without poisons: devising and testing nontoxic marine coatings. *J Coat Technol* 72:44–56
Brady RF (2003) Antifouling coatings without organotin. *J Protect Coat Linings* 20(1):33–37
Brady RF, Singler IL (2000) Mechanical factors favoring release from fouling release coatings. *Biofouling* 15(1–3):73–81
Brady RF, Bonafede SJ, Schmidt DL (1999) Self-assembled water-born fluoropolymer coatings for marine fouling resistance. *JOCCA-Surf Coat Int* 82(12):582–585
Brennan AB, Baney RH, Carman ML, Estes TG, Feinberg AW, Wilson LH, Schumacher JF (2005) Surface topography for non-toxic bioadhesion control. USA Patent 20060219143
Brusscher HJ, Bos R, van der Mei HC (1995) Initial microbial adhesion is determinant for the strength of biofilm adhesion. *FEMS Microbiol Lett* 128:229–234
Bunyard WC, Romack TJ, DeSimone JM (1999) Perfluoropolyether synthesis in liquid carbon dioxide by hexafluoropropylene photooxidation. *Macromolecules* 32:8224–8226
Callow ME, Fletcher RL (1994) The influence of low surface energy materials on bioadhesion: a review. *Int Biodeterior Biodegradation* 34:333–343
Callow ME, Callow JA (2000) Substratum location and zoospore behavior in the fouling alga Enteromorpha. *Biofouling* 15:49–56

Callow ME, Callow JA (2006) Biofilms. In: Fusetani N, Clare AS (eds.) Antifouling compounds. Progress in molecular and submolecular biology, vol 32. Springer, Berlin Heidelberg New York, pp. 141–169

Callow ME, Callow JA, Pickett-Heaps JD, Wetherbee R (1997) Primary adhesion of Enteromorpha propagules: quantitative settlement studies in video microscopy. *J Phycol* 33:938–947

Callow ME, Callow JA, Ista LK, Coleman SE, Nolasco AC, Lopez GP (2000) The use of self-assembled monolayers of different wettabilities to study surface selection and primary adhesion process of green algal (Enteromorpha) zoospores. *Appl Environ Microbiol* 66:3249–3254

Callow ME, Jennings AR, Brennan AB, Seegert CE, Gibson A, Wilson L, Feinberg A, Baney R, Callow JA (2002) Micro topographic cues for settlement of zoospores of the green fouling alga Enteromorpha. *Biofouling* 18:229–236

Carman ML, Estes TG, Feinberg AW, Schumacher JF, Wilkerson W, Wilson LH, Callow ME, Callow JA, Brenan AB (2006) Engineered antifouling microtopographies – correlating wettability with cell attachment. *Biofouling,* 22:11–21

Casse F, Swain GW (2006) The development of microfouling on four commercial antifouling coatings under static and dynamic immersion. *Int Biodeterior Biodegradation* 57:179–185

Chan CM (1993) Polymer surface modification and characterization, Chapters 5–7. HanserGardner, Brookfield

Chan CM, Ko TM, Hiraoka H (1996) Polymer surface modification by plasmas and photons. *Surf Sci Rep* 24(1–2):1–54

Chaudhury MK, Finlay JA, Chung JY, Callow ME, Callow JA (2005) The Influence of elastic modulus and thickness of the release of the soft-fouling green alga Ulva linza from PDMS model networks. *Biofouling* 21(1):41–48

Cooksey KE, Wigglesworth-Cooksey B (1995) Adhesion of bacteria and diatoms to surfaces in the sea – a review. *Aquat Microb Ecol* 9:87–96

Corpe WA (1970) Attachment of marine bacteria to solid surfaces. In: Manly S (ed.) Adhesion in biological systems.Academic, New York, pp. 73–87

Costerton JW, Cheng KJ, Geesey GG, Ladd TI, Nickel JC, Dasgupta M, Marrie TJ (1987) Bacterial biofilms in nature and disease. *Ann Rev Microbiol* 41:435–464

Cunliffe D, Smart CA, Alexander C, Vulfson EN (1999) Bacterial adhesion at synthetic surfaces. *Appl Environ Microbiol* 65 (11):4995–5002

Dahlström M, Jonsson H, Jonsson PR, Elwing H (2004) Surface wettability as a determinant in the settlement of the barnacle Balanus improvisus (DARWIN). *J Exp Mar Biol Ecol* 305:223–232

Derjaguin BV (1955) Theory of the heterocoagulation, interaction and adhesion of dissimilar particles in solutions of electrolytes. *Discuss Faraday Soc* 18:85–86

Dexter SC (1979) Influence of substratum critical surface tension on bacterial adhesion in situ studies. *J Coll Interface Sci* 70:346–354

Dunne WM (2002) Bacterial adhesion: seen any good biofilms lately? *Clin Microbiol Rev* 15:155–166

Elbert DL, Hubbel JA (1996) Surface treatments of polymers for biocompatibility. *Annu Rev Mater Sci* 26:365–394

Finlay JA, Callow ME, Ista LK, Lopez GP, Callow JA (2002) The influence of surface wettability on the adhesion strength of settled spores of the green alga Enteromorpha and the diatom Amphora. *Integ Comp Biol* 42:1116—1125

Flammang P, Jangoux M (2004) Bioadhesion models from marine invertebrates: an integrated study – biomechanical, morphological, biochemical, molecular – of the processes involved in the adhesion of Cuvierian Tubules on sea cucumbers (Echiodermata, Holothuroidea). Mons-Hainaut University, Belgium http://www.stormingmedia.us/86/8607/A860724.html. Last accesses 14 July 2008

Flemming H-C, Greenhalgh M (2008) Concept and consequences of the EU biocide guideline. Springer Ser Biofilms. doi: 10.1007/7142_2008_12

Gan D, Mueller A, Wooley KL (2003) Amphiphilic and hydrophobic surface patterns generated from hyberbranched fluoropolymer/linear polymer networks: minimally adhesive coatings via the crosslinking of s. *J Polym Sci Part A: Polym Chem* 41:3531–3540

Ghatak A, Chaudhury MK, Shenoy V, Sharma A (2000) Meniscus instability in a thin elastic films. *Phys Rev Lett* 85:4329–4332

Gerbig YB, Phani AR, Haefke H (2005) Influence of nanoscale topography on the hydrophobicity of fluoro-based polymer thin films. *Appl Surf Sci* 242:251–255

Good RG (1992) Contact angle, wetting and adhesion: a critical review. *J Adh Sci Techn* 6:1269–1302

Gölander C-G (1986) Preparation and properties of functionalised polymer surfaces. PhD thesis, Royal Institute of Technology, Stockholm

60. Gölander C-G, Jönsson S-E, Vladkova T (1984) A surface coated article, process and means for the preparation of thereof and use of thereof. Sweden patent no 8404866–9/28.09.1984; Bulgarian Patent 67997/28.09.1984; European Patent 022966/28.09.84; PCT SE85/00376/28.09.84

Gölander C-G, Jönsson S-E, Vladkova T, Stenius P, Eriksson J-C (1986) Preparation and protein-adsorption properties of photo-polymerized hydrophilic films coating N-vinyl pyrollidone (NVP), acrylic acid (AA) or ethylene oxide (EO) units as studied by ESCA. *Coll Surf* 21:149–165

Gölander C-G, Jönsson S-E, Vladkova T, Stenius P, Kisch E (1987) Protein adsorption on some hydrophilic films. Presented at 31st IUPAC, 13–18 July 1987, Sofia

Griffith AA (1921) The phenomena of rupture and flow in solids. *Phil Trans R Soc London A* 221:163–198

Grinnell F, Milam M, Spree P (1972) *Arch Biochem Biophys* 153:193

Griesser HJ, Hartley PG, McArthur SL, McLean KM, Meagher L, Thissen H (2002) Interfacial properties and protein resistance of nano-scale polysaccharide coatings. *Smart Mater Struct* 11:652–661

Groll J, Amirgoulova EV, Ameringer T, Heyes CD, Röcker C, Nienhaus GU, Möller M (2004) Biofunctionalized ultrathin coatings of cross-linked star-shaped poly(ethylene oxide) allow reversible folding of immobilized proteins). *J Am Chem Soc* 126:4234–4339

Grunlan MA, Lee NS, Gai G, Gedda T, Mabry JM, Mansfeld F, Kus E, Wendt DE, Kowalke GL, Finley JA, Callow JA, Callow ME, Weber WP (2004) Synthesis of a,w-bis epoxy oligo (1'H,1'H,2'H, 2'H-perfluoroalkyl siloxane)s and properties of their photo-acid cross-linked films. *Chem Mater* 16:2433–2441

Grunlan MA, Lee NS, Mansfeld F (2006) Minimally adhesive polymer surfaces prepared from star oligosiloxanes and star oligofluorosiloxanes. *J Polym Sci Part A: Poly Chem* 44:2551–2566

Gudipati CS, Greenlieaf CM, Johnson JA, Pornpimol P, Wooley KL (2004) Hyperbranched fluoropolymer and linear PEG based amphiphilic crosslinked networks as efficient anti-fouling coatings: an insight into the surface compositions, topographies and morphologies. *J Polym Sci Part A: Poly Chem* 42:6193–6208

Gudipati CS, Finlay JA, Callow JA, Callow ME, Wooley KL (2005) The antifouling and foulin-release performance of hyperbranched fluoropolymer (HBFP)-poly(ethylene glycol) (PEG) composite coatings evaluated by adsorption of biomacromolecules and the green fouling alga Ulva. *Langmuir* 21:3044–3053

Hamza A, Pham VA, Matsuura T, Santerre JP (1977) Development of membranes with low surface energy to reduce fouling in ultrafiltration applications. *J Membr Sci* 131:217–223

Harder T (2008) Marine epibiosis – concepts, ecological consequences and host defense. Springer Ser Biofilms. doi: 10.1007/7142_2008_16

Harris JM (1992) Poly(ethylene glycol) chemistry. Biotechnical and biomedical applications. Plenum , New York

Hester JF, Banerjee P, Won YY, Akthakul A, Acar MH, Mayers AM (2002) ATRP of amphiphilic graft copolymers based on PVDF and their use as membrane additives. *Macromolecules* 35:7652–7661

Hillborg H, Gedde UW (1999) Hydrophobicity changes in silicone rubbers. *IEEE Trans Dielect Elect Insul* 6:703–717

Hlady V, Van Vagenen RA, Andrade JD (1985) In: Andrade JD (ed.) Surface and interfacial aspects of biomedical polymers, vol 2. Plenum , New York, p. 81
Holland R, Dugdale TM, Wetherbee R, Brennan AB, Finlay JA, Callow JA, Callow ME (2004) Adhesion and motility of fouling diatoms on silicone elastomer. *Biofouling* 20:323–329
Holm ER, Kavanagh CJ, Meyer AE, Wiebe D, Nedved BT, Wendt D, Smith CM, Hadfield MG, Swain G, Wood CD, Truby K, Stein J, Montemarano J (2006) Interspecific variation in patterns of adhesion of marine fouling to silicone surfaces. *Biofouling* 22(3–4):233–243
Homma H, Kuroyagi I, Izumi K, Murley CL, Ronzello J, Boggs SA (1999) Diffusion of low molecular weight siloxane from bulk to surface. *IEEE Trans Dielect Elect Insul* 6:370–375
Humphrey AJ, Finlay JA, Pettitt ME, Stanley MS, Callow JA (2005) Effect of Ellman's reagent and dithiothreitol on the curing of the spore adhesive glycoprotein of the green alga Ulva. *J Adhesion* 81:791–803
Ikada Y, Suzuki M, Tamada Y (1984) Polymer surfaces possessing minimal interaction with blood components. In: Shalaby SW, Hoffman AS, Ratner BD, Horbett TA (eds.) Polymers as biomaterials. Plenum, New York.
Ista LK, Callow M, Finlay S, Coleman E, Nolasco AC, Simons RH, Callow JA, Lopez GP (2004) Effect of substratum surface chemistry and surface energy on attachment of marine bacteria and algal spores. *Appl Environ Microbiol* 70(7):4151–4157
Johston E, Bullock S, Uilk J, Gatenhohnm P, Wynne KJ (1999) Networks from a,w-dihydroxypoly(dimethylsiloxane) and (tridecafluoro-1,1,2,2-tetrahydrooctyl)triethoxysilane: surface microstructures and surface characterization. *Macromolecules* 32:8173–8182
Kamino K, Inoue K, Maruyama T, Takamatsu N, Harayama S, Shizuri Y (2000) Barnacle cement proteins: importance of disulfide bonds in their insolubility. *J Biol Chem* 275:27360–27365
Kavanagh CJ, Swain GW, Kovach BS (2003) The effect of silicone fluid additives and silicone matrices on the barnacle adhesion strength. *Biofouling* 19 (6):381–390
Kendall K (1971) The adhesion and surface energy of elastic solids. *J Phys D: Appl Phys* 4:1186–1195
Kendall K (1994) Adhesion: molecules and mechanics. *Science* 263:1720–1725
Kiaie D, Hoffman AS, Horbett TA, Lew KR (1995) Platelet and monoclonal antibody binding to fibrinogen adsorbed on glow discharge deposited polymers. *J Biomed Mat Res* 29:729–739
Kicheva J, Kostov V, Mateev M, Vladkova T (1992) Evaluation of the in vitro and in vivo biocompatibility of PVC materials with modified surfaces. In: Proceedings VI colloquium on biomaterials, Aachen, 24–25 Sept 1992, pp. 24–39
Kingshott P, Griesser HJ (1999) Surfaces that resist bioadhesion. *Curr Opin Solid State Mater Sci* 4 (4):403–412
Kinloch AJ, Young RJ (1983) Fracture behavior of polymers. Applied Science, London.
Klebe R (1974) Isolation of collagen-dependant cell attachment factor. *Natura* 250:248–252
Kohl JG, Singler IL (1999) Pull-off behaviour of epoxy bonded to silicone duplex coatings. *Prog Org Coat* 36:15–20
Krishnan S, Callow JA, Fischer DA (2006) Anti-fouling properties of comb-like block copolymers with amphiphilic side chains. *Langmuir* 22 (11):5075–5086
Kuhl TL, Leckband DE, Lasic DD, Israelachvili JN (1994) Modulation of interaction forces between bi-layer exposing short-chained ethylene oxide head groups. *Biophys J* 66:1479–1488
Li X, Logan BE (2004) Analysis of bacterial adhesion using a gradient force analysis and colloid probe atomic force microscopy. *Langmuir* 20(20):8817–8822
Linder E (1992) A low surface energy approach in the control of marine biofouling. *Biofouling* 6:193–205
Loeb GI, Neihof RA (1975) Marine conditioning films. *Adv Chem* 145:319–335
Malmsten M (1998) Biopolymers at interfaces. Marcel Dekker, New York.
Mark JE (1979) Interpretation of polymer properties in terms of chain conformations and spiral configurations. *Acc Chem Res* 12:49–55
McGuire J, Swartzel KR (1987) Proceedings National Meeting American Institute Chemical Engineers, Minneapolis, p. 31

Mera AE, Goodwin M, Pike JK, Wynne KJ (1999) Fluorinated silicone resin fouling release composite. *Polymer* 40:419

Meyer A, Baier R, Wood CD, Stein J, Truby K, Holm E, Montemarano J, Kavanagh C, Nedved B, Smith C, Swain G, Wiebe D (2006) Contact angle anomalies indicate that surface-active eluates from silicone coatings inhibit the adhesive mechanisms of fouling organisms. *Biofouling* 22(6):411–423

Milne A (1977a) Coated marine surfaces. UK Patent 1470465

Milne A (1977b) Antifouling marine compositions. US Patent 4025693

Mori Y, Nagaoka S (1982) A new antithrombogenic material with long poly(ethylene) oxide chains. *Trans Am Soc Artif Intern Organs* 28:459

Morra M, Cassinelli C (1997) Bacterial adhesion to polymer surfaces: a critical review of surface thermodynamic approaches. *J Biomat Sci Polymer Ed* 9:55–74

Nedved BT, Hadfield MG (2008) *Hydroides elegans* (Annelida: Polychaeta): a model for biofouling research. Springer Ser Biofilms. doi: 10.1007/7142_2008_15

Newby BZ, Chaudhury MK (1997) Effect of interfacial slippage on viscoelastic adhesion. *Langmuir* 13:1805–1809

Newby BZ, Chaudhury MK, Brown HR (1995) Macroscopic evidence of effect of interfacial slippage on adhesion. *Science* 269:1407–1409

Oliviera R (1997) Understanding adhesion: a means for preventing fouling. *Exp Thermal Fluid Sci* 14:316–322

Ostuni E, Chen CS, Ingber DE (2001) Selective deposition of proteins and cells in arrays of microwells. *Langmuir* 17:2828–2834

Ostuni E, Grzybowski BA, Mrksich M, Roberts CS, Whitesides GM (2003) Adsorption of proteins to hydrophobic sites on mixed self-assembled monolayers. *Langmuir* 19(5):1861–1872

Owen MJ (1990) Silicon surface reactivity. In: Zeigler JM, Fearon FWG (eds.) Silicon-based polymer science: a comprehensive resource. ACS Symposium Series 223. American Chemical Society, Washington DC, pp. 709–717

Pasche S (2004) Mechanisms of protein resistance of adsorbed PEG-graft copolymers. DSc thesis, Swiss Federal Institute of Technology, Zurich

Pasmore M (2008) Biofilms in hemodialysis. Springer Ser Biofilms. doi: 10.1007/7142_2008_5

Pasmore M, Todd P, Smith S, Baker D, Silverstein J, Coons D, Bowman CN (2001) Effect of ultrafiltration membrane surface properties on Pseudomonas aeruginosa biofilm initiation for the purpose of reducing biofouling. *J Membr Sci* 194:15–21

Pedry L (2005) Interaction of bacteria with hydrophobic and hydrophilic interfaces. PhD thesis, Stanford University

Pike JK, Ho T, Wynne KJ (1996) Low surface energy fluorinated poly(amide urethane) block copolymers and other low surface energy polymers. *Chem Mater* 8:856–860

Ratner BD, Chilkoti A, Lopez GP (1990) Plasma deposition and treatment for biomedical applications. In: d'Agustino R (ed.) Plasma deposition, treatment and etching of polymers. Academic, San Diego, pp. 463–516

Russell TP (2002) Surface responsive materials. *Science* 279:964–967

Satriano C, Conte E, Marletta G (2001) Surface chemical structure and cell adhesion onto ion beam modified polysiloxane. *Langmuir* 17:2243–2250

Satriano C, Carnazza S, Guglielmino S, Marletta G (2002) Differential cultured fibroblast behavior on plasma and ion-beam-modified polysiloxane surfaces. *Langmuir* 18(24):9469–9475

Scardino A, de Nys R, Ison O, O'Connor W, Steinberg P (2003) Microtopography and antifouling properties on the shell surface of the bivalve mollusks *Mytilus galloprovincialis* and *Pictada imbriticata*. *Biofouling* 19:221–230

Schackenraad JM, Stokroos I, Bartels H, Busscher HJ (1992) Patency of small caliber, superhydrophobic E-PTFE vascular grafts: a pilot study in rabbit carotid artery. *Cells Mater* 2:193–199

Scheuerman TR, Camper AK, Hamilton MA (1998) Effects of substratum topography on bacterial adhesion. *J Coll Interface Sci* 208:23–33

Schmidt DL, Coburn CE, DeKoven BM, Potter GE, Meyers GF, Fischer DA (1994) Water-based non-stick hydrophobic coatings. *Nature* 368:39–41

Schmidt DL, Brady RF, Lam K, Schmidt DC, Chaudhury MK (2004) Contact angle hysteresis, adhesion and marine biofouling. *Langmuir* 20(7):2830–2836

Sheu MS, Chen JY, Wang LP (1995) Biomaterials surface modification using plasma gas discharge processes. In: Wise DL et-al. (eds.) Encyclopedic handbook of biomaterials and bioengineering. Part A: Materials, vol 1. Marcer Dekker, New York, pp. 865–894

Sigal GB, Mrksich M, Whitesides GM (1998) Effect of surface wettability on the adsorption of proteins and detergents. *J Am Chem Soc* 120:3464–3473

Silberzan P, Perutz S, Kramer EJ, Chaudhury MK (1994) Study of the self-adhesion hysteresis of a siloxane elastomer using the JKR method. *Langmuir* 10:2466–2470

Sinde E, Carballo J (2000) Attachment of *Salmonella* and *Listeria monocytogenes* to stainless steel, rubber and PTFE: the Influence of the free energy. *Food Microbiol* 17:439–447

Smeltzer MS (2008) Biofilms and aseptic loosening. Springer Ser Biofilms. doi: 10.1007/7142_2008_1

Speranza G, Gottardi G, Pederzolli C, Lunelli L, Carli E, Lui A, Brugnara M, Anderle M (2004) Role of chemical interactions in bacterial adhesion to polymer surfaces. *Biofouling* 25(11):2029–2037

Stanley MS, Callow ME, Callow JA (1999) Monoclonal antibodies to adhesive cell coat glycoproteins secreted by zoospores of the green alga Enteromorpha. *Planta* 210:61–71

Stein J, Truby K, Wood CD (2003) Silicon foul release coatings: effect of interaction of oil and coating functionalities on the magnitude of macro fouling attachment strengths. *Biofouling* 195:71–82

Sun Y, Akhremitchev B, Walker GC (2004) Using the adhesive interaction between atomic force microscopy tips and polymer surfaces to measure the elastic modulus of compliant samples. *Langmuir* 20:5837–5845

Swain GWJ, Schultz MP (1996) The testing and evaluation of non-toxic antifouling coatings. *Biofouling* 10:187–197

Tang Y, Finlay JA, Kowalke GL (2005) Hybrid xerogel films as novel coatings for antifouling and fouling release. *Biofouling* 21(1):59–71

Tidball JG, Albrecht DA (1998) Regulation of apoptosis by cellular interactions with the extracellular matrix. In: Lockshin RA, Zakeri Z, Tilly JL (eds.) When cells die: a comprehensive evaluation of apoptosis and programmed cell death. Wiley-Liss, New York, pp. 411–427

Truby K, Wood C, Stein J, Cella J, Carpenter J (2000) Evaluation of the performance enhancement of silicone biofouling-release coatings by oil incorporation. *Biofouling* 15(1–3):141–150

Uilk J, Johnston EE, Bullock S, Wynne KJ (2002) Surface characterization, microstructure and wetting of networks from a,w-dihydroxy(polydimethylsiloxane) and 1,1,2,2-tetrahydrotridecafluoro octyltriethoxysilane. *J Macromol Chem Phys* 203:1506–1511

Van deVivere P, Kirchman DL (1993) Attachment stimulates exopolysaccharide synthesis by a bacterium. *Appl Environ Microbiol* 59:3280–3286

Van Loosdrecht MCM, Lyklemam J, Norde W, Zehnder AJB (1990) Hydrophobic and electrostatic parameters in bacterial adhesion. *Aquat Sci* 52:103–113

Velegol SB, Logan BE (2004) Correction to: "Contributions of bacterial surface polymers, electrostatics and cell elasticity to shape of AFM force curves". *Langmuir* 20:3820

Verwey EJW, Overbeek JTG (1948) Theory of stability of lyophobic colloids. Elsevier, Amsterdam

Vladkova TG (1995) Modification of polymer surfaces for medical application. Presented at XIII conference on modification of polymers, Kudowa Zdroj, Poland, 11–15 Sept 1995

Vladkova TG (2001) Some possibilities to polymer surface modification. UCTM, Sofia

Vladkova T, Krasteva N, Kostadinova A, Altankov GP (1999) Preparation of PEG-coated surfaces and a study for their interaction with living cells. *J Biomat Sci* 10(6):609–615

Vladkova TG, Keranov Il, Dineff PD, Avramova IA, Altankov GP (2005) Plasma based Ar+ beam assisted PDMS surface modification. *Nucl Instrum Methods Phys Res B* 236:552–562

Vladkova TG, Dineff PD, Zlatanov I, Katirolyi S, Venkatesan R, Murthy S (2006) Composition coating for biofouling protection. Bulgarian Patent Appl no 109779; WO 2008/074102 A1

Vreeland V, Waite JH, Epstein L (1998) Polyphenols and oxidases in substratum adhesion by marine algae and molluscs. *J Phycol* 34:1–8

Wagner VE, Koberstein JT, Bryers JD (2004) Protein and bacterial adhesion. *Biomaterials* 25:2247–2263

Waite JH (1999) Reverse engineering of bioadhesion in marine mussels. *Ann N Y Acad Sci* 18:301–309

Walker GC, Sun Y, Guo S, Finlay JA, Callow ME, Callow JA (2005) Surface mechanical properties of the spore adhesive of the green alga Ulva. *J Adhesion* 81:1101–1118

Wang J, Mao GP, Ober CK, Kramer EJ (1997) Liquid crystalline, semifluorinated side group block copolymers with stable low energy surfaces: synthesis, liquid crystalline structure, and critical surface tension. *Macromolecules* 30:1906–1914

Wynne KJ, Swain GW, Fox RB, Bullock S, Uilk J (2000) Two silicone nontoxic fouling release coatings: hydrosilation cured PDMS and CaCO3 filled ethoxysiloxane cured RTV11. *Biofouling* 16:277–288

Xu L-C, Logan BE (2005) Atomic force microscopy colloid probe analysis of interactions between proteins and surfaces. *Environ Sci Technol* 39(10):3592–3600

Youngblood JP, Andruzzi L, Ober CK, Hexemer A, Kramer EJ, Callow JA (2003) Coatings based on side-chain ether-linked poly(ethylene glycol) and fluorocarbon polymers for the control of marine biofouling. *Biofouling* 19:91–97

Zhao Q, Wang S, Müller-Steinhagen H (2004) Tailored surface free energy of membrane diffusers to minimize microbial adhesion. *Appl Surf Sci* 230:371–378

Zhili L, Brokken-Zijp JCM, de With G (2004) Determination of the elastic moduli of silicone rubber coatings and films using depth-sensing indentation. *Polymer* 45:5403–5406

A Strategy To Pursue in Selecting a Natural Antifoulant: A Perspective

K.E. Cooksey (✉), B. Wigglesworth-Cooksey, and R.A. Long

Abstract With the ban of tributyl tin and its analogs and the fact that copper and its derivatives are under legislative pressure, we must consider alternatives for the control of biofouling. So-called fouling-release coatings are one potential solution, but they do not control microfouling well. Compounds derived from macrobiota that appear to be fouling-free appear to be suitable molecules for investigation as components of antifouling coatings. If a molecule is to be active against a variety of organisms, it is important that it inhibit some universal metabolic process. Ideally, such a molecule would interfere with the surface sensing process itself and the events that result from the reception of that signal. Cell signaling in all eukaryotes is mediated by changes in the internal Ca^{2+} concentration. Therefore, a molecule that interferes with Ca-mediated events would be an ideal candidate to inhibit cellular adhesion and thus fouling. Using an image analysis-directed assay with diatoms and Ca-fluorophores to detect Ca fluxes, we report how 2-*n*-pentyl-4-quinolinol, D-600,and *trans*, *trans*-2,4-decadienal influence diatom adhesion and motility. All three molecules show activity as antifoulants.

1 Rationale

Here we report strategies to select anti-diatom agents since these organisms form a major portion of the initial fouling biomass (Marszalek et al. 1979; Sieburth 1979; Wetherbee et al. 1998). Trialkyl tins are no longer available for use in marine antifoulant coatings and the days of copper-based paints are numbered (Burgess et al. 2003) Thus it behooves the marine antifouling coating community to search for alternate strategies to control the deterioration of marine structures. One such strategy is to use coatings that have a low surface energy, which allow organisms to settle, but not to adhere strongly. The rationale for this approach is that poorly adhered organisms can

K.E. Cooksey
Department of Microbiology, Montana State University, Bozeman, MT 59717, USA and Environmental Biotechnology Consultants, Manhattan, MT 59741, USA
e-mail: umbkc@gemini.msu.montana.edu

Springer Series on Biofilms, doi: 10.1007/7142_2008_11

then be removed by shear forces generated by a vessel when underway. Recent work (Holm et al. 2006) showed that in a series of coatings, the most efficient in releasing an invertebrate fouling burden depends on which organism is chosen as the test organism. Thus it is not likely that a coating with a particular surface energy will control all types of macrofouling. In any event, it is well known that such coatings do not control microfouling well. The problem is obviously exacerbated on slow-moving vessels, stationary marine structures such as buoys, or optical surfaces. For instance, it has been noted (Holland et al. 2004) that ships treated with a silicone elastomer surface coating accumulated diatom fouling which was not released, even at 30 knots. Terlizzi et al. (2000) and Jelic-Mrcelic et al. (2006) came to a similar conclusion, but the observations from these latter authors were for classical antifouling paints.

Diatoms demonstrate the so-called Baier curve (after Robert E. Baier who first published this information for Ehrlich ascites cells; Baier 1980). The curve in Fig. 1 indicates that the adhesion of diatoms on surfaces with surface energies of 15–70 dynes cm^{-1} shows a minimal adhesion value (~25 dynes cm^{-1}), but that value is not zero and represents about 30% of the maximal adhesion value (35 dynes cm^{-1}) (Characklis and Cooksey 1983). It is possible that the reason for this adhesive adaptability in diatoms is because they secrete a glycoproteinaceous adhesive, which will have hydrophilic and hydrophobic domains (Cooksey and Cooksey 1986). In more recent work using atomic force microscopy (AFM), Arce et al. (2004) showed that the works of removal of a species of *Navicula* from freshly-cleaved mica (very hydrophilic) and Intersleek 425 paint (hydrophobic) were statistically equivalent. In this experiment the AFM was operated in the bioprobe mode using a single diatom attached to the cantilever surface of the AFM, so measurements of the adhesive power of the diatom could be measured directly. Thus it is our opinion that no coating which depends *only* on a particular surface energy will be successful in controlling microfouling.

The question then arises whether control of microfouling is really necessary when classical thought is that invertebrates are the major culprits in producing hydrodynamic drag on a vessel. We believe that it is not yet a settled question whether microbes always prepare a surface for invertebrate settlement. In some cases microbiota stimulate invertebrate larval adhesion, and in other cases, they inhibit it (Dahms et al. 2004). We will not deal with that topic here since it has been reviewed recently (Maki and Mitchell 2002; Dahms et al. 2004). However, it has been shown by Bohlander (1991) that microfouling alone on a ship (US Navy Frigate, *USS Brewton*) can cause a considerable fuel penalty. The fuel saving for the ship at 26 knots was that required to produce the extra 4,500 HP needed for a fouled ship. Thus there are at least two reasons to prevent microfouling: possibly to break the settlement succession and to reduce biofilm-generated drag.

2 Approach

There are many marine organisms that appear not to support a surface layer of microorganisms. Materials extracted from these organisms (both plant and animal) can be tested to determine if they contain antifouling activity. It is possible to screen

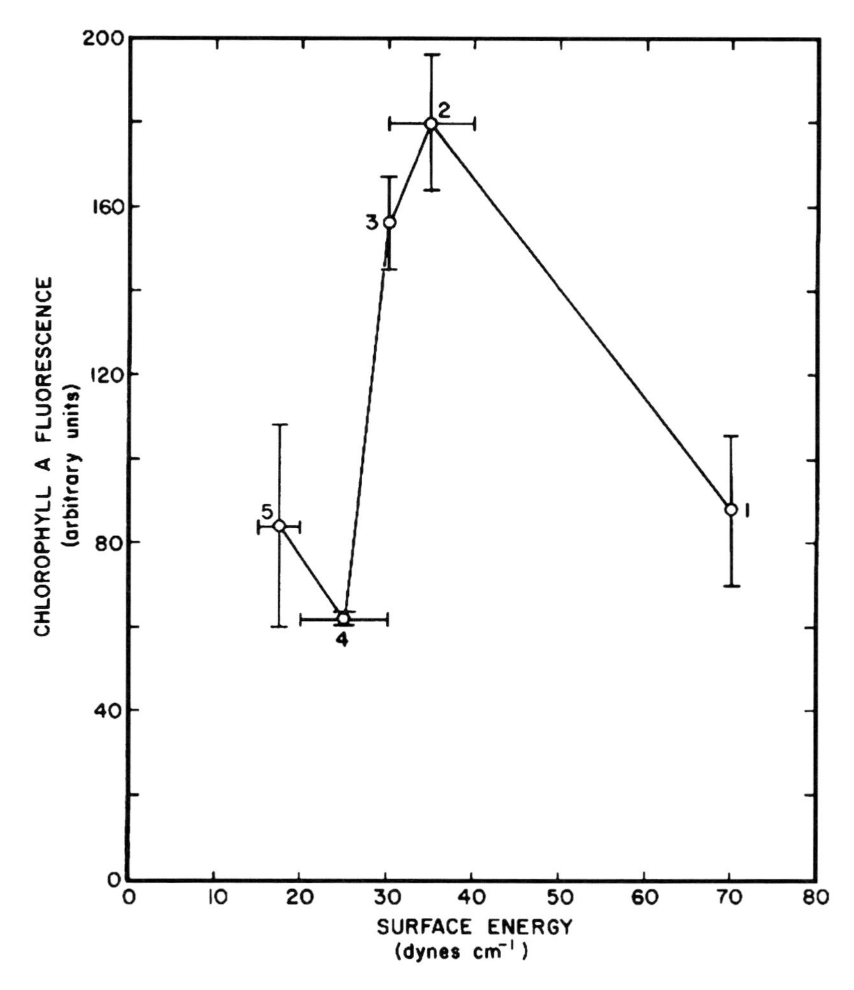

Fig. 1 Adhesion of *A. coffeaeformis* on glass surfaces as measured by the chlorophyll content of the attached cells. The surfaces were treated to modify their surface energy as follows: *1* radio-frequency discharge cleaned and stored under sterile water; *2*, as 1, but stored in air; *3* chloropropyl trichlorosilane; *4* dichloromethyl silane; *5* perfluorinated silane (from Characklis and Cooksey 1983, with permission)

these extracts for the presence of compounds that interfere directly with cellular adhesion, for processes related to adhesion such as motility, and with fundamental physiological processes such as energy generation. Another physiological process well worth consideration is the ability for one cell to communicate with another, so-called quorum sensing. There is however a drawback to these approaches: often the compounds responsible for the antifouling activity are available only from the organism from which they were extracted originally and economic synthesis is unlikely. The pioneering work of Rittschof and colleagues is a good example of this

(Rittschof et al. 1986). Renillafoulin from Sea Pansies provided insights into the adhesion process in barnacles, but as yet, there is no means to synthesize this molecule, nor is there a possibility to harvest this molecule on a grand scale from the marine environment. Thus it is unlikely that this molecule will be used in coatings for aircraft carriers and the like.

3 Development of an Assay Capable of Detecting and Quantifying the Response of Diatoms to Compounds with Potential Antifouling Properties

Since diatoms form a major (perhaps *the* major) component of the initial foulingfilm (Marszalek et al. 1979), it is pertinent to use these organisms as model fouling microorganisms. In fact they may well be the *prime* indicator to use as their physiology is both plant- and animal-like (Webster et al. 1985; Wigglesworth-Cooksey and Cooksey 1992; Armbrust et al. 2004; Vardi et al. 2006). Almost all of our work has been with two species of diatoms, both of which can be found on fouled surfaces in the sea (Callow 1986). At this point it is as well to consider just what we want an antifoulant molecule to do. Do we wish to kill all cells arriving at a surface and if so what time scale will we use? Should the active material allow initial adhesion-but prevent cellular growth and thus colonization of the surface? Is it possible for the potential antifoulant to prevent initial adhesion, or to allow initial adhesion but promote cellular removal by engineering the properties of the surface? These questions must be taken into account in assessing potential activities of candidate molecules. A bioassay should provide this information but, in practical terms, it is just not possible to screen compounds for all of these activities at one time in one type of bioassay. In our assay, diatoms are allowed to attach to a clean microscope slide cover glass, unattached cells are removed and the number of cells remaining is determined by their chlorophyll fluorescence (Cooksey 1981). The assay also allows an assessment of diatom motility (Cooksey and Cooksey 1980). An indication of the reproducibility of the adhesion assay is shown in Table 1. There are

Table 1 Adhesion of *Amphora coffeaeformis* to glass: a test of the reproducibility of the assay

Unattached cell removal	Cells remaining/mm^{2a}	% Remaining[b]
Not washed	841 ± 54 (6.4)	95
Washed ×1	824 ± 51 (6.2)	93
Washed ×2	806 ± 35 (4.3)	91
Washed ×3	841 ± 85 (10.1)	95

Modified from Wigglesworth-Cooksey et al. (1999)
[a]n = 4 ± SD (coefficient of variation). Figures in brackets show % variation
[b]Calculated from the cell concentration in the assay medium and their fluorescence

many variables in this assay which must be optimized. The level of the cell concentration used is critical. If this is too large (>10^5 cells mL^{-1}), the diatom biofilm will slough from the glass substrate as a cohesive film at very low shear velocities. The experimental surfaces must be reproducibly clean. Glow discharge cleaning is not recommended as it produces a chemically unstable surface which quickly adsorbs airborne contaminants (see Fig. 1). Various methods of removing poorly attached cells have been investigated, but the shear produced by dipping the glass substrate into medium was found to be the least invasive and thus the most reproducible. The time frame in which the measurements are made is also critical and must be kept constant as cells condition the surface, i.e., change its surface energy with time due to the adsorption of materials from the aqueous milieu (Wigglesworth-Cooksey et al. 1999). The way such a "conditioning film"influences the performance in the marine environment of a coating with a low surface energy has been reviewed by Maki and Mitchell (2002). A coating with a second strategy, such as the incorporation of a molecule which inhibits some essential metabolic process, may be less susceptible to the influence of a conditioning film.

It can be seen that adhesion and motility (Table 2) are dependent on respiratory-derived energy, not photosynthesis. This has been confirmed by Smith and Underwood (1998). Both adhesion and motility are Ca-dependent (Cooksey and Cooksey 1980; Cooksey 1981). Assays such as this are suitable for laboratory screening studies and do not necessarily represent the conditions existing on the hull of a ship that is underway. Adhesion assays using calibrated flumes are expensive, time consuming, and do not facilitate testing of multiple samples simultaneously. Findlay et al. (2002) make the point that assays involving only the measurement of diatom motility are not suitable for testing the efficacy of fouling release coatings. We are proposing to test the activity of compounds that influence diatom metabolic activity, so this observation does not apply here.

As we learned more about the adhesion of diatoms to surfaces, we realized that adhered cells always were motile, at least initially, and loss of motility was usually a precursor of cell detachment. Thus if we could measure motility quickly, we could screen more efficiently. We had also noticed that cells that were compromised

Table 2 Influence of selected inhibitors on diatom adhesion and motility

Compound, concentration	Motility	Adhesion	Site of action
DCMU, 2 μM	0[a]	0	Photosynthesis, PSII
Darkness	0	0	Photosynthesis
CCCP, 1.25 μM	–[b]	–	All energy generation
Cycloheximide, 3.6 μM	NT	–	Protein synthesis
Tunicamycin, 0.5 μg mL^{-1}	–	–	Glycoprotein synthesis
Ca in medium reduced from 5 to 0.25 mM	–	–	Secretion and all signaling processes

DCMU 3-(3,4-dichlorophenyl)-1,1-dimethylurea, *CCCP* carbonylcyanide *m*-chlorophenylhydrazone, *NT* not tested

[a]0 indicates no effect

[b]– indicates that the compound causes inhibition

physiologically by the inclusion of some test substance in the medium appeared to move more slowly. Initially, in our research, measurements were made manually using a video camera, a video recorder, and a TV monitor (Cooksey and Cooksey 1988). Now all experiments are scored using an image analysis computer system (Cell Trak, Motion Analysis Corp., Santa Rosa, CA). Even though the instrument is capable of real-time assessments of diatom motility, it is convenient to preserve all motility data using a video recorder and use the computer to assess the results at a later time. The equipment allows measurement of speed in micrometers per second, direction of travel as a compass bearing, turning velocity in degrees per second, and changes in speed as micrometers per second per second of up to 80 cells in a single field simultaneously. These parameters make it possible to measure cell behavior in response to a chemical challenge.

Using this equipment we determined that the speed of a diatom over a surface was not constant, but varied from second to second. This is difficult to see with a naked eye. Figure 2 shows the speed with time of *Amphora coffeaeformis* on clean glass. Furthermore, we have observed cells that are in a medium with a low level of toxicant "shunt," i.e., they move backwards and forwards with no significant change in their overall position. Tunicamycin causes this response at <0.5 μg mL^{-1} (Cooksey and Cooksey 1986). Others have proposed using diatom motility as an indicator of toxicity of sediment elutriates (Cohn and McGuire 2000).

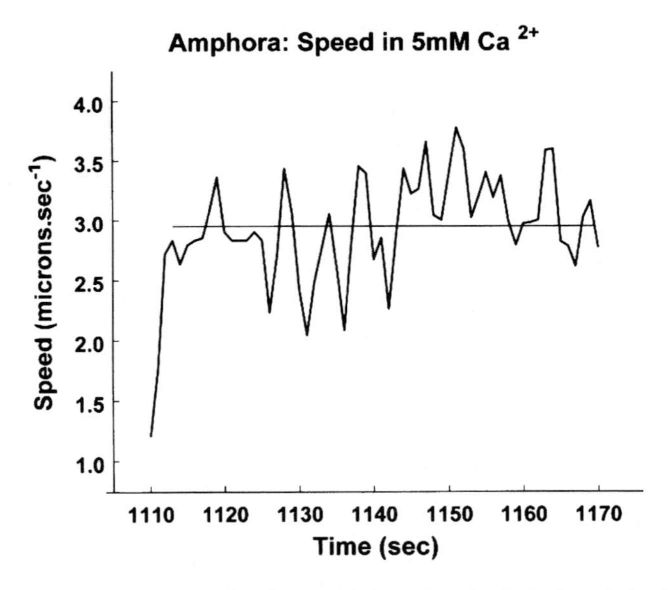

Fig. 2 Speed of a single cell of *A. coffeaeformis*with time. Note the rhythmic variation in speed over 60 s. The horizontal line represents the average speed of the cell, i.e., 2.9 μm s^{-1} (±0.5 μm s^{-1})

4 Why a Calcium Antagonist Can be a Potential Antifoulant

The use of compounds that interfere with Ca homeostasis as active components of antifouling coatings has not, to our knowledge been suggested previously and therefore requires some justification. The impetus for this idea came from our finding that the initial adhesion of diatoms to surfaces and their subsequent motility requires a certain level of Ca in the external milieu. The Ca was shown to act both externally to the cell as well as internally (Cooksey and Cooksey 1980; Cooksey 1981). Intracellular Ca concentration (Ca_i) regulates most responses to external signals in all eukaryotes studied (Berridge et al. 1998). Although initial studies in this field concerned excitable cells, it is now known that the responses of non-excitable cells are also Ca controlled (e.g., Dolle and Nultsch 1988 – *Chlamydomonas*; Andrejaustkast et al. 1985 – higher plants). In fact Ca has been termed the "life or death signal" (Berridge et al. 1998) because not only does its intracellular concentration modulate cellular processes such as movement, it also controls cell death. To carry out these regulatory activities, the Ca_i needs to be set within narrow limits. There are two sources of Ca_i: (a) Ca that arises from outside the cell and crosses the plasma membrane to reach the cytoplasm, and (b) Ca that is released from internal stores. These signals have two functions. They can either activate localized cell processes or they can involve channels throughout the cell, which in turn gives rise to waves of increased Ca_i. Further examples relevant to diatom biology include vesicle secretion (Webster et al. 1985) and, because of its ability to release bound intracellular Ca, the fact that a chemotactic effector facilitates temporary cell motility in a Ca environment that is otherwise insufficient to support motility (Cooksey and Cooksey 1988). It should be mentioned that the increases in Ca_i referred to above are of the order of tenfold, i.e., 10^{-7}–10^{-6} M Ca. Organisms living in seawater, which is 10 mM Ca, must control Ca_i levels closely or become inactive and/or die because elevated Ca_i is lethal. Major contributors to this process in this environment are the membrane Ca-ATPases,which export Ca from the cell. Cells avoid long periods of elevated Ca_i by delivering the Ca signals as transient increases in Ca_i rather than one signal over a long period. These are seen as Ca_i oscillations (see Fig. 2). In some cases, cells exhibiting Ca waves in the proximity of other cells cause those cells also to exhibit Ca waves. This ensures a tissue-like response. Examples from the diatom literature include the directed cellular dispersal from a colony (Wigglesworth-Cooksey et al. 1999) and Ca wave propagation in *Phaeodactylum tricornutum* (Vardi et al. 2006). Thus we can conclude that any compound that interferes with the delicately poised Ca homeostasis in a diatom is likely to reduce its ability to colonize a surface. Such molecules would act as specific antifoulants for diatoms. Since the role played by Ca_i in biological systems is universal, Ca homeostasis antagonists may act generally on all fouling organisms.

5 Some Results

It is possible to gauge changes in internal Ca concentration if diatom cells are loaded with a fluorophore, the fluorescence of which depends on the intracellular free Ca^{2+} concentration. We loaded cells of *A. coffeaeformis*with the fluorophore Ca-Green (Molecular Probes, Eugene, OR). The cells were washed and allowed to attach to a glass surface, which was then placed in minimal medium containing 0.25 mM Ca^{2+} (Cooksey and Cooksey 1980) contained in a fluorometer cuvette. Then, 0.1 M Ca^{2+} was added incrementally to a final concentration of 5 mM and changes of cellular fluorescence were recorded with time (Fig. 3). It can be seen from Figs. 2 and 3 that the changes in motility and Ca^{2+} fluorescence temporally coincide, suggesting that these phenomena are linked and that Ca^{2+} waves, as described for mammalian cells (Berridge et al. 1998), could be the signal system controlling diatom motility. Incremental additions of Ca^{2+} synchronized the cellular response; otherwise the signal would not have been so strong. From this and from earlier work (Cooksey and Cooksey 1988) it is clear that signal response in diatoms is controlled by internal Ca^{2+} levels, which have to arise from the concentration of Ca^{2+} in the external milieu. We have investigated two examples of compounds that

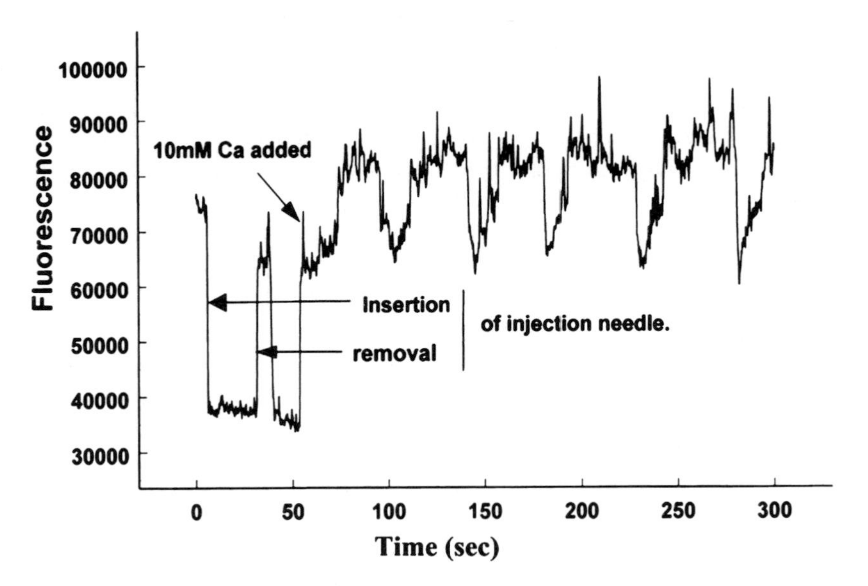

Fig. 3 Fluorescence of synchronized population of *A. coffeaeformis* loaded with Calcium Green. Note the oscillation of the fluorescence signal, which can be compared to that of the speed variation in a single cell (see Fig. 2)

are known to interfere with Ca homeostasis. The first of these D-600, or gallopamil, is an anti-hypertensive drug related to the more commonly used verapamil. Its action in mammalian cells is to prevent Ca^{2+} uptake via Ca^{2+} channels. Binding of ^{3}H-verapamil to plasmamembranes in *Chlamydomonas reinhardtii*(Dolle and Nultsch 1988) and other plants (Andrejaustkast et al. 1985) has also been demonstrated. D-600 also causes apoptosis in adenocarcenoma cells (Fleckenstein 1977). At a concentration of 25 μM, D-600 prevented adhesion of *Amphora* and reduced motility to 0–5% of the population in 2 h.

A second compound is from a natural source, although it is available in synthetic form commercially (Sigma-Aldrich, St. Louis, MO). *Trans*, *trans*-2,4-decadienal (DD) is produced by planktonic diatoms (Miralto et al. 1999). The compound has been implicated as a chemical defense molecule as it impairs invertebrate grazing of planktonic diatoms (Ianora et al. 2004; Romano et al. 2003; Ianora et al. 2006). Vardi et al (2006) showed that treatment of the planktonic organisms *Phaeodactylum tricornutum* and *Thalassiosira weissflogii*, with DD caused intracellular Ca^{2+} transients similar to those we have seen with *Amphora*. Furthermore, DD generated nitric oxide bursts, which produced a phenomenon akin to programmed cell death (apoptosis). The same groups have proposed DD as an infochemical implicated in phytoplankton bloom collapse. Caldwell et al. (2004) found that DD is toxic to the developmental stages of a range of invertebrate species and microfilament and microtubule related events are at the center of its activity. Since these same events are also central to diatom adhesion and motility (Webster et al. 1985), it seemed reasonable to us that DD could be the cellular dispersal agent (i.e., a negative chemotactic effector) described previously (Wigglesworth-Cooksey et al.1999). Although there has been considerable investigation of DD (see references in Vardi et al. 2006), there are no studies on the use of DD as antifoulant, in spite of the fact that it is commercially available. It appeared to us that a natural compound that promotes programmed cell death and interferes with events involving the cytoskeletoncould be an ideal candidate for such a purpose. It may prevent the onset of cell signaling and thus prevent the metabolic actions which take place subsequent to arrival of a cell on a surface. In a preliminary study, we found that 66 μM DD caused a loss in motilityin both *Amphora* and a species of *Navicula* in 2 h. In 20 h all cells became permeable to Sytox Green (1 μM, 15 min) showing that their cell membranes had become compromised (Figs 4a,b). Further work with more organisms, including bacteria, is needed to establish that, most, if not all, cells are indeed sensitive to this compound.

A natural compound that causes similar effects to DD in both diatoms and bacteria is 2-*n*-pentyl-4-quinolinol(PQ). This was isolated from an *Aeromonas* sp. (Long et al. 2003) and has been synthesized (Long et al. 2003; Wratten et al. 1977). Long et al. (2003) found that PQ was active in reducing growth and respiration in planktonic bacteria and growth in planktonic diatoms. We have measured the effects of PQ on three pennate biofilm-forming diatoms, i.e., *A. coffeaeformis, Navicula* sp.,and an *Auricula* sp. (Wigglesworth-Cooksey et al. 2007). This group of diatoms represent three types of adhesion mechanism: (a) secreted adhesive via the raphe slit (*Amphora*, Webster et al. 1985), (b) adhesive secreted through pores in

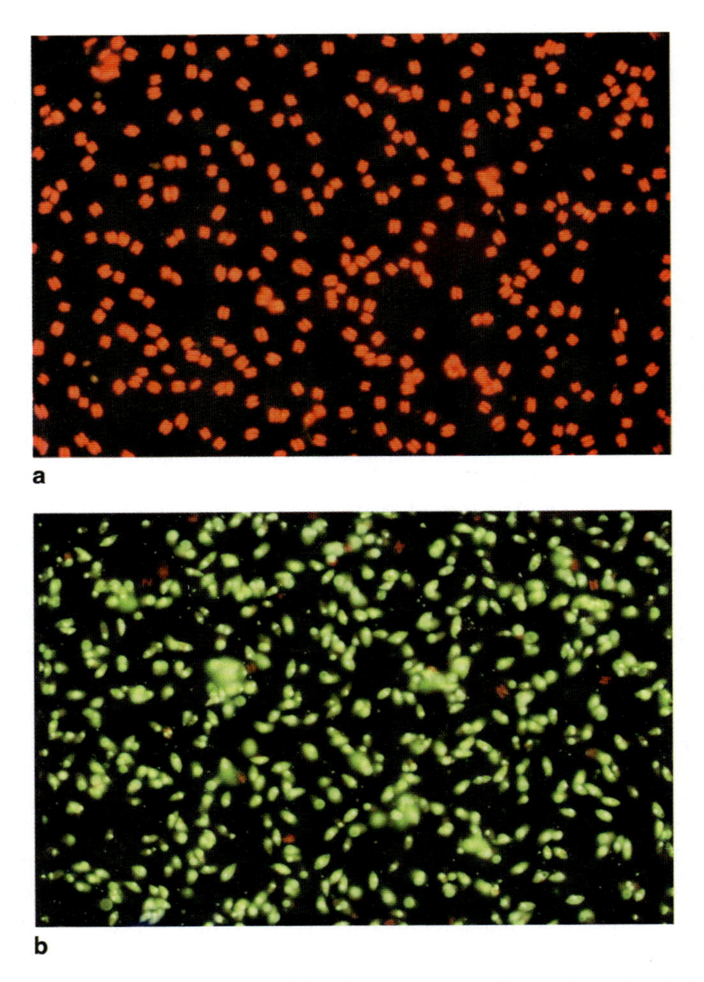

Fig. 4 Influence of decadienalon the proclivity of an *Amphora* biofilmto take up the vital stain Sytox Green. **a** Control, untreated cells are *red* from chlorophyll fluorescence. The cell membrane is intact. **b** Treated cells fluoresce *green/yellow* showing that their plasma membranes are compromised

the frustule (*Navicula*, Chamberlain 1976), and (c) adhesive secreted through a raphe which is on a keel (A*uricula*, unpublished observation). In general, PQ inhibited growth, adhesion, and motility in all three organisms, but the results were species-specific. Whereas PQ prevented initial adhesion in *Amphora* (ED_{50} = 120 μM), it did not do so in *Navicula.* PQ caused cell membrane damage in most (94–97%) cells (result similar to that depicted in Figs 4a,b for DD). Burgess et al. (2003) investigated a similar molecule (a quinone, rather than a quinolinol). It showed activity against barnacle larvae, macro-algal spores and some species of bacteria. It was not tested against fouling diatoms.

Table 3 Effect of 100 μM PQ on diatom motility

Diatom	Treatment	Motile cells[a]
Amphora	Control	51 ± 3
	+ PQ	0
Navicula sp.1	Control	60 ± 13
	+ PQ	5 ± 9
Auricula sp.	Control	67 ± 1
	+ PQ	0

[a] Motile cells as a percentage of the population. Cells moving at less than 0.25 μm s^{-1} were considered non-motile

Motility in all three diatoms was reduced to near zero by 100 μM PQ (Table 3). The difference between the adhesion results for *Amphora* and *Navicula* underscore the need for care when generalizing from results with one organism. In this case motility was a better indicator of inhibitory activity than the adhesion of the cell to a surface. Thus it is important to use several organisms in the same type of assay when screening potential antifoulants.

6 Conclusions and Thoughts for the Future

The motility of a diatom is a sensitive indicator of its metabolic health and can be used to screen compounds for activity as antifoulants. Because of their animal-like behavior, it is possible that results obtained with diatoms can be expanded to include all fouling organisms. Although a fully automated diatom motility assessment is convenient, purely manual measurements are possible. Compounds isolated from nature are not likely to accumulate in the environment. Materials with potential for commercial use should be able to be synthesized economically. Because of this, compounds of fairly low molecular weight are likely to be of the most use. There are many laboratories, both industrial and academic, and at least one national program seeking to produce an environmentally benign antifouling coating. We suggest that if the search to find new ways to protect structures in the sea is to accelerate and we believe that it must, a forum for the exchange of ideas is needed.

References

Andrejaustkast E, Hertel R, Marme D (1985) Specific binding of the calcium antagonist[3H] verapamil to membrane fractions from plants. J Biol Chem 260:5411–5414

Armbrust EV (2004) The genome of the diatom *Thalassiosira pseudonana*: ecology, evolution and metabolism. Science 306:79–86

Arce FT, Avci R, Beech IB, Cooksey KE, Wigglesworth-Cooksey B (2004) A live bioprobe for studying diatom-surface interactions. Biophys J 87:4284–4297

Baier RE (1980) Substrata influences on adhesion of microorganisms and their resultant new surface properties. In: Bitton G, Marshall KC (eds.) Adsorption of microorganism to surfaces. Wiley, New York, pp. 59–104

Berridge MJ, Bootman MD, Lipp P (1998) Calcium – a life or death signal. Nature 395:645–648

Bohlander GS (1991) Biofilm effects on drag: measurements on ships. Trans Inst Marine Engineers (C) 103:135–138

Burgess JG, Boyd KG, Armstrong E, Jiang Z, Yan L, Berggren M, May U, Pisacane T, Adams DR (2003) The development of a marine natural product-based antifouling paint. Biofouling 10:197–205

Caldwell GS, Bentley MG, Olive PJW (2004) First evidence of sperm motility inhibition by the diatom aldehyde 2*E*,4*E* decadienal. Mar Ecol Progr Series 273:97–108.

Callow ME (1986) Fouling of "in service" ships. Botanica Marina 29:351–357

Chamberlain AHL (1976) Algal settlement and secretionof adhesive materials. In: Sharpley JM, Kaplan AM (eds.) Proceedings of the third international biodegradation and biodeterioration symposium, Applied Science, London, pp. 417–432

Characklis WG, Cooksey KE (1983) Biofilms and microbial fouling. Adv Appl Microbiol 29:93–138

Cohn SA, McGuire JR, (2000) Using diatom motility as an indicator of environmental stress: effects of toxic sediment elutriates. Diatom Res 15:19–29

Cooksey KE (1981) Requirement for calciumin adhesion of a fouling diatomto glass. Appl Environ Microbiol 41:1378–1382

Cooksey B, Cooksey KE (1980) Calcium is necessary for motility in the diatom *Amphora coffeaeformis*. Plant Physiol 65:129–131

Cooksey KE, Cooksey B (1986) Adhesion of fouling diatomsto surfaces: some biochemistry. In: Evans LV, Hoagland KD (eds.) Algal biofouling. Elsevier, Amsterdam, pp. 41–53

Cooksey B, Cooksey KE (1988) Chemical signal-response in diatomsof the genus *Amphora*. J Cell Biol 91:523–529

Dahms HU, Dobretsov S, Quian PY (2004) The effect of bacterial and diatom biofilms on the settlement of the bryozoan Bugula neritina. J Exp Mar Biol Ecol 313:191–209

Dolle R, Nultsch W (1988) Specific binding of the calcium blocker[^{3}H] verapamil to membrane fractions of *Chlamydomonas reinhardtii*. Arch Microbiol 49:451–456

Findlay JA, Callow ME, Ista LK, Lopez GP, Callow JA (2002) The influence of surface wettabilityon the adhesion strength of settled sporesof the green alga *Enteromorpha* and the diatom *Amphora*. Integr Comp Biol 42:1116–1122

Fleckenstein A (1977) Specific pharmacology of calcium in myocardium, cardiac pacemakers and vascular smooth muscle. Annu Rev Pharmacol 17:149–166

Holland R, Dugdale TM, Wetherbee R, Brennan AB, Finlay JA, Callow JA, Callow ME (2004) Adhesion and motility of fouling diatoms on a silicone elastomer. Biofouling 20:323–329

Holm ER, Kavanagh CJ, Meyer AE, Wiebe D, Nedved BT, Wendt D, Smith CM, Hadfield GM, Swain G, Wood CD, Truby K, Stein J, Montemarano J (2006) Interspecific variation in patterns of adhesion of marine fouling to silicone surfaces. Biofouling 22:233–243

Ianora A, Miralto A, Poulet SA, Carotenuto Y, Buttino I (2004) Aldehyde suppression of copepod recruitment in blooms of a ubiquitous planktonic diatom. Nature 429:403–407

Ianora A, Boersma M, Casotti R, Fontana A, Harder J, Hoffman F, Pavia H, Potin P, Poulet SA, Toth G (2006) New trends in marine chemical ecology. Estuaries Coasts 29:531–551

Jelic-Mrcelic G, Sliskovic M, Antolic B (2006) Biofouling communities on test panels coated with TBTand TBT-free copper-based antifouling paints. Biofouling 22:293–302

Long RA, Qureshi A, Faulkner DJ, Azam F (2003) 2-*n*-Pentyl-4-quinolinol produced by a marine *Alteromonas* sp. and its potential ecological and biogeochemical roles. Appl Environ Microbiol 69:568–576

Maki JS, Mitchell R (2002) Biofouling in the marine environment. In: Bitton G (ed.) Encyclopedia of environmental microbiology. Wiley-Interscience, New York, pp. 610–619

Marszalek DS, Gerchacov SM, Udey LR (1979). Influence of substrate composition on marine microfouling. Appl Environ Microbiol 38:987–995

Miralto A, Barone G, Romano G, Poulet SA, Ianora A (1999) The insidious effect of diatoms on copepod reproduction. Nature 402:173–176

Rittschof D, Clare AS, Costlow JD (1986) Barnacle settlement inhibitors from Sea Pansies, *Renilla reniformis*. Bull Mar Sci 39:376–382

Romano G, Russo GL, Buttino I, IanoraA, Mirealto A (2003) A marine diatom-derived aldehyde induces apoptosis in copepod and sea urchin embryos. J Exp Biol 206:3484–3494

Sieburth JMc (1979) Sea microbes. Oxford University Press, New York, p. 489

Smith DJ, Underwood GJC (1998) Exopolymer production by intertidal epipelic diatoms. Limnol Oceanogr 43:1578–1591

Terlizzi A, Conte E, Zupo V, Mazzella L (2000) Biological succession on silicone fouling-release-surfaces: long-term exposure tests. Biofouling 15:327–342

Vardi A, Formiggini F, Casotti R, De Martino A, Ribalet F, Miralto A, Bowler C (2006) A stress surveillance system based on calcium and nitric oxide in marine diatoms. PLOS Biol 4:411–419

Webster DR, Cooksey KE, Rubin RW (1985) An investigation of the involvement of cytoskeletal structures and secretionin gliding motilityof the marine diatom *Amphora coffeaeformis*. Cell Motility 5:103–122

Wetherbee R, Lind JL, Burke J, Quatrano RS (1998) The first kiss: establishment and control of the initial adhesion of raphid diatoms. J Phycol 23:9–15

Wigglesworth-Cooksey B, Cooksey KE, (1992) Can diatoms sense surfaces? State of our knowledge. Biofouling 5:227–238

Wigglesworth-Cooksey B, van der Mei H, Busscher HJ, Cooksey KE (1999) The influence of surface chemistry on the control of cellular behavior: studies with a marine diatom and a wettability gradient. Colloids Surf B Biointerfaces 15:71–79

Wigglesworth-Cooksey B, Long RA, Cooksey KE (2007) An antibiotic from the marine environment with antimicrobial fouling activity. Environ Toxicol 22:275–280

Wratten SJ, Wolfe MS, Anderson RJ, Faulkner DJ (1977) Antibiotic metabolites from a marine pseudomonad. Antimicrob Agents Chemother 11:411–414

Novel Antifouling Coatings: A Multiconceptual Approach

D. Rittschof

Abstract The development of novel antifouling and foul release coatings must be considered in the context of business, government, and academic research. Existing antifouling technology is based upon the use of broad-spectrum biocides. Foul release technology is partially developed, has incompletely understood mechanisms and unknown long term fates and effects. Business is structured to register, generate, deliver, apply, and remove antifouling coatings based upon broad-spectrum biocides. Business is weak in biology and study of fates and effects beyond those required for registration. Government is structured to regulate, respond to, and support basic research. Government is not proactive or cooperative. Academics are highly competitive, still relatively isolated from reality, strong in basic research, and not well versed in business or government. Rapid progress in novel coatings technology is unlikely in this environment. Business responses to regulations and awareness of environmental responsibility will drive the process.

1 Introduction

The intent of this discourse is to provide a stationary target to promote conversations on how best to make substantial progress toward environmentally benign antifouling and foul release coatings. It is presented from the perspective of a researcher who has dedicated 25 years to the development of environmentally benign antifouling coatings. From that perspective, understanding four related topics is central to thinking and planning new approaches. The four topics are:

D. Rittschof
Duke University Marine Laboratory, Nicholas School of the Environment, 135 Duke Marine Lab Road, Beaufort, NC 28516, USA
e-mail: ritt@duke.edu

Springer Series on Biofilms, doi: 10.1007/7142_2008_21

1. Fouling, antifouling, and foul release concepts and technology (Costlow and Tipper 1984; Kavanagh et al. 2003)
2. The role of business models and issues as they relate to antifouling and foul release coatings(Champ 2000; Rittschof and Parker 2001)
3. The role of government and regulatory agencies and laws in novel coatings development (Champ 2000; Rittschof 2000; International Maritime Organization 2003)
4. The role of academic research in the process (Clare 1997; Rittschof and Holm 1997; Rittschof and Parker 2001)

I argue that progress requires that all four topics be addressed simultaneously. However, it is likely that other equally important yet to be conceived topics will be added to this list.

I will make the argument that progress in developing novel coatings is unlikely because the existing global, political, business, and research structures inhibit sharing of expertise and the necessary cooperation between business, government, and academia. Finally, although I am pessimistic about any hope of rapid progress due to my perception and growing cynicism of human nature, I will suggest ways to facilitate progress in assessing developing environmentally benign solutions to fouling.

2 Fouling

In our context, fouling is the attachment and/or growth of undesirable molecules and organisms to submerged surfaces. Fouling includes molecules, microbes, and macroorganisms and spans the spectrum from true temporal succession (each subsequent stage requiring a prior stage) to essentially random events determined by availability of foulers. The actual process of fouling ranges from passive mechanisms comparable to dust settling on a surface to highly specific mechanisms that require active behavior (Fig. 1; Clare et al. 1992; Hadfield 1998). Biological fouling impacts the esthetics, performance, and economics of stationary and mobile platforms.

3 Antifouling Coatings

Classically, antifouling is any of a large number of control measures that use broad spectrum biocides to control fouling (Fisher et al. 1984; Preiser et al. 1984). Antifouling coatings are coatings that control fouling by releasing broad-spectrum biocides. Antifouling coating technology is closely tied to anticorrosion coating technology. The two technologies are generally present on hulls in a multilayered coating system. Development of these systems preceded global awareness of environmental degradation due to build up and impact of biocides. Regulations have been in general reactive rather than proactive (Brancato et al. 1999; Champ 2000; International Maritime Organization 2003). Global business models, rules, regulations,

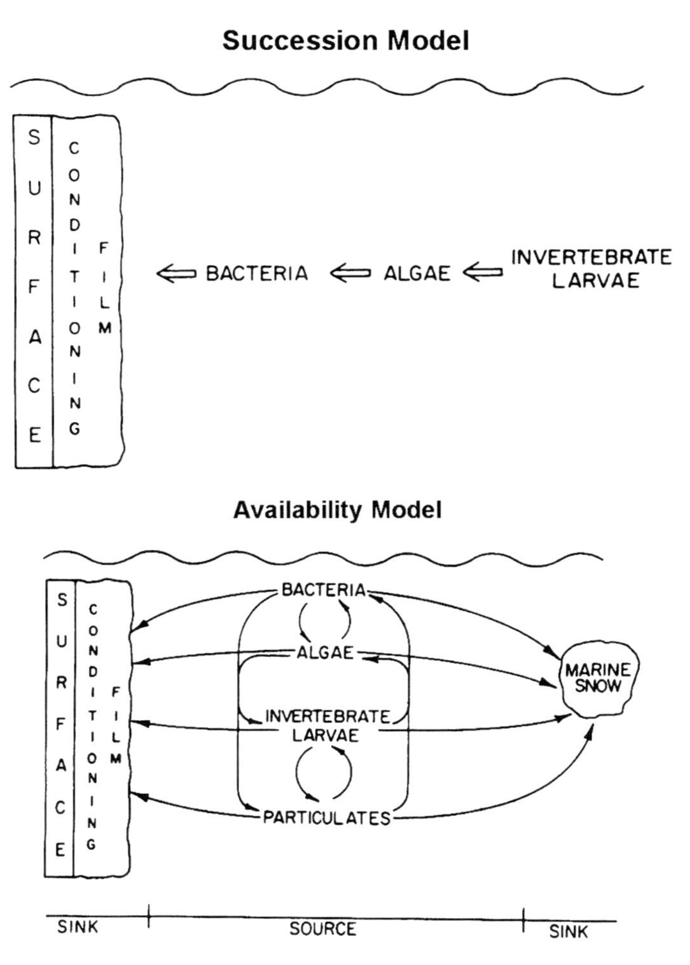

Fig. 1 The spectrum of biological fouling (from Clare et al. 1992). Classically, biofouling was thought to be an exclusively successional process (**a**). However, since the early 1990s evidence has accumulated that supports the availability model (**b**), especially for the majority of the cosmopolitan fouling organisms now found in the worlds ports. These foulers have been introduced through shipping and dominate because they are analogous to terrestrial weeds

and performance expectations are tied to economics and performance of toxic coatings (Champ 2000; Rittschof 2000).

4 Foul Release Coatings

The concept of developing foul release surfaces, surfaces that are easy to clean because biological adhesives adhere to them poorly, has been popular for over three decades. Originally, the concept was based upon fluoropolymers (Bultman et al. 1984).

In the last several decades foul release coatings have been based upon silicone polymers, which are much easier to clean than fluoropolymers (cf Brady 2000; Stein et al. 2003). The physical/chemical mechanisms resulting in low adhesion of silicones are incompletely understood. Poor physical properties of foul release coatings are a major stumbling block. When physical properties are improved, the ease of cleaning is reduced. Because foul release coatings are based upon coating technology that is different from antifouling coatings, facilities that can apply either kind of coating are unlikely at this time.

5 Antifouling–Foul Release Coatings

The notion of coatings that are hybrids of antifouling and foul release technology is beginning to be a popular one. The basic concept is to deliver antifoulants through foul release coatings. One popular idea is to improve the physical characteristics of the foul release coating and then to add antifoulants that target pernicious foulers that are hard to clean. The overall concept may be based upon observations that nontoxic silicone coatings containing organotin or other toxic catalysts and additives such as silicone oils (Kavanagh et al. 2003) alter macrofouling larval behavior (Afsar et al. 2003), inhibit fouling, and are routinely easy to clean. Some of these coatings release sufficient toxic species that they may perform as antifouling coatings for months (Rittschof and Holm 1997; Holm et al. 2005).

6 Business Models

Business approaches are harder to document. This section is based upon my experience in working with a variety of businesses for over 20 years. The antifouling coating industry has the mission of making money. Products are designed for specific performance niches within the broader market. For example, the small pleasure boat antifouling coating niche is for products that can be self-applied and that are effective for 3–6 months, while the niche for coatings for large ships is for products that are applied in specialized facilities and are expected to maintain physical and antifouling properties up to for 10 years. The yacht and intermediate-sized vessel market is for coatings with expectations that are intermediate to the extremes.

The antifouling coatings business has three major components:

1. Antifouling additive suppliers, which develop and register biocides
2. Coatings manufacturers, which develop polymer systems, mix, deliver, and develop protocols for product application
3. Coatings appliers, which apply and remove anticorrosive and antifouling coatings

The technical infrastructure is strong in protection of intellectual property, polymer chemistry, anticorrosion chemistry, registration of toxic compounds, and customer support. It is relatively weak in fouling biology and in environmental stewardship.

The business of antifouling coatings is based upon coatings that control fouling with broad-spectrum biocides. The biocides are delivered either by coatings (resin–rosin systems) that act as slow release reservoirs from the bulk coating or by toxin release that is controlled hydrolysis of the polymer (usually acrylates), which expose and release new toxin over time. The highly effective but environmentally damaging organotinpolymer films, which were voluntarily removed from global markets in 2003, were of the ablative (hydrolytic) or self-polishing type. The base polymers used in commercial antifouling coatings meet a spectrum of important physical and anticorrosion characteristics (Preiser et al. 1984).

The coatings industry has a history of modifying its products when they are shown to have unacceptable environmental impacts. More recently, environmental concerns have resulted in regulatory pressure to reduce the release of copper from antifouling coatings. The industry response to restriction of use of copper has been to reduce amounts of copperand to supplement the coatings with broad-spectrum organic biocides(Gough 1994; Liu et al. 1999; Readman 1996; Tolosa et al. 1996; Rittschof 2000).

7 Government

The role of government in regulating antifouling technology is centered on registration of additives (national governments) and in making and enforcing laws and regulations (cf International Maritime Organization 2003). Governments are, in general, reactive and adversarial rather than proactive and cooperative. Even when government is well informed by stakeholders such as business, scientists, and environmental groups, political solutions are inadequate, expensive, and slow (cf Champ 2000). For example, US registration of a new compound for use as an antifouling biocide can take over a decade and cost over US$10,000,000 (Rittschof 2000).

However, governments also have a dramatic positive impact on fouling control technology. Due to defense and economic and environmental considerations, governments have a major role in supporting research in antifouling, foul release, and environmental impacts(cf. Exploratory Research for Advance Technology, ERATO, Biofouling Project Japan, 1990–1995; US Office of Naval Research Antifouling program, Alberte et al. 1992; Nordic Council of Ministers, Dahllöf et al. 2005; AMBIO 2006). Government involvement is central to funding of basic research; training of government, industry and academic researchers; and development of assessment techniques and concepts.

8 Academia

Academia and government researchers associated with antifouling are highly competitive, strong in basic research and relatively weak in applied research. Academic researchers are generally uninterested or uniformed about the workings of business and government. In general, academics are isolated from business and from government agencies charged with regulation. Historically, in the USA and many other developed countries, academics from upper tier research universities were encouraged to avoid interests in societally relevant research. More recently, there have been gradual changes in attitudes in major research universities, associated with changes in the scope of funding opportunities. Researchers have a role in initiating new avenues of research. As ideas mature, often over the course of several decades, potential productive avenues are identified and new research structures are generated to move the concepts toward products. In the case of antifouling, these changes and advances are on the horizon

9 Synopsis

Existing antifouling technology is based upon the use of broad-spectrum biocides. Foul release technology is partially developed, has incompletely understood mechanisms and unknown long-term fates and effects. Business is structured to protect intellectual property and to register, generate, deliver, apply, and remove antifouling coatings based upon broad spectrum biocides. Business is weak in biology and in the study of fates and effects beyond those required for registration. Government is structured to regulate broad spectrum biocides and to enforce laws such as the US Clean Water Act. Government is not proactive or cooperative. Academics are highly competitive, still relatively isolated from reality, strong in basic research, and not well versed in business or government. It is in this context that novel antifouling coatings development should be considered.

10 Goals of Environmentally Benign Antifouling Coatings

Environmentally benign antifouling is a possibility that could be efficiently approached if one could generate the necessary list of goals and their associated assumptions and prioritize them. Given the context provided I suggest the following list of goals and assumptions that all should be met by the first generation of environmentally benign coatings:

1. Coating should be compatible with existing business models
2. Coating should use existing polymers, production, delivery, application methods, and application facilities
3. Compatible with existing anticorrosion systems
4. Function comparably to coatings that are presently on the market
5. Deliver compounds with known minimal environmental impact

To make a long story short, in the context provided, the only solution that fits the context established is a short-lived broad spectrum biocide. If one looks at the more recent products on the market one can see that there are two business strategies that meet all but the assumption of known minimal environmental impact. Both strategies use organic toxins to enable reduction in the level of copper released by antifouling coatings:

1. Use of clearly long lived broad spectrum biocides such as Irgarol (Gough 1994; Liu et al. 1999; Readman 1996; Tolosa and Readman 1996; Tolosa et al. 1996; New York State Department of Environmental Conservation 1996) and Diuron (PAN 2008)
2. Use of shorter lived broad spectrum biocides, copper pyrithione (Dahllöf et al. 2005), and Sea 9–211 (Willingham and Jacobson 1996). These strategies are clearly in line with the argument developed that business models can tolerate only minor changes.

There are two related possibilities for the rationale of replacement of heavy metal biocideswith long lived organic biocides: (1) Following the classical business model to the letter by replacing one long lived biocide with another. Perhaps, it is possible that these businesses could claim ignorance as an excuse for lack of concern for environmental consequences. (2) Following the classical business model by replacing one long lived biocide with another with knowledge of consequences, but without legal, moral, or ethical responsibility for protecting the environment. Independent of which possibility is correct, the result is the same high potential for environmental degradation.

Similarly, using short lived broad spectrum biocides to replace long lived biocides, could be unintentional or it could be a carefully considered decision that reflects a company with a forward looking business model that includes responsibility for environmental impacts. In these case there is documentation that the environment was considered (cf. Callow and Willingham 1996; Galvin 1998; Willingham and Jacobson 1996; Harrington et al. 2000; Dahllöf et al. 2005). Independent of which option is correct, the end result is companies positioned to move more quickly away from the old business model. It is these companies that should have a competitive edge in a global community that will eventually recognize that business must take responsibility for the environmental impacts of its products (Rittschof 2000).

In addressing development of novel antifoulant compounds from this perhaps naive academic point of view, I came to understand one important concept. It is not the half-life of the toxin that is important, it is whether or not the toxin will build-up in the environment to deleterious levels. Although this kind of analysis is sufficiently well developed to be standard fare in environmental chemistry textbooks (cf Schwarzenbach et al. 2003), this kind of analysis is not to my knowledge part of the regulatory structure. My sense is that business (most easily because they track customer use and sales and know trade secrets) or academics could develop models that could be used to predict conditions of use where there would be environmental impacts. From an environmental perspective this would be a preferred alternative to reporting dangerous build-ups (Gough 1994; Liu et al. 1999; Readman 1996; Tolosa and Readman 1996; Tolosa et al. 1996) after the fact. Perhaps this will be the next step in the process of developing environmental responsibility.

One inevitable conclusion from this intellectual exercise is that in the existing global regulatory and business and research structure, novel antifouling technology will evolve slowly from small changes in existing art. If one asks the question, “Could the process be accelerated?” the answer is straightforward, but not easily implemented. That answer is restructure business, government, and academia to enable these sectors to work cooperatively by sharing expertise in working toward a common goal (Rittschof and Parker 2001).

Such utopian restructuring is unlikely at the national level, but might be possible at the level of an international organization such as the European Union or the United Nations. The EU has a research and development structure, the Advanced Nanostructured Surfaces for Control of Biofouling (AMBIO) project(AMBIO 2006), which meets many of these objectives. In reality, development of a multi-functional cooperative structure would be a novel infrastructure that could be charged with addressing a variety of global problems.

Acknowledgements Thanks to Murthy and Venkat and RAMAT for the stimulation and travel funds to generate this document. The ideas represented were generated while working on projects funded by the Office of Naval Research, agencies of the Government of Singapore, and several chemical and coatings companies. LN polished the manuscript.

References

Afsar A, de Nys R, Steinberg P (2003) The effects of foul-release coatings on the settlement and behaviour cyprid larvae of the barnacle Balanus amphitrite amphitrite Darwin. *Biofouling* 19(Suppl):105–110

Alberte RS, Snyder S, Zahuranec B (1992) Biofouling research needs for the United States Navy: program history and goals. *Biofouling* 6:91–95

AMBIO (2006) Advanced nanostructured surfaces for the control of biofouling http://www.ambio.bham.ac.uk/. Last accessed 14 July 2008

Brady RF, Singer IL (2000) Mechanical factors favouring release from fouling release coatings. *Biofouling* 15(1–3):73–81

Brancato MS, Toll J, DeForest D, Tear L (1999) Aquatic ecological risks posed by tributyltin in United States surface waters: pre-1989 to 1996 data. *Environ Toxicol Chem* 18:567–577

Bultman JD, Griffith JR, Field DE (1984) Fluopolymer coatings for the marine environment. In: Costlow JD, Tipper RC (eds.) Marine biodeterioration: an interdisciplinary study. US Naval Institute, Annapolis, MD, pp. 237–242

Callow ME, Willingham GL (1996) Degradation of antifouling biocides. *Biofouling* 10:239–249

Champ M (2000) A review of organotin regulatory strategies, pending actions, related costs and benefits. *Sci Total Environ* 258:21–71

Clare AS (1997) Towards nontoxic antifouling (mini-review). *J Mar Biotechnol* 6:3–6

Clare AS, Rittschof D, Gerhart DJ, Maki JS (1992) Molecular approaches to nontoxic antifouling. *J Invert Reprod Dev* 22:67–76

Costlow JDC, Tipper RC (1984) Marine biodeterioration: an interdisciplinary approach. US Naval Institute, Annapolis, MD

Dahllöf I, Grunnet K, Haller R, Hjorth M, Maraldo K, Petersen DG (2005) Analysis, Fate and toxicity of zinc and copper pyrithione in the marine environment. TemaNord 550–583

Fisher EC, Castelli VJ, Rodgers SD, Bleile HR (1984) Technology for control of marine biofouling – a review. In: Costlow JD, Tipper RC (eds.) Marine biodeterioration: an interdisciplinary study. US Naval Institute Press, Annapolis, MD, pp. 261–299

Galvin RM, Mellado JMR, Neihof RA (1998) A contribution to the study of the natural dynamics of pyrithione (ii): deactivation by direct chemical and adsorptive oxidation. *Eur Water Manag* 4:61–64

Gough MA, Fothergill J, Hendrie JD (1994) A survey of southern England coastal waters for the *s*-triazine antifouling compound Irgarol 1051. *Mar Pollut Bull* 28:613–620
Hadfield MG (1998) Research on settlement and metamorphosis of marine invertebrate larvae: past, present and future. *Biofouling* 12:9–29
Harrington JC, Jacobson A, Mazza LS, Willingham G (2000) Designing an environmentally safe marine antifoulant. Presented at the10th international congress on marine corrosion and fouling, DSTO, Melbourne
Holm ER, Orihuela B, Kavanagh CJ, Rittschof D (2005) Variation among families for characteristics of the adhesive plaque in the barnacle *Balanus amphitrite*. *Biofouling* 21:121–126
International Maritime Organization (IMO) (2003) International Convention for the Prevention of Pollution from Ships, 1973, as modified by the Protocol of 1978 relating thereto (cited 6 April 2003). http://www.imo.org/home.asp. Last accessed 14 July 2008
Kavanagh CJ, Swain GW, Kovach BS, Stein J, Darkangelo-Wood C, Truby K, Holm E, Montemarano J, Meyer A, Wiebe D (2003) The effects of silicone fluid additives and silicone elastomer matrices on barnacle adhesion strength. *Biofouling* 19:381–390
Liu D, Pacepavicius GJ, Maguire RJ, Lau YL, Okamura H, Aoyama I (1999) Survey for the occurrence of the new antifouling compound Irgarol 1051 in the aquatic environment. *Water Res* 33:2833–2843
New York State Department of Environmental Conservation (1996) NYSDEC Registration of Irgarol algicide. http://pmep.cce.cornell.edu/profiles/herb-growthreg/fatty-alcohol-monuron/irgarol/new-act-ing-irgarol.htmlLast accessed 14 July 2008
Pesticide Action Network (2008) Diuron identification, toxicity, use, water pollution potential, ecological toxicity and regulatory information. PAN pesticides database. http://www.pesticideinfo.org/Detail_Chemical.jsp?Rec_Id=PC33293 Last accessed 14 July 2008
Preiser HS, Ticker A, Bohlander GS (1984) Coating selection or optimum ship performance. In: Costlow JD, Tipper RC (eds.) Marine biodeterioration: an interdisciplinary study. US Naval Institute Press, Annapolis, MD, pp. 223–228
Readman JW (1996) Antifouling herbicides – a threat to the marine environment? *Mar Pollut Bull* 32:320–321
Rittschof D (2000) Natural product antifoulants: one perspective on the challenges related to coatings development. *Biofouling* 15:199–207
Rittschof D, Holm ER (1997) Antifouling and foul-release: a primer. In: Fingerman M, Nagabhushanam R, Thompson MF (eds.) Recent advances in marine biotechnology, vol I. Endocrinology and reproduction. Oxford and IBH, New Delhi, pp. 497–512
Rittschof D, Parker KK (2001) Cooperative antifoulant testing: a novel multisector approach. In: Fingerman M, Nagabhushanam R (eds.) Recent advances in marine biotechnology, vol VI. Science, Einfield, pp. 239–253
Stein J, Truby K, Wood CD, Gardner M, Swain G, Kavanagh C, Kovach B, Schultz M, Wiebe D, Holm E, Montemarano J, Wendt D, Smith C, Meyer A (2003) Silicone foul release coatings: effect of the interaction of oil and coating functionalities on the magnitude of macrofouling attachment strengths. *Biofouling* 19(Suppl):71–82
Schwarzenbach RP, Gschwend PM, Imboden DM (2003) Environmental organic chemistry, 2nd edn. Wiley, Hoboken
Tolosa I, Readman JW (1996) Simultaneous analysis of the antifouling agents: tributyltin, tripheyltin and Irgarol 1051 used in antifouling paints. *Mar Pollut Bull* 335:267–274
Tolosa IJ, Readman W, Blaevoet A, Ghilini S, Bartocci J, Horvat M (1996) Contamination of Mediterranean (Cote d'Azur) coastal waters by organotins and Irgarol 1051 used in antifouling paints. *Mar Pollut Bull* 32:335–341
Willingham GL, Jacobson AH (1996) Designing an environmentally safe marine antifoulant. *ACS Symp Ser* 640224:–233

Concept and Consequences of the EU Biocide Guideline

H.-C. Flemming (✉) and M. Greenhalgh

Abstract The Biocide Product Directive (BPD) of the European Union is intended to balance the efficacy of biocides in their intended use with their impact on human and animal health and the environment. It legally organizes the process of putting a biocide on the market and harmonizes the regulation of the EU Member States. Biocides have to be subjected systematic tests for efficacy and risk before approval. The BPD achieves its aims using a two-stage regime of rigorous evaluation of biocidal active substances and products, to ensure they pose no unacceptable risks to people, animals or the environment.

Ultimately only those biocidal products that contain an active substance that is included in Annex I of the Directive will be authorized for use. Active substances have to be evaluated to ascertain whether or not they will be included in Annex I. This requires industry to submit data, which is evaluated by Member States with decisions over Annex I inclusion being taken at the European level. Industry is charged a fee for this process. Once an active substance has been included in Annex I, national Competent Authorities can authorize products containing it in individual Member States (providing that any necessary data have been supplied and any conditions put on Annex I inclusion are met). Once a product has been authorized in the first Member State, it will be possible for it to be mutually recognized and therefore authorized by other Member States (providing relevant conditions are similar). However, there will have to be an application from other Member States, and again there will be a fee for these processes.

1 Brief Historical Outline

By nature, biocidal products are directed against living organisms, frequently not really restricted to "target organisms". This implies that they inevitably also can harm the health of non-target organisms such as humans or animals. An example is DDT,

H.-C. Flemming
Biofilm Centre, University of Duisburg-Essen, Geibelstrasse 41, D-47057, Duisburg, Germany
e-mail: hanscurtflemming@compuserve.com

Springer Series on Biofilms, doi: 10.1007/7142_2008_12

which was very effective against mosquitoes spreading malaria but was bio-accumulated and spread into the environment to an extent that made the further use of this substance unacceptable. Furthermore, a wide variety of substances for biocidal use have been developed and applied worldwide. They are authorized by regulations that are very different in different countries, frequently incompatible, and based on (partially insufficient) systems of risk assessment. In order to overcome this situation, the European Commission drafted legislation for harmonizing provisions for biocidal products, and to ensure a more uniform and higher level of health and environmental protection throughout the EU without compromising the internal market. This was (and still is) a very challenging approach and the European Commission has gone further by co-operating with other non-EU countries through the auspices of the Organisation for Economic Co-operation and Development (OECD) to try to harmonize biocide regulations on a global basis. The basic idea is that it is not enough just to develop a new and more effective biocide but that the health and environmental issues involved in its application are also considered. The concept of the authors of the guideline was to implement a system that would force the chemical industry to behave in an ethically acceptable way. In 1993, the first draft of a Directive concerning the placing of biocidal products on the market was submitted by the European Commission. It was found that risk assessment should be an integral part of the Directive. The Commission therefore submitted a revised version in 1995. After long and controversial discussions, a final text was adopted by the Council in 1998 as "Biocidal Products Directive" (BPD) (European Parliament 1998). Each Member State appointed an agency to deal with the new legislation: the so-called Competent Authority. The biocides industry transformed dramatically with the implementation of the BPD. There has been much debate amongst industry and the national Competent Authorities on how to operate the scheme, within the idealistic legal framework of the Directive. Some Competent Authorities themselves are considered by some to take an idealistic approach, whereas others are more pragmatic and give higher priority to the needs of industry and biocide users. Nevertheless, the BPD is significantly more stringent than any previous legislation, either within Europe or in the rest of the world. Knight and Cooke (2002) assume that it may cost the industry well over 500 million €. Much of the information provided in this chapter is based on their excellent work.

From the point of view of health and environment, it clearly is a major step forward to better protection, although the direction is very complicated in many details and it is not free from discrepancies.

2 Scope of the Guideline

The BPD defines biocidal products as "preparations containing one or more active substances that are intended to control harmful organisms by either chemical or biological, but not physical, means". This encompasses a wide range of products including disinfectants, insect repellents, and anti-fouling paints. Despite the name "biocide", a biocidal product does not actually have to kill. If it is used to destroy, deter, make harmless, or control a harmful organism by chemical or biological means, it maybe considered to be a biocide. For example a repellent used to "deter" a mosquito could be considered to be a biocidal product.

The Directive will not apply to certain products already subject to European legislation, including plant protection products, human medicines, veterinary medicines, medical devices or cosmetics. Article 1 of the BPD lists those Directives that are not covered within the scope of the BPD. The legislation also excludes the non-biocidal uses of products and active substances

The official objective of the guideline is laid down in the Foreword, Chap. 1, of the guideline:

> "Whereas, in their resolution of 1 February 1993 on a Community programme of policy and action in relation to the environment and sustainable development ([4]), the Council and the representatives of the Governments of the Member States, meeting within the Council, approved the general approach and strategy of the programme presented by the Commission, in which the need for risk management of non-agricultural pesticides is emphasised;"

In Chap. 3 of the Foreword the EU commits that biocides are necessary for the control of organisms dangerous for the health of humans and animals:

> "Whereas biocidal products are necessary for the control of organisms that are harmful to human or animal health and for the control of organisms that cause damage to natural or manufactured products; whereas biocidal products can pose risks to humans, animals and the environment in a variety of ways due to the intrinsic properties and associated use patterns;"

These are the official scientific reasons for implementing the BPD.

The biocidal products guideline is divided into three parts. The third and last one, the Annexes, is potentially the most important because it lists the conditions for placing biocidal products on the market. Annex V of the BPD classifies biocidal products into four main groups: disinfectants and general biocides, preservatives, pest controls, and other biocides as shown in Table 1 (Knight and Cooke 2002).

Ultimately only those biocidal products that contain an active substance that is included in Annex I of the Directive will be authorized for use. An area of dispute concerning scope is the regulation of in-situ generated biocides. These include substances that are mixed together or otherwise generated by the consumer to create the biocidal active ingredient. The EU Commission and Member States have agreed that for example the in-situ generation of ozone is not covered.

When the Directive has been fully implemented in all Member States, existing and new active substances will have to be evaluated to ascertain whether or not they will be included in Annex I. Both processes will require industry to submit data, with a system of data protection. The data will be evaluated by Member States with decisions over Annex I inclusion being taken at the European level. Industry will be charged a fee for this process.

Once an active substance has been included in Annex I, national Competent Authorities can authorize products containing within individual Member States (providing that any necessary data have been supplied and any conditions put on Annex I inclusion are met).

Table 1 Products defined as biocides within the BPD

Main group 1: disinfectants and general biocides
1. Human hygiene products
2. Private and public health area disinfectants
3. Veterinary hygiene biocides
4. Food and feed area disinfectants
5. Drinking water disinfectants
Main group 2: preservatives
6. In-can preservatives
7. Film preservatives
8. Wood preservatives
9. Preservatives for fibre, leather and polymerized materials
10. Masonry preservatives
11. Preservatives for liquid cooling systems
12. Slimicides
13. Metal-working fluid preservatives
Main group 3: pest control
14. Rodenticides
15. Avicides
16. Molluscicides
17. Piscicides
18. Insecticides, acaricides and products to control other anthropods
19. Repellants and attractants
Main group 4: other biocides
20. Preservatives for food or feedstocks
21. Anti-fouling products
22. Embalming and taxidermist fluids
23. Control of vertebrates

Following product authorization in a first Member State, it is then possible for the product to be mutually recognized and therefore authorized by other Member States (providing relevant conditions are similar). However, there will have to be an application to other Member States, and once again there will be a fee for these processes.

Each European Union Member State is responsible for implementing the BPD. In Great Britain the Directive was implemented through the Biocidal Products Regulations 2001, which came into force on 6 April 2001. The Directive was implemented in Northern Ireland through the Biocidal Products Regulations (Northern Ireland) 2001 on 16 January 2002. The Biocidal Products (Amendment) Regulations came into force on 1 April 2003 and there has been a further amendment, the Biocidal Products (Amendment) Regulations 2005, which came into force on 1 October 2005.

3 Approval Systems

To obtain authorization for the marketing of a biocide, the applicant must submit at least two data packages: the first on the active substance, and the second on the formulated product relating to the product type. For each additional product type for which authorization is sought, a further dossier is required at additional cost. The BPD makes a pragmatic but arbitrary distinction between those biocidal active substances on the market before 14 May 2000 ("existing" active substances) and those placed on the market for the first time after this date ("new" active substances). The Member States and the Scientific Committee on Biocidal Products review the scientific content of the dossier and make an appropriate recommendation to the Commission. If the recommendation is favourable, the Commission will enter the active substance in an approved list (Annex I of the BPD). A review programme is established in the BPD to assess systematically during a 10-year period all the existing active substances (European Commission 2006).

Standard biocidal products containing active substances in Annex 1 of the BPD require a full dossier of information, and applications for authorization are evaluated by the national Competent Authority without undue delay. Biocidal products containing a new active substance for which a decision for Annex 1 listing is pending may be provisionally authorized for up to 3 years. New products containing existing active substances can be authorized under existing national schemes for up to 10 years during the review programme. Under the BPD, applicants may use the concept of "frame formulation" to facilitate authorization of re-branded biocidal products. Frame formulations are groups of biocidal products with the same active substance of the same technical specification and use, which differ only in minor details of the formulation composition such as colour or perfume ingredients, and hence have the same risk and efficacy. Once a Member State approves a biocidal product, all other Member States must approve the product, according to the principle of mutual recognition. If there are disputes between Member States by some reason, the Standing Committeeon Biocidal Products will have to resolve them. Member States may opt out of the mutual recognition procedure for avicide, piscicide and vermin-control biocidal products.

4 Risk Assessment

Risk assessment is an important aspect of the regulatory process. It is performed for both the intended use and a reasonable worst-case situation. The risk from a chemical substance is determined from its intrinsic hazardous properties and the likely exposures of humans and the environment throughout its life-cycle. The intrinsic chemical, health, and environmental hazardous properties can be quantified as a hazard assessment. The hazard of the biocide is assessed predominantly through toxicological testing in animal models (Annex IIA and IIB). Good quality human data may also be available, perhaps from epidemiological studies. The hazard assessment is combined with an exposure assessment to produce a risk assessment. If the outcome is favourable, the substance will be recommended for Annex 1 listing. If not, further information on toxicity or exposure in order to refine the risk assessment may be demanded. If the risk remains unfavourable, a regulatory decision may be taken to implement risk management requirements, such as additional labelling or restrictions to use, to permit product approval.

Exposure assessment is a more complex issue. There are two basic options: measuring or modelling. Modelling can be carried out using generic data for chemical release. Estimates of environmental release are improved by gathering information on the release of biocides from specific processes to develop emission scenarios. Risk characterization is also conducted regarding animals kept and used by humans. The humaneness of biocidal products targeted at vertebrates is also considered, e.g. for biocides directed against rats.

The rule is that biocidal products can only be approved if, when used as prescribed, they do not present unacceptable risks to man, animals or the environment, are efficacious and use permitted active substances. Approval of biocidal products requires that they are used properly at an effective but minimized application rate. The regulatory authority also assesses the packaging, labelling and accompanying safety data sheet.

Acute and repeat-dose toxicity, irritation and corrosivity, sensitization, mutagenicity, carcinogenicity, toxicity for reproduction and the physicochemical properties of each active substance in the biocidal product are considered. If possible, they are also quantified, preferably as a dose–response effect. This includes the exposure of professionals, non-professionals, and those exposed indirectly via the environment to each active substance in the biocidal product during its lifestyle. Only as a last resort is the use of personal protective equipment taken into account to enable a biocidal product to be used safely. Replacement of hazardous substances by non-hazardous ones is preferred. Biocidal products containing category 1 or 2A or 2B carcinogens, mutagens or substances toxic to reproduction cannot be approved for use by the general public. Carcinogens are defined after the International Agency for Research on Cancer (1987):

Category 1 is for substances for which there is sufficient evidence for a causal relationship with cancer in humans (confirmed human carcinogen)

Category 2A is for substances for which there is a lesser degree of evidence in humans but sufficient evidence in animal studies, or degrees of evidence considered appropriate to this category, e.g. unequivocal evidence of mutagenicity in mammalian cells (probable human carcinogen)

Category 2B is for substances for which there is sufficient evidence in animal tests, or degrees of evidence considered appropriate to this category (possible human carcinogen)

5 Costs for Authorization of a Biocidal Product

The Health and Safety Executive of the United Kingdom (2008) gives an interesting look at the current costs for authorization of a biocidal product, which is available from its website (http://www.hse.gov.uk/biocides/index.htm).

For Annex I inclusion the HSE currently charges £10,000 (approximately US$ 20,000) for a completeness check of a dossier prior to full evaluation, in addition to an evaluation fee of £84,000 –89,000 for the full evaluation. The fees are based on actuals, meaning that if it was calculated that more time was spent on an evaluation, a further fee to cover the work may be charged; consequently if less time was spent on an evaluation the calculated difference could be refunded.

The UK figures quoted here, are estimates of the likely costs involved in evaluating a product dossier. At the time of writing there are three active agents authorized in Annex 1, i.e. dichlofluanid, difethialone and sulphuryl fluoride, while carbon dioxide has been authorized in Annex 1A. Actual costs are unknown, but costs for product authorizations will vary depending on how relevant the data (that was supplied for inclusion into Annex I of the BPD) is to a product. In the UK, the estimate for the first product authorization after Annex inclusion could cost between £8,500 and £20,000. The fee will be based on the actual work done and will depend on how much of the work was done at the Annex I inclusion stage. Once the initial product has been authorized, fees will be lower. It is expected that products can be authorized that contain the same active ingredient and are the same formulation etc. as the original product, providing the company seeking authorization holds the relevant letters of access. Also it is expected that the cost for authorization of subsequent products that have the same use, user type and contain the same active substance, but with differences in composition from a previously authorized product that do not affect the level of risk or efficacy associated with the product, is to be in the region of £500.

6 Some Problems with the BPD

The given definition for biocides as given in the BPD is very broad. It includes chemical compounds, formulations of compounds and also microoganisms and viruses, which strictly are not chemical substances. The entire legal work is not without inconsistencies and discrepancies. Thus, organisms such as *Bacillus thuringiensis* are dealt within a guideline that was created to reduce the usage of hazardous substances such as the "Seveso toxin" dioxin (2,3,7,8-tetrachlordibenzo-*p*-dioxin and 2,3,7,8-tetrachlor-dibenzofuran).

Even substances that are commonly accepted as non-toxic are enclosed into this definition: Ethanol for instance is not poisonous according to the guideline 67/EWG. However, if a technical product (e.g. a cleaner) contains sufficient amounts of ethanol, it will be preserved against microbiological growth just by that ethanol. So it is debatable whether ethanol is added to improve the cleaning results when applying the product or whether ethanol is added to inhibit microbiological growth.

The problem leads to the transfer of judicial power to the public administration. The administration may decide on the basis of the form sheet of the biocidal product whether an ingredient is in fact a biocidal substance or not. The administration uses the information provided when other users of biocidal products registered their active ingredients before. When the notification process is completed, the public administration will have had to deal with many more borderline cases like that outlined above. So, the public administration will have to build up the capability to evaluate biocidal compounds and biocidal products as well as decide which biocidal compound or which biocidal product may be placed on the market and which may not. Thus, the biocidal product guideline 98/8/EG not only imposes a heavy burden on the chemical industry but also on the public administration simply because of the immense administrative time and effort. The process of implementation is designed in a way that the public authorities must learn and understand how biocides act in a scientific way as well as the benefits they give to the user and the damage they may potentially cause to human health and the environment. This all takes a considerable time to accomplish.

The BPD guideline differentiates between biocidal substances and biocidal products. Biocidal products are materials that are used to control harmful microorganisms or wildlife, such as rodents. Biocidal substances are chemical substances that have the ability to kill or inactivate target organisms and are used in biocidal products for control of microorganisms or wildlife.

Biocidal substances (active agents) have to be listed in Annex 1 to be allowed to be used in biocidal products. Biocidal substances that are not dangerous substances according to the guideline 67/EWG may be listed in Annex IA. To have a substance listed in Annex IA is more difficult than to have it listed in Annex 1, but the listing of a biocidal product that contains the substance is easier. Biocidal substances that are so common that their usage cannot be controlled by the administration will be listed in Annex IB. The guideline gives some examples for such substances. Quoted in the guideline itself are ethanol and carbon dioxide, used so widely that they are listed in Annex IB.

To be placed in Annex 1, IA or IB, the items laid down in Annex 2 have to be determined by tests and their results have to be disclosed to the public administration. To ensure that the results of the different tests from different laboratories are comparable, test methods have to be developed and rules for the interpretation of the test results have been published. A list of properties of the substances, which have to be determined for substances to be listed in one of the chapters of Annex I is given in Annex IIA. The Annex IA gives a list of data that have to be specified by test results or by other data of equivalent significance and reliability. A major weighting is placed upon data relating to the fate of the biocidal compound in the environment. Data have to be submitted on the toxicity against aquatic organisms, the fate of the compound in soil, in addition to data about the biological degradation. Biodegradation is especially important as biocidal compounds must not be allowed to build up in the environment. The types of tests are legally determined, for example OECD biodegradation study protocols. Some tests that have been undertaken to allow chemical compounds to be placed on the US market will have to be repeated as the US tests use different vertebrate species.

Annex IB lists the data necessary to place a biocidal product on the market, which consists of biocidal compounds plus additional substances that might influence the activity of the biocidal substance.

The type of data necessary to be allowed to place a compound or a product on the market depends on the intended use or the purpose of the biocidal compound or biocidal product. Annex V lists 23 different purposes for biocidal compounds or products (product types). By legal definition this list of purposes is considered to be complete so that all notified biocidal compounds or products have to be listed within one of the defined 23 product classes (see Table 1). The decision whether a biocidal substance may be placed on the market is made by one of the national public administrations of the Member States of the EU. Currently most substances that have been filed under the notification scheme have been on the market prior to the introduction of the biocidal product Directive. All existing biocidal substances (i.e. those placed on the market before 14 May 2000), which are to be notified and supported, have been placed onto four lists with dossier call-in dates of 28 March 2004, 30 April 2006, 31 July 2007 and 31 October 2008. Once dossiers are received the review process begins. The reviews have been distributed among the national public administrations (Competent Authorities) according to their capabilities. The amount of data that has to be controlled by the public administration to be able to decide whether a substance may be placed on the market is so large that the administrations have yet to finish their work. One important date was 1 September 2006, at which time all identified existing actives that were not notified as being supported through the authorization process *must* be withdrawn from the European Union.

Once a decision is taken by one of these national public administrations it has to be accepted by the other Member States, as long as there are no cogent scientific reasons against it. It could be for instance, that one Member State rejects the decision of the administration of another one because the target organism, which should be controlled by the compound, does not occur in the Member State, which then rejects the decision. A compound can be notified for a period of not more than 10 years for non-toxic substances and not more then 5 years for toxic substances. After that period the notification has to be renewed. If new data or new scientific knowledge becomes known to the administration that are in contrast to the data presented for the notification and make it desirable to withdraw the substance from the market for safety or environmental reasons, the public administration has to withdraw the notification and thus the substance from the market.

7 Consequences of the Guideline

In general, the BPD acts as a threshold for the development and diversification of the biocide market. This is intended to control developments that might otherwise lead to unwanted effects on health and the environment and takes responsibility for preventing such effects. It represents a real threshold because generating the data necessary to notify a compound or a product is very expensive and time-consuming. Estimates range from about 2–10 million € (approximately US$ 15 million) for testing and generation of the data in addition to the costs of the administration process. Therefore, only highly profitable companies are able to afford the time-consuming process and the costs required to notify a substance. The lower the data requirements, the lower the risk of losing money by the notification process due to insufficient sales after putting the compound or the product on the market. This doesn't take into account the additional time and costs associated with the invention and development of new active agents. Companies have tended to notify compounds and products that are have already been on the market for a long time and of which most of the data required by the administration have been generated and paid for already. Consequently, the system of notification implemented by the biocidal products guideline 98/8/EC is advantageous to old products. Given the market size for the various biocidal product types and the barrier of the BPD (including the substitution principle) most EU-based companies have withdrawn their new active research programmes.

Initially more than 1,500 active substances were identified, of which around 800 were notified to the EU Commission as being supported through the authorization process. Only approximately half of these dossiers have been submitted, leading to the conclusion that only a total of approximately 400 will go through the review process. As a consequence, the EU's Competent Authorities have had to rearrange their dossier review process.

Despite its complexity, the problems of internal coherence, and the political problems within the EU the BPD still represents a serious and responsible compromise to balance the intended effect of biocides (i.e. killing organisms) and the protection of humans and the environment against the non-intended effects. On the other hand, the BPD has certainly slowed down the development and implementation of new biocides.

Additional EU Directives such as the so-called VOC (volatile organic compounds) Directive (European Parliament 1999), and REACH (registration, evaluation, authorization and restriction of chemicals) (European Parliament 2006) have complicated the issue of biocidal products. The aim of the VOC Directive is to significantly reduce the use of VOCs within the EU. However, this has resulted in more water-based products, which will require preservation. With REACH, aspects of this Directive will impact on components of biocidal products. This Directive seeks to regulate a further 30,000 chemicals found within the EU. It is estimated that it will result in around 120,000 dossiers, something that will pose significant strain on both industry and national authorities.

For the development of new biocides, the BPD represents a significant barrier concerning the European market. This includes many chemical anti-fouling approaches, which have to be thought over on this background. It is a wake-up call against the unregulated dispersion of chemical agents, which tend not to be considered sufficiently. Therefore, alternative anti-fouling strategies such as nutrient limitation, cleaning-friendly design and surfaces gain a specific advantage over conventional chemical treatment.

References

Commission of the European Communities (2007) Commission Regulation (EC) no 1451/2007 of 4 December 2007 on the second phase of the 10-year work programme referred to in Article

16(2) of Directive 98/8/EC of the European Parliament and of the Council concerning the placing of biocidal products on the market. Off J Eur Union L325:11.12.2007

European Parliament and the Council of the European Union (1998) Directive 98/8/EC of the European Parliament and of the Council of 16 February 1998 concerning the placing of biocidal products on the market. Off J Eur Commun L123:24.04.1998 http://ecb.jrc.it/legislation/1998L0008EC.pdf. Last accessed 13 July 2008

European Parliament and the Council of the European Union (1999) Council Direcive 1999/13/EC of 11 March 1999 on the limitation of emissions of volatile organic compounds due to the use of organic solvents in certain activities and installations. Off J Eur Commun L85:29.3.1999

European Parliament and the Council of the European Union (2006) Regulation (EC) No 1907/2006 of the European Parliament and of the Council of 18 December 2006 concerning the registration, evaluation, authorisation and restriction of chemicals (REACH). Off J Eur Union L396:30.12.2006

Health and Safety Executive of the United Kingdom (2008) Biocides and pesticides. http://www.hse.gov.uk/biocides/index.htm. Last accessed 13 July 2008

International Agency for Research on Cancer, World Health Organisation (1994) IARC monographs on the evaluation of carcinogenic risks to humans, vols 1–60, 1972–1994 and Suppl 7, 1987

Knight, DJ, Cooke, M (2002) Regulatory control of biocides in Europe. In: Knight, DJ, Cooke, M (eds.) The biocide business. Wiley, Weinheim, pp. 45–74

Part II
Macrofouling

Hydroides elegans (Annelida: Polychaeta): A Model for Biofouling Research

Brian T. Nedved and Michael G. Hadfield(✉)

Abstract The small serpulid polychaete *Hydroides elegans* is a problem fouling organism in warm water marine harbors around the world. Often the first significant animal biofouler on newly submerged surfaces, its calcareous tubes can accumulate rapidly and create serious problems for ships. *H. elegans* is easily adapted for laboratory biofouling research because of its rapid generation time (~3 wks) and ease of propagation. The dioecious adult worms spawn readily in the laboratory, and their metamorphically competent larvae develop in ~5 d at 25 °C. The larvae of *H. elegans* settle in response to natural biofilms or films formed by many, but not all, single marine bacterial species. Tubes of *H. elegans* adhere very tightly to surfaces and are more resistant to dislodgement than many barnacles. Thus, *H. elegans* is an excellent model organism for experimental studies, including tests of newly formulated marine coatings.

1 Introduction

The fouling communities that occur on ships and other man-made structures submerged in the sea are diverse assemblages of organisms (Carlton and Hodder 1995; Gollasch 2002; Godwin 2003). Due to that diversity, the variety of adhesives that fouling organisms utilize to cement themselves to settlement substrata are equally diverse (Naldrett and Kaplan 1997; Brady and Singer 2000; Wiegemann 2005; Smith and Callow 2006), posing a significant challenge for the development of new coatings to combat biofouling processes (Holm et al. 2006). Minimizing fouling on ship hulls is important because of the negative influence fouling has on hull performance (Woods Hole Oceanographic Institution 1952), expenses associated with dry-docking, scraping and re-painting hulls, and the substantial costs from propulsive fuel losses required to overcome the increased drag created by hull fouling (Townsin 2003). Research to find new coatings to combat biofouling has

M.G. Hadfield
Kewalo Marine Laboratory, University of Hawaii, 41 Ahui Street, Honolulu, HI 96813, USA
e-mail: hadfield@hawaii.edu

Springer Series on Biofilms, doi: 10.1007/7142_2008_15

two major thrusts, one in chemically formulating experimental coatings and another in testing these coatings in both field and laboratory settings. The serpulid polychaete *Hydroides elegans* (Haswell 1883) has proven to be an excellent organism for testing experimental coatings under both field and laboratory conditions.

Hydroides elegans is a common member of fouling communities throughout tropical and subtropical seas (ten Hove 1974; Hadfield et al. 1994; Unabia and Hadfield 1999; Bastida-Zavala and ten Hove 2002, 2003). *H. elegans* is a problematic fouling organism because: (1) it quickly colonizes newly submerged surfaces (Unabia and Hadfield 1999; Holm et al. 2000); (2) it grows as much as 1.5 mm day^{-1} (Paul 1937); (3) it reaches sexual maturity in as short a time as 9 days in a tropical harbor (Paul 1937); (4) it has a short larval period (Hadfield et al. 1994; Carpizo-Ituarte and Hadfield 1998); and (5) aggregations of its calcified tubes can accumulate to several centimeters thick on submerged surfaces in as short a time as 1–2 months in Pearl Harbor, Hawaii (Edmondson 1944) (Fig. 1).

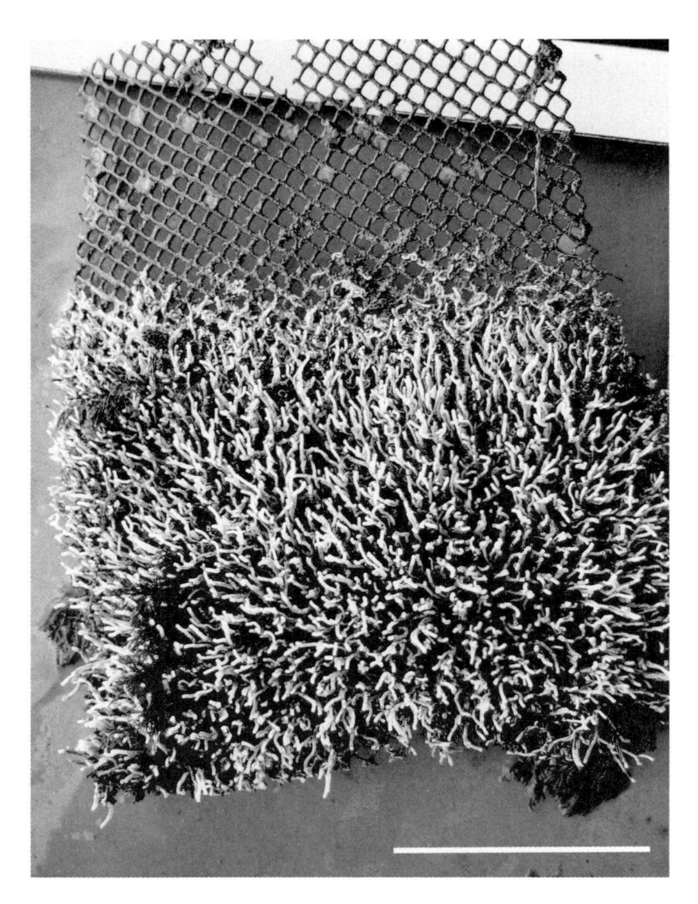

Fig. 1 Dense accumulation of tubes of *Hydroides elegans* on Vexar after a 1-month submersion in Pearl Harbor, HI

Due to its importance as a fouler of ship hulls, there is a growing body of research concerned with the natural inductive cues for recruitment of *H. elegans* (Hadfield et al. 1994; Hadfield and Strathmann 1996; Walters et al. 1997; Unabia and Hadfield 1999; Hadfield and Paul 2001; Lau and Qian 2001; Harder et al. 2002; Lau et al. 2002; Huang and Hadfield 2003; Lau et al. 2005; Shikuma and Hadfield 2006), as well as the metamorphic cascades that are triggered by this process (Carpizo-Ituarte and Hadfield 1998; Holm et al. 1998; Carpizo-Ituarte and Hadfield 2003). Additionally, it has been employed in studies of neurogenesis (Nedved and Hadfield 1998, and unpublished), muscular development (Nedved and Hadfield 2001), segment development (Seaver and Kaneshige 2006; Seaver et al. 2005), and the heritability of egg size (Miles et al. 2007). *H. elegans* will undoubtedly find use as a research model in many other types of studies due to the ease with which is can be maintained and reared in the laboratory.

Upon submersion in seawater, surfaces undergo a well-characterized progression from initial coating of adsorbed organic molecules (Zobell and Allen 1935), to the formation of biofilms (Marshall 1981; Baier 1984) composed of a wide variety of microorganisms that form highly organized communities (Costerton et al. 1999). These complex biofilms provide settlement cues for larvae of many sessile marine invertebrate species (reviewed by Hadfield and Paul 2001). Biofilm bacteria produce the inductive cue for settlement of competent larvae of *H. elegans* (Hadfield et al. 1994; Unabia and Hadfield 1999; Lau and Qian 2001; Lau et al. 2002; Huang and Hadfield 2003; Lau et al. 2005; Shikuma and Hadfield 2006). Laboratory evidence that other biofilm organisms may produce inductive cues for larvae of *H. elegans* (e.g. diatoms: Harder et al. 2002) remain provisionary, given the difficulty of producing absolutely axenic cultures of such organisms for testing.

Larvae of *H. elegans* require a minimum bacterial density for the induction of metamorphosis, and increased larval settlement positively correlates with the density of bacteria in a biofilm (Hadfield et al. 1994; Huang and Hadfield 2003). Settlement by *H. elegans* is greatly reduced or eliminated when multi-species biofilms are treated with a variety of agents that act either as fixatives or antiseptics, demonstrating that the microorganisms within the biofilms must also be alive for induction to occur (Unabia and Hadfield 1999). Recent studies by Lau et al. (2005) and Shikuma and Hadfield (2006) using denaturing gradient gel electrophoresis (DGGE) have examined the effect that changes in the bacterial assemblages of biofilms have on the induction of metamorphosis of *H. elegans*, both demonstrating a stronger positive correlation between settlement of *H. elegans* and bacterial density than between settlement and differences in natural community composition. However, the effectiveness of a bacterial biofilm as an inducer of metamorphosis of *H.elegans* is not solely due to the sheer number of bacteria residing in it. Huang and Hadfield (2003) demonstrated that single-strain, low-density biofilms of *Pseudoalteromonas luteoviolacea* induced metamorphosis of *H. elegans* (Fig. 2), while mono-specific biofilms of *Flexibacter* sp. and *Cytophaga* sp. were non-inductive even though the cell densities of these biofilms were 7–12 times greater (Fig. 3). These data indicate that induction of metamorphosis is due to specific chemical characteristics of *P. luteoviolacea* (Huang and

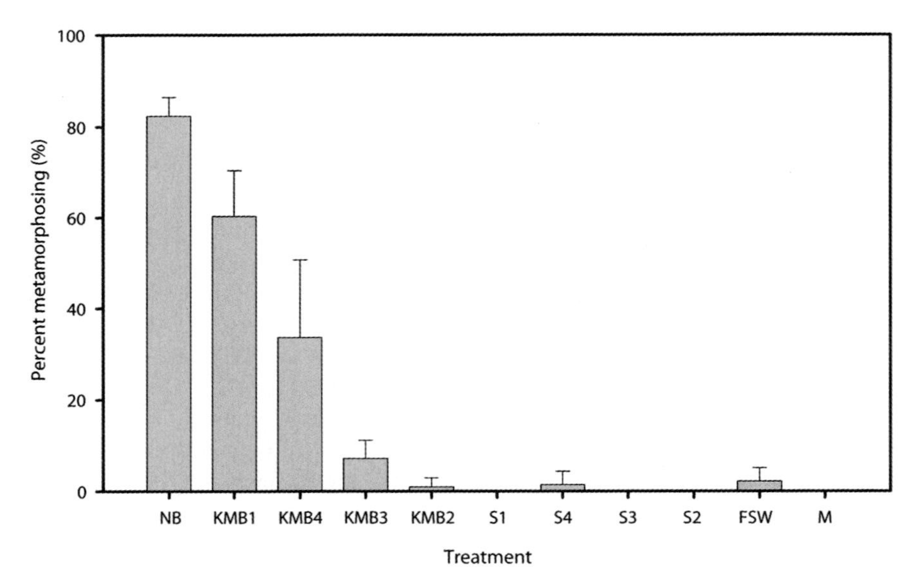

Fig. 2 Induction of metamorphosis of *Hydroides elegans* by mono-specific biofilms on plastic Petri dishes prepared from bacterial strains KMB1, KMB2, KMB3 and KMB4. Controls include: (1) dishes similarly treated with filter-sterilized culture medium from each bacterial strain (*S1*, *S2*, *S3*, *S4*); (2) natural biofilms (*NB*) allowed to accumulate on Vexar mesh in flowing seawater and placed in a Petri dish of filtered seawater (*FSW*); (3) untreated Petri dishes filled with FSW; and (4) dishes rinsed with fresh culture medium (*M*). *KMB1 Pseudoalteromonas luteoviolacea*, *KMB2 Flexibacter* sp, *KMB3 Cytophaga* sp, *KMB4 Cytophaga lytica*. Inoculation density of all strains was approximately 10^{-8} cells mL^{-1}. *Bars* represent mean percent of larvae that metamorphosed in 24 h + SD (n = 5 replicates per treatment) (reproduced from Huang and Hadfield 2003)

Hadfield 2003). Furthermore, production of this metamorphic cue is strain-specific; a different strain of *P. luteoviolacea* obtained from the American Type Culture Collection (Manassas, VA) does not induce settlement of larvae of *H. elegans* (unpublished personal observations).

Hydroides elegans is particularly well-suited for use in testing of experimental coatings. The adhesive that secures the calcareous tubes of *H. elegans* appears to be stronger than that of the balanoid barnacles *Balanus eburneus* and *B. amphitrite*, two species often employed in testing of marine coatings. The mean removal force for *H. elegans* that had settled on six different silicone coatings in Pearl Harbor was nearly three times greater than the mean removal force required to remove *B. eburneus* from replicate panels immersed in the Indian River Lagoon, FL (Fig. 4A, Stein et al. 2003). Additionally, more spat of *B. amphitrite* than newly settled juveniles of *H. elegans* are removed from the silicone coating RTV11 (General Electric, New York) by a 4-min exposure to a wall-shear force equivalent to 100 Pa (unpublished personal observations presented in Fig. 4B) in a turbulent flow apparatus (described in Schultz et al. 2003).

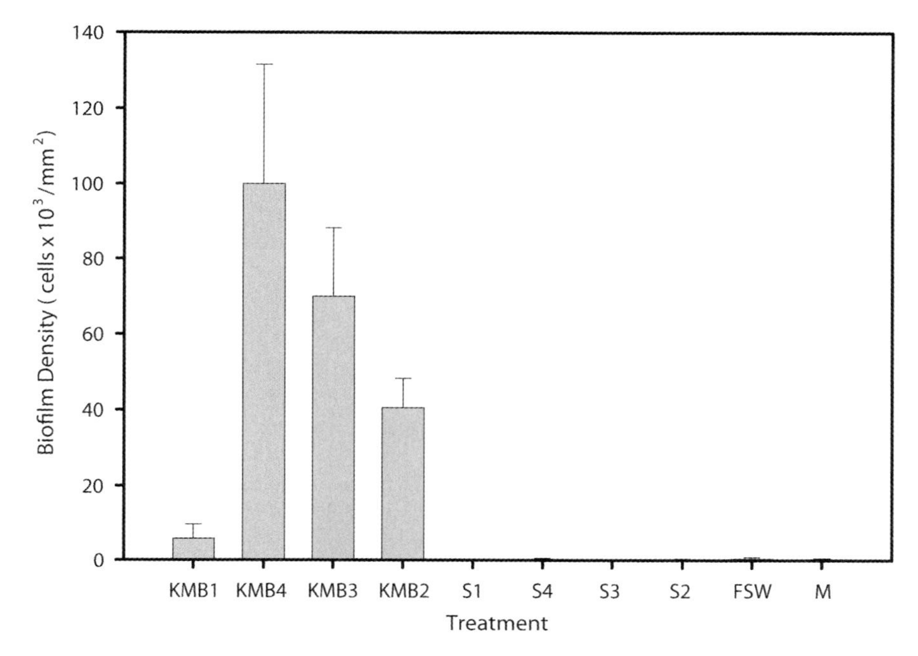

Fig. 3 Bacterial densities in single-species biofilms (see Fig. 2), counted under fluorescence microscopy after formalin fixation and DAPI staining. KMB1, KMB2, KMB3 and KMB4 are the bacterial strains studied; *S1–S4* are dishes treated with the supernatants from each strain, respectively; *FSW* and *M* are filtered seawater and culture-medium-only controls, respectively. *KMB1 Pseudoalteromonas luteoviolacea*, *KMB2 Flexibacter* sp, *KMB3 Cytophaga* sp, *KMB4 Cytophaga lytica*. Inoculation density of all strains was approximately 10^{-8} cells mL^{-1}. *Bars* represent bacterial cell numbers × 10^3 mm^{-2} + SD (n = 25, consisting of five area counts per replicate and five replicate dishes per treatment; replicate effects were not significant (reproduced from Huang and Hadfield 2003)

Information on the occurrence and biology of *H. elegans* has been published under several taxonomic names (e.g. Edmondson 1944; Wisely 1958). This confusion has been resolved in taxonomic reviews by Zibrowius (1971), ten Hove (1974) and Bastida-Zavala and ten Hove (2003), who concluded that *H. norvegica* is a species of the northern Atlantic Ocean and the Mediterranean Sea and that the similar species in warm seas around the world should be referred to as *H. elegans*. There are, of course, other tropical species of *Hydroides*, and they may easily be confused with *H. elegans* without careful observation of the operculum and setae, which are well illustrated in ten Hove (1974), Bailey-Brock (1987) and Bastida-Zavala and ten Hove (2003). According to ten Hove (1974), *H. elegans* is the only *Hydroides* species that forms dense aggregations in warm water bays and estuaries worldwide.

The remainder of this chapter provides a concise summary of methods developed in our laboratory for the culture of *H. elegans* for use in biofouling testing. Our techniques have been used to successfully culture *H. elegans* elsewhere (e.g. Bryan et al. 1997), including areas that do not have access to coastal waters.

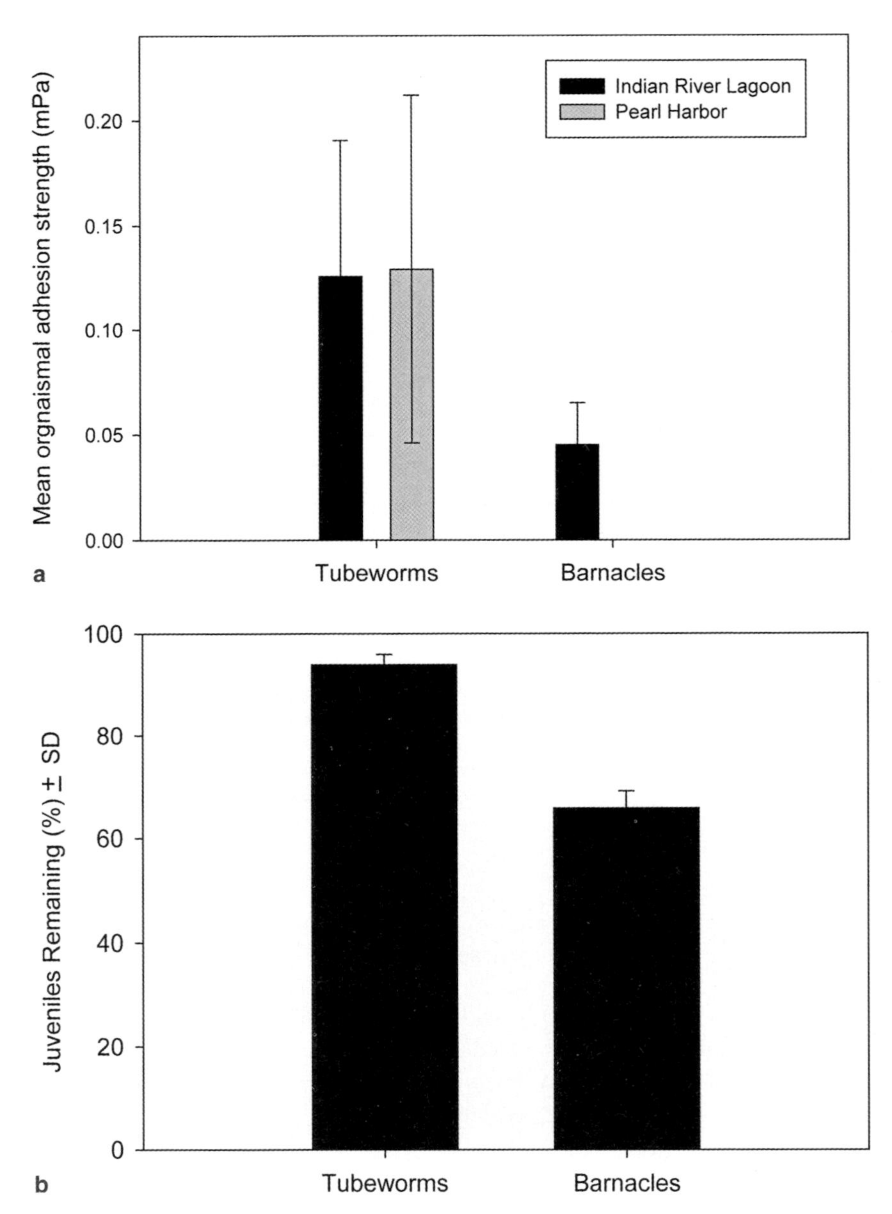

Fig. 4 Strength of adhesion of tubeworms and barnacles in the field (**a**) and a laboratory trial (**b**). **a** Mean attachment strength of barnacles and tubeworms on test coatings. Data are for *Hydroides dianthus* at Indian River and *H. elegans* at Pearl Harbor, and *Balanus eburneus* at Indian River (reproduced with permission from Stein et al. 2003). **b** Percent of juveniles remaining after exposure to a wall shear stress equivalent to 100 Pa for 4 min. Data are for *H. elegans* and *Balanus amphitrite* (previously unpublished data from our laboratory)

2 Collection and Care of Adults

In Pearl Harbor, larvae of *Hydroides elegans* settle on biofilmed surfaces throughout the year and can reach high densities on both natural and man-made surfaces (Fig. 1). Several different materials have been used as artificial settlement substrata for the field collection of *H. elegans* (Walters et al. 1997; Lau and Qian 2001; McEdward and Qian 2001; Lau et al. 2002; Walters et al. 2003). We prefer to use small pieces of extruded plastic mesh (Vexar) as settlement substrata for collecting *H. elegans* in Pearl Harbor. These screens are hung from a pier approximately 1 m below the mean low tide line, and within 3–4 weeks thousands of recruits have settled on them and grown to reproductive maturity (Fig. 1). These dense populations of worms are then transported back to the laboratory and kept in continuously flowing, unfiltered seawater for several weeks without a noticeable decrease in fecundity.

Vexar is preferred over solid substrata, because: (1) it greatly increases the surface area of the material; (2) it provides crevices that may entrain larvae near the surface of screen facilitating settlement on its surface; and (3) it allows water flow through the mesh to bring food, oxygen, and remove wastes from settled worms (Walters et al. 1997).

Large populations of adult *H. elegans* can be kept in closed-system aquaria containing either natural or artificial seawater. Additionally, individual worms can be reared separately by allowing larvae to settle on small (1 × 1 cm) biofilmed chips of polystyrene, removing all but one juvenile worm from each chip, and maintaining the chips in individual wells of plastic ice cube trays (Eric Holm, personal communication; Miles et al.2007). In both settings, adult worms survive and continue producing gametes when fed *Isochrysis galbana* (6×10^4 cells mL^{-1}).

3 Spawning

Spawning of *H. elegans* can be achieved using either destructive or non-destructive methods. We typically use the destructive spawning method, because large numbers of gravid worms are available to us throughout the year in Hawaii. When using this method, we remove 30–40 worms from a piece of Vexar and place them in a small glass dish containing 100 mL of 0.22 μm Milipore-filtered seawater (FSW). To induce release of gametes, the calcareous tubes of the worms are broken in half using forceps, and the abdominal segments of the worms are exposed. This process causes release of thousands of small (~45 μm diameter) orange eggs or clouds of sperm from the abdominal segments of the worm. We then repeat this process until all the worms have been removed from their tubes. After 15–20 min, the fertilized eggs are separated from the adult worms and debris by passing them through a 200 μm sieve (Nitex) into a 500-mL beaker. Filtered seawater is added to achieve a volume of 200 mL. This addition of seawater dilutes the sperm concentration to prevent polyspermy. Fertilization occurs within minutes of exposure of eggs to sperm, and first cleavage occurs approximately 1 h after fertilization (23–26 C). Using this method it is easy to obtain tens of thousands of embryos at a time.

If it is not possible to sacrifice large numbers of adult worms to obtain gametes, a non-destructive method may be used. In this method, worms whose tubes are still attached to their substratum are placed in dishes containing FSW, and the aperture of the worm's tube is gently broken with fine forceps. This mechanical disturbance causes release of eggs and sperm into the tube, and the worm then expels the gametes from the tube by muscular peristaltic action. Generally, females induced to spawn using this method release fewer eggs than can be obtained with the destructive method. After the worms have spawned, they can be placed back into their individual containers where they will repair the apertures of their tubes, and can be induced to spawn again in 2–3 days (personal observations). We have also noted that when adult worms are kept individually in ice cube trays, they occasionally release gametes spontaneously after the FSW is changed in the wells of the trays.

4 Feeding and Care of Larvae

4.1 Seawater

In our laboratory, natural coastal seawater (salinity 35 ‰) filtered through a 0.22 μm Millipore filter (FSW) is used in larval culture to minimize bacterial contamination. We have raised larvae of *H. elegans* in "MBL artificial seawater" (Cavanaugh 1975; Bidwell and Spotte 1985; Strathmann 1987) with no deleterious effects. Antibiotics (60 μg mL^{-1} penicillin G and 50 μg mL^{-1} streptomycin sulfate) may be used if larval mortality is high, but this mixture is generally not required to maintain the larvae through metamorphic competence.

4.2 Temperature and Light

Larval cultures of *H. elegans* are maintained on the bench-top at room temperature (23–26 C) and kept under the ambient lighting regime of the laboratory. However, maintaining larval cultures at 25 C in an incubator provides a greater degree of synchrony of early larval stages. At 24–25°C, larvae attain metamorphic competence after 5 days in culture. Lower temperatures increase time to competence, at 20 C, larvae become competent to metamorphose after 8–10 days in culture (Wisely 1958, as *H. norvegica*).

4.3 Culture Vessels

All glassware used for larval culture in our laboratory is scrubbed under running tap water, rinsed several times with deionized water, and allowed to air dry prior to use.

Additionally, all of our glassware is periodically soaked in a strong acid solution (25% HCl and 25% H_2SO_4) to destroy any organic films that may have developed on the interior surfaces of the containers. Adsorbed organic films hasten the development of a biofilm in the culture vessels, which can provide a metamorphic cue for competent larvae and may cause a substantial number of larvae to metamorphose on the walls of the glassware and the surface film. In our laboratory, larvae are cultured in 1- or 2-L beakers at an initial density of 5–10 larvae mL^{-1}. Cultures of larval *H. elegans* are maintained without stirring or aeration with high levels of larval survival through attainment of competence.

In order to prevent evaporation, beakers are covered with plastic wrap.

4.4 Changing Water in Cultures

After the second day, the larvae are transferred to clean beakers with fresh FSW. To do this, each larval culture is poured into a small plastic beaker whose bottom has been replaced by a piece of 41 μm Nitex sieve. The beaker is placed in a small bowl of seawater in a sink and, as the larval culture is poured through the sieve, the old culture water is allowed to run over the top of the bowl, and the larvae are confined above the screen (see Strathmann 1987). The concentrated larvae are then gently washed from the sieve into clean, acid-washed beakers containing fresh FSW and phytoplankton. Care is taken to retain a small volume of water in the sieve to prevent larvae from being crushed against it. This procedure is subsequently repeated daily until larvae attain competence. Once attaining competence, larvae can be maintained in this manner for several weeks (Unabia and Hadfield 1999).

4.5 Larval Food and its Culture

The unicellular alga *Isochrysis galbana* (Tahitian strain) is the most commonly used food source for larvae of *H. elegans* (Hadfield et al. 1994; Carpizo-Ituarte and Hadfield 1998; Holm et al. 1998; McEdward and Qian 2001; Carpizo-Ituarte and Hadfield 2003; Huang and Hadfield 2003; Lau et al. 2005; Shikuma and Hadfield 2006). However, other algal species have been utilized to raise larvae of this species through metamorphosis (Wisely 1958 as *H. norvegica*; Hadfield et al. 1994). We use *I. galbana* at a density of 6×10^4 cells mL^{-1}. Larval cultures of *H. elegans* are fed *I. galbana* daily by adding aliquots from our working alga cultures; we do not attempt to separate the alga from its culture media.

In our laboratory, *I. galbana* is grown in a commercially produced, modified Guillard's f/2 media (Micro Algae Grow, Florida Aqua Farms, Dade City, FL). We syringe filter (0.22 μm) Micro Algae Grow and use it at a working concentration of 1:1,000 in autoclaved seawater (salinity 25‰). A stock culture of *I. galbana* is maintained in 50-mL screw-top Erlenmyer flasks and recultured bi-weekly. Algal cultures for larval cultures are started every 3 days and maintained in autoclaved culture containers described in Switzer-Dunlap and Hadfield (1981). These cultures are then used as a larval food source when the algal populations are in the later portion of their growth phase. All cultures are bubbled and kept in continuous light supplied by 20-W cool white fluorescent bulbs at room temperature.

4.6 Larval Development

Although the embryonic and larval development of *H. elegans* has been previously described (Wisely 1958 as *H. norvegica*), the timing of larval development is highly dependent on the culture conditions. Larvae cultured in our laboratory develop more rapidly due to the higher ambient temperatures of Hawaii. Cell division proceeds rapidly after first cleavage, and larvae hatch after about 4 h. Larvae of *H. elegans* begin feeding as early trochophores approximately 9 h after fertilization (25°C), and by 12 h they are well developed trochophores with an apical sensory organ (ASO), a single eyespot on the right side of the larva (ES), a prototroch (Pro), and a metatroch (Met) (Fig. 5a).

Larvae remain in the unsegmented prototroch stage for approximately 60 h longer. Three days after fertilization, larvae have developed into metatrochophores with a second eyespot on the left side of the larval episphere, rudiments of the

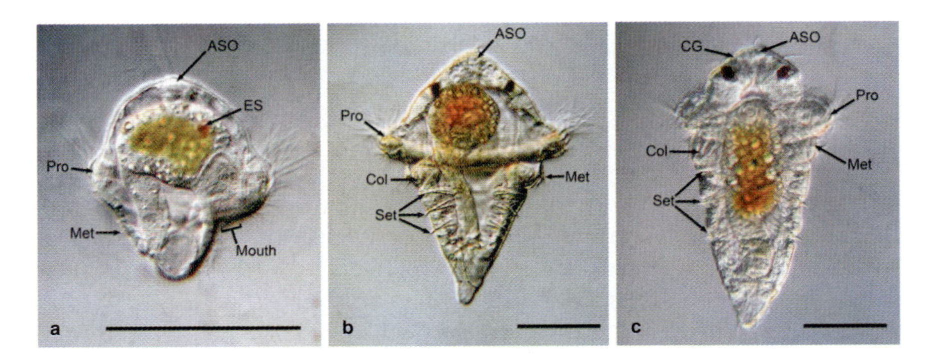

Fig. 5 Differential interference contrast microscopy images of larval development in *Hydroides elegans*. **a** Lateral view of trochophore-stage larva (12 h post-fert at 25°C). **b** Ventral view of metatroch-stage larva (72 h post-fert). Notice the appearance of a second eyespot in the episphere of the larva, and the precocious development of the collar and the first three abdominal segments. The segments are identifiable by the positioning of the paired setigers within each segment. **c** Ventral view of a competent larva (~120 h post-fert). The larva has grown considerably since the trochophore stage, and the hyposphere has become considerably longer. The horse-shoe shaped cerebral ganglion have become quite developed, and the mid-gut of the larva has almost been entirely displaced from the episphere of the larva. Larval mid-gut is easily visualized due to the pigmented algal cells in its lumen. *ASO* apical sensory organ; *CG* cerebral ganglia; *Col* collar; *ES* eyespot; *Met* metatroch; *Pro* prototrochal band; *Set* setae. *Scale bars:* 50 µm in all panels

collar (Col), and elements of the first three abdominal segments including setae (Set) (Fig. 5b).

Larvae of *H. elegans* become competent to metamorphose 5 days after fertilization (25°C). The hyposphere of the larvae has lengthened considerably and the gut has shifted posteriorly. The larval midgut (discernable by the algal cells within it, visible in Figs. 5a–c) has been almost entirely displaced from the episphere by the differentiating cerebral ganglia (CG, Fig. 5c). In competent larvae, the growth and differentiation of the cerebral ganglia is accompanied with a change in a shape of the larval episphere. The lateral margins of the episphere appear to constrict in the region immediately anterior of the prototroch (compare Figs. 5b and 5c), so that the previously hemispherical episphere becomes conical and provides a morphological landmark for the development of competence.

4.7 Metamorphosis

In addition to exposure to biofilm bacteria, larvae of *Hydroides elegans* can be artificially induced to metamorphose by the bath application of the cations K^+ and Cs^+ and the phosphodiesterase inhibitor 3-isobutyl-1-methylxanthine (IBMX). IBMX (0.1 mM) induces 80% of the larvae exposed to it to undergo metamorphosis (Holm et al. 1998). Potassium ions (50 mM excess in FSW) typically induce metamorphosis in over 70% of larvae, but the response is much slower than the rate of metamorphosis induced by biofilms (Carpizo-Ituarte and Hadfield 1998). Maximal induction by caesium ions occurs when 10 mM excess Cs^+ is applied in a 3.0–4.5 h pulse.

Carpizo-Ituarte and Hadfield (1998) described the morphogenic events associated with metamorphosis in larvae of *H. elegans* (Fig. 6). Competent larvae of *H. elegans* may initiate metamorphosis almost immediately after contacting inductive surfaces and begin the process by excreting a sticky thread from their posterior end that serves to tether the larvae to the substratum. Almost immediately after, larvae lie flat on the surface and begin secreting a primary tube from most or all of the segments. They shape the newly secreted tube by rotating within it as they erect their setae to push the primary tube away from their bodies (Carpizo-Ituarte and Hadfield 1998). The secretion of the primary tube, which can be completed in as little as 10 min after contact with a surface, is an irreversible process that permanently attaches a larva to the substratum. As the primary tube is secreted, the collar is everted, the area immediately surrounding the collar constricts, the larval body elongates, and the metatroch is lost. Simultaneously, the pair of lobes that are the precursors to the branchial radioles become apparent on the anterolateral margins of the episphere of the juvenile. The primary tube is never calcified.

Secretion of the calcified secondary tube begins at the anterior margin of the primary tube approximately 2 h after the commencement of metamorphosis, after which new material is added to the secondary tube continuously. As the secondary tube is secreted, the prototroch is resorbed, and the branchial radioles begin to differentiate from the anterior lobes. Metamorphosis is complete and juvenile development has commenced by 11–12 h post-settlement (Carpizo-Ituarte and Hadfield 1998). Because both the primary tube and the early portions of the secondary tube are transparent, the events of metamorphosis and early juvenile development are easily observed with relatively low power microscopy (Fig. 6).

In summary, the major tropical marine fouler *H. elegans* has proven to be a near perfect laboratory-animal model for studies of biofouling processes.

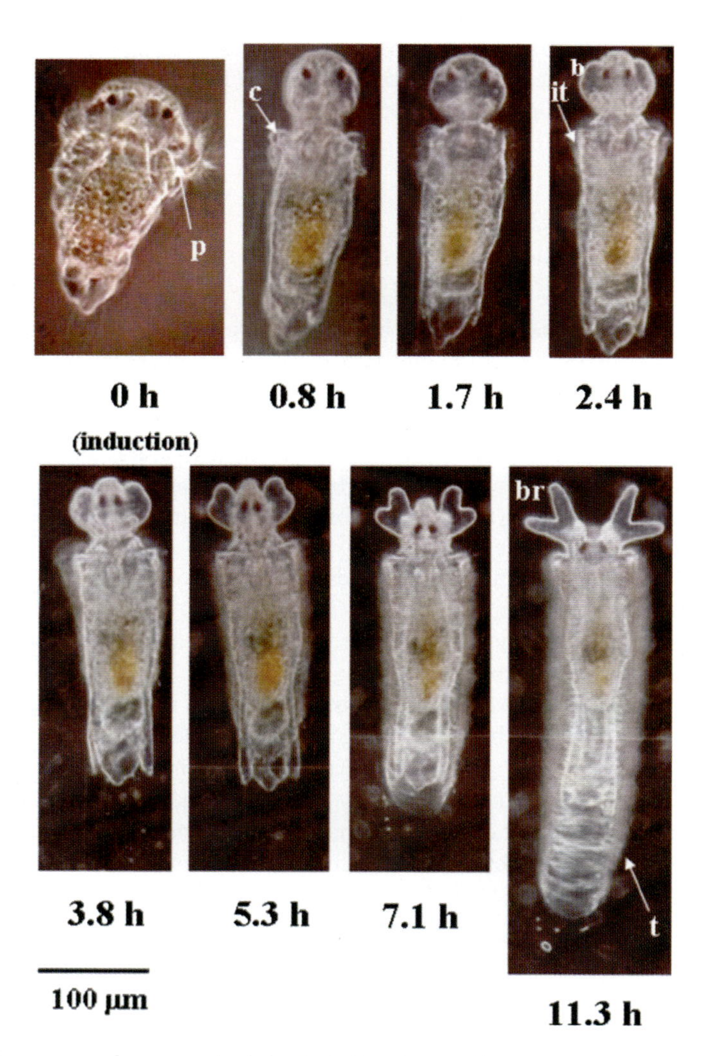

Fig. 6 Time-course of metamorphosis in *Hydroides elegans*. Frames represent a competent larva at the moment of induction to metamorphosis (0 h) and selected stages for the first 11.3 h after induction: *p* prototroch; *c* collar; *b* branchial lobes; *it* initiation point of calcareous tube; *br* branchial radioles; *t* calcareous tube covering the worm (Carpizo-Ituarte and Hadfield 1998)

The information provided above should make it possible to employ this organism for studies of biofouling or questions involving the development of polychaete worms in most laboratories. If scientists attempting to use *H. elegans* in their research find problems in its culture, we would be happy to communicate with them to find solutions.

Acknowledgments We are grateful to many former graduate students and postdoctoral fellows in the Hadfield laboratory who have contributed significantly to the development of the methods described here, especially E. J. Carpizo-Ituarte, E. R. Holm, S. Huang, Y. Huang, N. Shikuma and C. Unabia. The authors are grateful to M. J. Huggett and T. DuBuc for their assistance in creating the figures in this manuscript. Research reported here has been supported by the US Office of Naval research, currently grant no. N00014-05-1-0579 to MGH.

References

Baier RE (1984) Initial events in microbial film formation. In: Costlow JD, Tipper RC (eds.) Marine biodeterioration: an interdisciplinary study. US Naval Institute, Annapolis, MD, pp. 57–62

Bailey-Brock JH (1987) Phylum Annelida. In: DM Devaney and LG Eldredge (eds.) Reef and shore fauna of Hawaii, Sections 2 and 3. Bishop Museum Press, Honolulu, pp. 213–454

Bastida-Zavala JR, ten Hove HA (2002) Revisions of *Hydroides* Gunnerus, 1768 (Polychaeta: Serpulidae) from the Western Atlantic region. *Beaufortia* 52(9):103–178

Bastida-Zavala JR, ten Hove HA (2003) Revisions of *Hydroides* Gunnerus, 1768 (Polychaeta: Serpulidae) from the Eastern Pacific region and Hawaii. *Beaufortia* 53(4):67–110

Bidwell JP, Spotte P (1985) Artificial seawaters: formulas and methods. Bartlett, Boston, p. 349

Brady Jr RF, Singer IL (2000) Mechanical factors favoring release from fouling release coatings. *Biofouling* 15:73–81

Bryan PJ, Qian PY, Kreider JL, Chia FC (1997) Induction of larval settlement and metamorphosis by pharmacological and conspecific associated compounds in the serpulid polychaete *Hydroides elegans*. *Mar Ecol Prog Ser* 146:81–90

Carlton JT, Hodder J (1995) Biogeography and dispersal of coastal marine organisms: experimental studies on a replica of a 16th century sailing vessel. *Mar Biol* 121:721–730

Carpizo-Ituarte E, Hadfield MG (1998) Stimulation of metamorphosis in the polychaete *Hydroides elegans* Haswell (Serpulidae). *Biol Bull* 194:14–24

Carpizo-Ituarte E, Hadfield MG (2003) Transcription and translation inhibitors permit metamorphosis up to radiole formation in the serpulid polychaete *Hydroides elegans*. *Biol Bull* 204:114–125

Cavanaugh GM (1975) Formulae and methods of the Marine Biological Laboratory Chemical Room, 6th edn. Marine Biological Laboratory, Woods Hole, MA, 84 pp

Costerton JW, Stewart PS, Greenberg EP (1999) Bacterial biofilms: a common cause of persistent infections. *Science* 284:1318–1322

Edmondson CH (1944) Incidence of fouling in Pearl Harbor. *Occas Pap Bernice P Bishop Mus* 18(1):1–34

Godwin LS (2003) Hull fouling of maritime vessels as a pathway for marine species invasions to the Hawaiian Islands. *Biofouling* 19(Suppl.):123–131

Gollasch S (2002) The importance of ship hull fouling as a vector of species introductions into the North Sea. *Biofouling* 18:105–121

Hadfield MG Paul VJ (2001) Natural chemical cues for settlement and metamorphosis of marine invertebrate larvae. In: McClintock JB, Baker W (eds.) Marine chemical ecology. CRC Press, Boca Raton, FL pp. 431–461

Hadfield MG, Strathmann M (1996) Variability, flexibility and plasticity in life histories of marine invertebrates. *Oceanologica Acta* 19(3–4):323–334

Hadfield MG, Unabia CC, Smith CM, Michael TM (1994) Settlement preferences of the ubiquitous fouler *Hydroides elegans*. In: Thompson MF, Nagabhushanam R, Sarojini R, Fingerman M (eds.) Recent developments in biofouling control. Oxford and IBH, New Delhi, pp. 65–74

Harder T, Lam C, Qian PY (2002) Induction of larval settlement in the polychaete *Hydroides elegans* by marine biofilms: an investagation of monospecific diatom films as settlement cues. *Mar Ecol Prog Ser* 229:105–112

Holm ER, Kavanagh CJ, Meyer AE, Wiebe D, Nedved BT, Wendt D, Smith CA, Hadfield MG, Swain G, Darkangelo Wood C, Truby K, Stein J, Montemarano J (2006) Interspecific variation in patterns of adhesion of marine fouling to silicone surfaces. *Biofouling* 22(4):233–243

Holm ER, Nedved BT, Carpizo-Ituarte E, Hadfield MG (1998) Metamorphic-signal transduction in *Hydroides elegans* (Polychaeta: Serpulidae) is not mediated by a G-Protein. *Biol Bull* 195:21–29

Holm ER, Nedved BT, Phillips N, DeAngelis KL, Hadfield MG, Smith CM (2000) Temporal and spatial variation in the fouling of silicone coatings in Pearl Harbor, Hawaii. *Biofouling* 15(1–3):95–107

Huang S, Hadfield MG (2003) Composition and density of bacterial biofilms determine larval settlement of the polychaete *Hydroides elegans*. *Mar Ecol Prog Ser* 260:161–172

Lau SCK, Mak KKW, Chen F, Qian PY (2002) Bioactivity of bacterial strains isolated from marine biofilms in Hong Kong waters for the induction of larval settlement in the marine polychaete *Hydroides elegans*. *Mar Ecol Prog Ser* 226:301–310

Lau SCK, Qian PY (2001) Larval settlement in the serpulid polychaete *Hydroides elegans* in response to bacterial films: an investigation of the nature of putative larval settlement cue. *Mar Biol* 138(2):321–328

Lau SCK, Thiyagarajan V, Cheung CK, Qian PY (2005) Roles of bacterial community composition in biofilms as a mediator for larval settlement of three marine invertebrates. *Aquat Microb Ecol* 38(1):41–51

Marshall KC (1981) Bacterial adhesion in natural environments. In: Berkeley RCW, Lynch JM, Melling J, Rutter PR, Vincent B (eds.) Microbial adhesion to surfaces. Ellis-Horwood, New York, pp. 187–196

McEdward LR, Qian PY (2001) Effects of the duration and timing of starvation during larval life on the metamorphosis and initial juvenile size of the polychaete Hydroides elegans (Haswell). *J Exp Mar Bio Ecol* 261(2):185–197

Miles CM, Hadfield MG, Wayne ML (2007) Estimates of heritability for egg size in the serpulid polychaete *Hydroides elegans*. *Mar Ecol Prog Ser* 340:155–162

Naldrett MJ, Kaplan DL (1997) Characterization of barnacle (*Balanus eburneus* and *B.crenatus*) adhesive proteins. *Mar Biol* 127:629–635

Nedved BT, Hadfield MG (1998) Neurogenesis in the larvae of the serpulid polychaete *Hydroides elegans*. *Am Zool* 38:167A

Nedved BT, Hadfield MG (2001) Fate of larval muscles during metamorphosis of *Hydroides elegans*. *Am Zool* 41:1536

Paul MD (1937) Sexual maturity of some organisms in the Mardras Harbor. *Curr Sci Bangalore* 5:478–479

Schultz MP, Finlay JA, Callow ME, Callow JA (2003) Three models to relate detachment of low form fouling at laboratory and ship scale. *Biofouling* 19(Suppl.):17–26

Seaver EC, Kaneshige LM (2006) Expression of 'segmentation" genes during larval and juvenile development in the polychaetes *Capitella* sp. I and *H. elegans*. *Dev Biol* 289:179–194

Seaver EC, Thamm K, Hill SD (2005) Growth patterns during segmentation in the two polychaete annelids, *Capitella* sp. I and *Hydroides elegans*: comparisons at distinct life history stages. *Evol Dev* 7:312–326

Shikuma NJ, Hadfield MG (2006) Temporal variation of an initial marine biofilm community and its effects on larval settlement and metamorphosis of the tubeworm *Hydroides elegans*. *Biofilms* 2(4):231–238

Smith AM, Callow JA (2006) Biological adhesives. Springer, Heidelberg Berlin New York, 284 pp
Stein J, Truby K, Wood CD, Stein J, Gardner M, Swain G, Kavanagh C, Kovach B, Schultz M, Wiebe D, Holm E, Montemarano J, Wendt D, Smith C, Meyer A (2003) Silicone foul release coatings: effect of the interaction of oil and coating functionalities on the magnitude of macrofouling attachment strengths. *Biofouling* 19(Suppl.):71–82
Strathmann MF (1987) Reproduction and development of marine invertebrates of the Northern Pacific Coast. University of Washington Press, Seattle, 670 pp
Switzer-Dunlap M, Hadfield MG (1981) Laboratory culture of *Aplysia*. In: Hinegardner RE, Fay R (eds.) Marine invertebrates, laboratory animal management. National Academy of Sciences, Washington, DC, pp. 199–216
ten Hove HA (1974) Notes on *Hydroides elegans* (Haaswell, 1883) and *Mercierella enigmatic* Fauvel, 1923, alien serpulid polychaetes introduced to the Netherlands. *Bull Zool Museum (University of Van Amsterdam)* 4(6):45–51
Townsin RL (2003) The ship hull fouling penalty. *Biofouling* 19(Suppl.):9–15
Unabia CRC, Hadfield MG (1999) Role of bacteria in larval settlement and metamorphosis of the polychaete *Hydroides elegans*. *Mar Biol* 133:55–64
Walters LJ, Hadfield MG, del Carmen K (1997) The importance of larval choice and hydrodynamics in creating aggregations of *Hydroides elegans* (Polychaeta: Serpulidae). *Invert Biol* 116(2):102–114
Walters LJ, Smith CM, Hadfield MG (2003) Recruitment of sessile marine invertebrates on Hawaiian macrophytes: do pre-settlement or post-settlement processes keep plants free from fouling? *Bull Mar Sci* 72(3):813–839
Wiegemann M (2005) Adhesion in blue mussels (*Mytilus edulis*) and barnacles (genus *Balanus*): mechanisms and technical applications. *Aquat Sci* 67:166–176
Wisely B (1958) The development of a serpulid worm, *Hydroides norvegica*, Gunnerus (Polychaeta). *Aust J Mar Freshwater Res* 9:351–361
Woods Hole Oceanographic Institution (1952) Marine fouling and its prevention. US Naval Institute, Annapolis, MD, 243 pp
Zibrowius H (1971) Les espèces Méditerranéennes du genre *Hydroides* (Polychaeta Serpulidae): remarques sur le prétendu polymorphisme de *Hydroides uncinata*. *Tethys* 2:691–746
Zobell CE, Allen EC (1935) The significance of marine bacteria in the fouling of submerged surfaces. *J Bacteriol* 29:239–251

Marine Epibiosis: Concepts, Ecological Consequences and Host Defence

T. Harder

Abstract The sessile mode of life is widespread in a variety of marine phyla. Sessile life requires a stable substratum. On the benthos, motile life stages and sessile adults compete for rigid surfaces making non-living, i.e. inanimate, hard substratum a limited resource. Epibiosis is a direct consequence of surface limitation and results in spatially close associations between two or more living organisms belonging to the same or different species. These associations can be specifically guided by host chemistry resulting in species-specific symbiotic or pathogenic assemblages. Most colonizers, however, are non-specific substratum generalists. The ecological consequences for the overgrown host (basibiont) and the colonizer (epibiont) can be positive and negative. The predominantly disadvantageous nature of epibiosis by microorganisms for the basibiont has resulted in a variety of defence mechanisms against microcolonizers, including physical and chemical modes of action. Besides antimicrobial effects of secondary metabolites emanating from the host, recent studies increasingly demonstrate that epibiotic bacteria associated with the host deter growth and attachment of co-occurring bacterial species or new epibiotic colonizers competing for the same niche.

1 Introduction

In the marine environment the sessile mode of life is dominant in the majority of phyla. Owing to their low specific weight not only microorganisms (e.g. bacteria, microalgae) but also the small motile life stages of macroorganisms (e.g. larvae and spores) behave like passive propagules in this viscous, hydrodynamically dominated environment. Clearly, for these species an attached filter-feeding mode of life is the energetically advantageous and favourable state, although the lack of locomotion necessitates a variety of new challenging survival strategies, such as reproduction and various defence forms against consumers and overgrowth.

T. Harder
Centre for Marine Bio-Innovation, University of New South Wales, Sydney, NSW 2052, Australia
e-mail: t.harder@unsw.edu.au

Springer Series on Biofilms, doi: 10.1007/7142_2008_16

Sessile life requires a stable substratum. On the benthos, motile life stages and sessile adult forms compete for rigid surfaces making non-living (i.e. inanimate) hard substratum a limited resource. Epibiosis (greek *epi* "on top" and *bios* "life") can be considered as a direct consequence of surface limitation and results in spatially close associations between two or more living organisms belonging to the same or different species. The substrate organism is considered the *basibiont*, while the organism(s) growing attached to the animate surface is referred to as the *epibiont*. Epibionts are further subdivided into *epizoans* (animals) and *epiphytes* (plants, algae). Epibiotic assemblages are rarely species-specific; on the contrary, numerous sessile organisms may live either as basibiont or as epibiont, or both simultaneously (Wahl 1997). Attachment and growth on inanimate surfaces is usually considered as *fouling*, although this term is frequently used synonymously in an epibiotic context in the literature.

Epibiosis is a typical aquatic phenomenon although many examples of terrestrial epibionts are known (e.g. algae, lichens). Depending on seasonality and location the average millilitre of seawater contains 10–100 microscopic larvae and spores, 10^3 fungal cells, 10^6 bacteria and 10^7 viruses. Thus, the colonization pressure exerted by meroplanktonic dispersal stages can be intense on submerged surfaces (Davis et al. 1989) with severe ecological consequences for basibionts and epibionts. The distinctive role of water as a food vector for sessile organisms is the main reason why surface attachment, fouling and hence epibiotic associations predominate in aquatic environments. A large variety of marine phyla have adopted the sessile mode of life for at least one ontogenetic phase. The list includes many bacteria, protozoa, diatoms, molluscs, tube-building polychaetes; most macroalgae, bryozoans, phoronids, cnidarians; some echinoderms, crustaceans; all sponges and tunicates.

2 Ecological Consequences for Epibionts

Marine fouling is an omnipresent phenomenon and the list of foulers is long. The different stages of the fouling process of dispersal stages on solid substrates have been described in successional (Davis et al. 1989) and probalistic models (Clare et al. 1992; Maki and Mitchell 2002) and are presented in detail in this volume. Irrespective of the fouling sequence, for most meroplanktonic larvae settlement is the ultimate prerequisite for successful metamorphosis into sedentary juveniles (Hadfield and Paul 2001). Also, microcolonizers such as bacteria, benthic diatoms and algal spores often proliferate more rapidly or sometimes exclusively when fixed to a solid substratum (Grossart et al. 2003). Once attached, microcolonizers are challenged by other members in the biofilm matrix (Fletcher and Callow 1992; Costerton et al. 1995). To successfully compete in biofilms, many representatives of the bacterial genus *Pseudoalteromonas* release anti-bacterial products that aid the cells in the colonization of host surfaces. Through the production of agarases, toxins, bacteriolytic substances and other enzymes, bacterial cells are assisted in their competition for nutrients and space as well as in their protection against predators grazing on surfaces (Holmström and Kjelleberg 1999).

In densely populated marine environments where competition for space is high, the advantage for colonizers in occupying empty animate surfaces is probably the main reason for epibiosis (Wahl 1989; Todd and Keough 1994) although a variety of other advantages for the colonizer may support specific host–epibiont associations. For instance, settlement on raised or elevated hosts results in a hydrodynamically favourable position of the epibiont (Keough 1986) as flow dynamics increase with distance from the benthos (Butman 1987). Increased flow ensures better supply of planktonic nutrients and more efficient removal of toxic excretory products such as ammonia. An exposed habitat supports phototrophic epibionts, especially in deeper or turbid waters where light penetration is weak (Brouns and Heijs 1986). While filter-feeding epibionts profit from nutrient currents created by the host (Laihonen and Furman 1986) deposit-feeding epibionts benefit from metabolites exuded by the basibiont (Harlin 1973). Regarding the fate of colonizers, epibionts either benefit from the host defence against consumers or other colonizers, a phenomenon termed "associational defence" (Hay 1986), or the fates of epibiont and host are closely interlinked and shared together, a phenomenon termed "shared doom" (Wahl and Hay 1995).

The predominantly advantageous associations of epibionts with host organisms indicate that the mere presence of a surface is often not the only criterion for successful colonization. Numerous colonizers are reported to be guided by specific "cues" mediating the suitability of the settlement site (Rodriguez et al. 1993; Wieczorek and Todd 1998; Steinberg et al. 2002). The recognition of appropriate cues activates the genetically scheduled sequence of behavioural and physiological processes during settlement (Morse 1990) and many larvae delay or even avoid settlement in the absence of appropriate settlement cues (Coon et al. 1990; Qian and Pechenik 1998).

3 Settlement Cues

There is clear experimental evidence for physical settlement cues, such as surface roughness (Berntsson et al. 2000) and wettability (Qian et al. 2000); environmental conditions in direct proximity to the surface, such as irradiation (Maida et al. 1994) and microhydrodynamics (Mullineaux and Butman 1991); and biogenic chemical signals emanating from the basibiont or other epibionts (e.g. bacteria) already present on the host surface (Johnson et al. 1991a,b; Krug and Manzi 1999). Several authors have presented experimental evidence for selective settlement of both generalist and specialist epibionts in response to invertebrate or plant host cues. In most of these studies, the host served as the obligate prey source for larvae or adults. This raises the question of how planktonically dispersed larvae locate their patchily distributed hosts. Given the large spatial scales that need to be screened by potential colonizers, one would expect either strong or very distinct cues that govern such host–epibiont associations. To address this question, a number of studies have focused on the selective response of sea slugs to host corals. For example, water-soluble cues from corals induce settlement and metamorphosis in larvae of the opisthobranchs *Phestilla sibogae* (Hadfield and Scheuer 1985), *Adalaria proxima* (Lambert and Todd 1997) and *Alderia modesta* (Krug and Manzi 1999). Other well-investigated host plants are coralline algae that govern larval settlement of taxonomically distinct invertebrates such as the sea urchin *Holopneustes purpurascens* (Williamson et al. 2000), the starfish *Acanthaster planci* (Johnson et al. 1991b; Johnson and Sutton 1994) and the mollusk *Haliotis* (Morse and Morse 1984; Hahn 1989). Other well-studied systems comprise obligate associations that seemingly benefit from close proximity of conspecifics to enhance reproductive output, such as in oysters (Tamburri et al. 1992; Turner et al. 1994) and barnacles (Clare and Matsumura 2000). However, the settlement cue(s) involved in the establishment of these systems were rarely identified at the molecular level.

In contrast to the numerous partially characterized inducers, only few settlement cues isolated from natural sources were in fact chemically identified, e.g. delta-tocopherols from *Sargassum tortile* that induce settlement of the hydroid *Coryne uchidai* (Kato et al. 1975); jacarone isolated from the red alga *Delesseria sanguinea* that induces settlement of the scallop *Pecten maximus* (Yvin et al. 1985); narains and anthosamines A and B isolated from marine sponges and lumichrome isolated from conspecifics that induce settlement of ascidian larvae (Tsukamoto et al. 1994, 1995, 1999); *N*-acylhomoserine lactone quorum sensing signal molecules that aid zoospores of the green macroalgae *Ulva* to exploit a bacterial sensory system and select permanent attachment sites by responding to bacteria already present on the surface (Joint et al. 2002). In most cases, the ecological relevance of these compounds in situ is not clear, either because the source of the settlement cue is not necessarily related to the recruitment patterns of the organism (Yvin et al. 1985; Tsukamoto et al. 1994, 1995), or because the availability of the cue to settling larvae has not been demonstrated unequivocally (Tsukamoto et al. 1999).

Interestingly, there is a high similarity in host recognition by pathogens in marine and terrestrial plants (Kolattukudy et al. 1995). For example, the pathogenic filamentous green alga *Acrochaete operculata* recognized its host, the red alga *Chondrus crispus*, by cell wall polysaccharides. *C. crispus* has an isomorphic life history, in which the gametophytic and sporophytic generations differ only in minor traits, such as sulfate-ester group distribution of their matrix polysaccharides, known as κ- and λ-carrageenans. Remarkably, the sporophytic generation is highly susceptible to infection whereas the gametophytic phase is naturally resistant. The virulence of the green algal endophyte is modulated by the presence of λ-carrageenan, which stimulates protein synthesis and elicits the production of specific polypeptides in the pathogen (Bouarab et al. 2001).

Only recent years have witnessed some complete characterizations of marine invertebrate larval settlement cues. In a series of investigations Matsumura et al. (1998) identified the key molecule responsible for gregarious settlement in the fouling barnacle *Balanus amphitrite* as a settlement-inducing protein complex (SIPC). This protein complex has now been fully elucidated as a α_2-macroglobulin-like glycoprotein (Dreanno et al. 2006). Although the SIPC is regarded as an adult cue that is recognized by the cyprid at settlement, it is also expressed in juveniles and in larvae, where it may function in larva–larva settlement interactions. In another series of investigations the structure and the different sources of coralline algae-derived settlement cues for two larval species of sea urchins, *Holopneustes purpurascens* and *Heliocidaris erythrogramma,* have been fully elucidated. The biogenic amine histamine was isolated from the red alga *Delisea pulchra* by bioassay-guided fractionation and identified as the inducer of settlement of *H. purpurascens* (Swanson et al. 2004). The alga still evoked larval settlement after antibiotic treatments, which effectively removed epiphytic bacteria on the algal surface, demonstrating that histamine was indeed an alga-derived cue. In contrast, the coralline alga *Amphiroa anceps,* which also stimulates larval settlement of *H. purpurascens,* lacked detectable amounts of histamine. Interestingly, antibacterial treatment of *A. anceps* removed the settlement cue, suggesting a bacterial origin of the cue from this alga; indeed bacterial films of two isolates from the surface of *A. anceps* induced settlement of *H. purpurascens* in laboratory assays (Swanson et al. 2006). The role of algae-associated bacteria as producers of settlement cues has been examined in more detail for the sea urchin *H. erythrogramma* (Huggett et al. 2006). The hypothesis of a bacterially derived settlement signal was supported by the fact that a variety of bacterial isolates from the surface of coralline algae triggered larval settlement at levels comparable to those of the positive control of coralline algae. One bacterial isolate from *A. anceps, Thallasomonas viridans*, is a known histamine producer. Given that larvae of both urchin species settle in response to histamine, these findings demonstrate a common settlement cue in coralline algae produced by the host alga and/or by associated bacteria.

4 Ecological Consequences for Basibionts

Any potential basibiont, i.e. the majority of sessile, relatively long-lived organisms, must either tolerate epibiosis or employ some sort of defence against this phenomenon. While epibiosis entails both benefits and disadvantages for epi- and basibionts the investment into defence depends on a finely tuned and often variable energy budget of the basibiont (Wahl 1989). Epibiosis causes a variety of beneficial effects to the basibiont, such as the induction of morphogenesis in macroalgae by symbiotic bacteria (Tatewaki et al. 1983; Nakanishi 1999), the interaction between macroalgae and nitrogen-fixing bacteria (Thevanathan et al. 2000), and the protection of seaweed surfaces from bacterial colonizers by associated bacteria (Lemos et al. 1985). A well-investigated example of a symbiotic association between host and epibiotic bacteria is the embryo of the American lobster, *Homarus americanus*, which is resistant to the fungus *Lagenidium callinectes*, a pathogen of many crustaceans. The surfaces of healthy lobster embryos are covered almost exclusively by a single, Gram-negative bacterium, which produces the antifungal substance 4-hydroxyphenethyl (Gil-Turness and Fenical 1992). Testing the effects of epibiosis on herbivory and predation, research by Wahl and colleagues suggested that epibionts on the blue mussel *Mytilus edulis* affected its susceptibility to predation by the shore crab *Carcinus maenas* (Wahl et al. 1997). Similarly, epibiosis by a variety of plants and animals altered the host susceptibility of the omnivorous sea urchin *Arbacia punctulata* (Wahl and Hay 1995). Furthermore, Wahl and Mark (1999) investigated the hypothesis that if the effects for epibiont and basibiont were predominantly beneficial then co-evolution would be expected to lead to some sort of associational specificity. However, by analyzing over 2000 patterns of epibiotic associations the authors concluded that many colonizers are non-specific substratum generalists and that epibiosis is predominantly facultative (Wahl and Mark 1999).

The adverse effects of epibiosis on the basibiont often outweigh the beneficial ones (Table 1). For instance, soft-bodied marine invertebrates and algae are susceptible to diseases and tissue necrosis induced by bacteria, fungi and microalgae (Mitchell and Chet 1975; Bouarab et al. 2001; Cooney et al. 2002). The sometimes drastic changes of pH and redox conditions created by microepibionts may attack chemically sensitive surfaces of the basibiont (Terry and Edyvean 1981). Importantly, the adverse effects of microbial epibiosis may reach beyond pathogenicity and virulence. Since microbial films are important sources of chemical cues for larval settlement in many benthic marine invertebrates (Lau et al. 2002; Harder et al. 2002), microbial epibiosis may promote subsequent colonization by rigid crustose epibiotic macroorganisms, which in turn significantly impair the basibiont's ability to exchange gases and nutrients (Jagels 1973), damage the tissue by increased weight, rigidity and drag (Dixon et al. 1981), and decrease the growth rate of photosynthetic basibionts by cutting surface irradiance levels (Sand-Jensen 1977; Silberstein et al. 1986). From a nutritional perspective it is evident that if the host and the epibiont share the same trophic requirements then planktonic nutrients reaching the basibiont may already be partially depleted after their passage through the epibiotic barrier. As epibionts may fall victim to predators of their hosts, so may basibionts suffer from

Table 1 Ecological consequences for epibiont and basibiont as a result of epibiotic associations (summarized from Wahl 1989)

	Advantages	Disadvantages
Epibiont	Colonization of new substrate	Unstable, non-durable substrate
	New surface due to growth of basibiont	Biologically variable substrate
	Nutrient flow from basibiont	Exposure to detrimental host defence
	Favourable hydrodynamic conditions	Shared doom
	Favourable exposure to light	
	Associational resistance	
Basibiont	Camouflage	Increased weight and drag
	Insulation against desiccation	Decreased elasticity
	Nutrient flow from epibiont	Increased surface roughness
	Associational resistance	Increased deposition of particulate material
		Insulation against exchange of gas and waste products
		Increased mechanical damage
		Increased chemical damage
		Decreased nutrient flow through epibiotic filter

"shared doom", i.e. damage due to grazers preying on epibionts (Dixon et al. 1981). Table 1 summarizes the advantages and disadvantages of epibiosis for epi- and basibionts.

5 Defence

Many marine invertebrates and plants have evolved a variety of physical and chemical defence mechanisms to suppress epibiosis and/or remove epibionts. Epibiont removal can be physically achieved by continuous or periodic surface renewal or by means of mucus secretion (e.g. in cnidaria, algae, molluscs, echinoderms and tunicates) and periodical shedding of the cuticula or epidermis (Sieburth and Tootle 1981; Littler and Littler 1999; Nylund and Pavia 2005) (see Fig. 1).

To create unfavourable or toxic conditions at or immediately above the living surface is a wide-spread adaptation of host organisms to cope with epibionts.

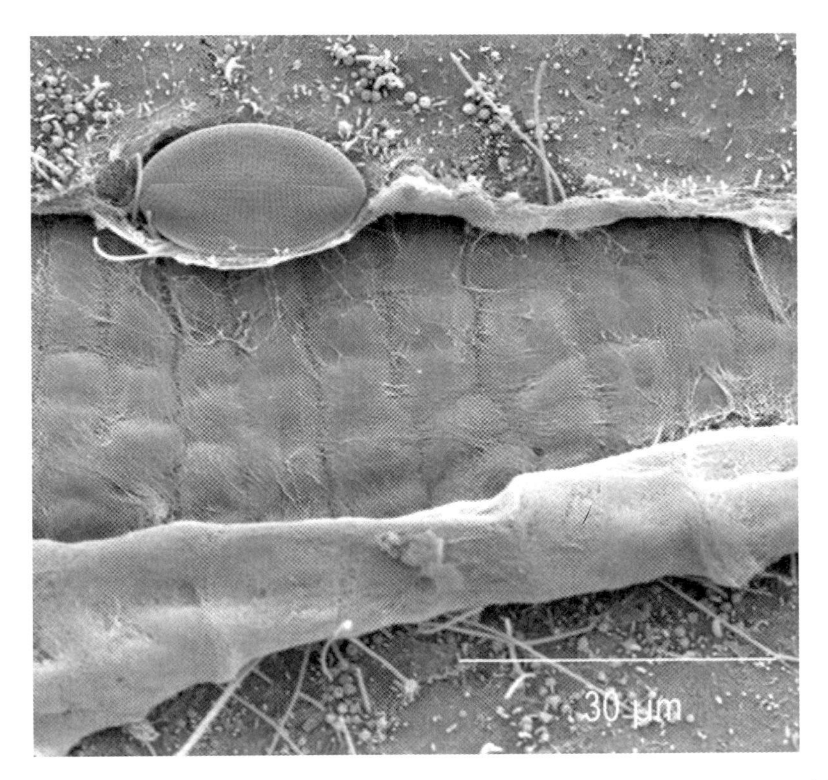

Fig. 1 Surface of the macroalga *Laminaria digitata* showing sloughing of the cuticle-containing bacteria and diatoms to reveal an uncolonized algal surface. *Scale bar*: 30 μm

The brown alga *Laminaria digitata* and the red alga *Gracilaria conferta* react with an oxidative burst to the presence of either alginate oligosaccharides or agar oligosaccharides, both of which are degradation products of their own cell walls (Küpper et al. 2001), resulting in the efficient elimination of bacterial epiflora (Weinberger et al. 2000). Moreover, sessile marine organisms feature a variety of chemical defence metabolites effective against different phyla of potential epibionts (reviewed by Clare 1996; Faulkner 2000). There are numerous studies on the inhibition of micro- and macroorganisms by extracts from diverse marine eukaryotes, such as corals, sponges, tunicates, ascidians and macrophytes (e.g. Michalek and Bowden 1997; Jensen et al. 1996; Wilsanand 1999; Slattery et al. 1995; Hellio et al. 2000; Dobretsov et al. 2006). Mostly, these investigations were descriptive and did not result in the purification and elucidation of inhibitory compounds. It remains unclear whether these extracts deter epibiosis at or near surfaces in situ and, if so, at what concentrations these effects are elicited. In this context, one of the better-studied models for algal secondary metabolism is the Australian red alga *Delisea pulchra*, which produces a range of structurally similar halogenated furanones (Steinberg et al. 2001). These metabolites are encapsulated in vesicles in the gland cells of *D. pulchra*, which provide a delivery mechanism to the surface of the alga at concentrations that deter a wide range of prokaryote and eukaryote epibionts (Maximilien et al. 1998). Being structurally related to acylated homoserine lactones (AHLs), halogenated furanones inhibit bacterial colonization through direct antagonism of bacterial cell-to-cell signalling. The AHL-mediated gene expression of bacteria is inhibited when halogenated furanones occupy the AHL-binding site of LuxR-like proteins, which represent the transcriptional activators in AHL regulatory systems (Manefield et al. 1999).

Information on the localization, identity and surface concentration of defence secondary metabolites is rapidly advancing (e.g. Salomon et al. 2001; Kubanek et al. 2002; Nylund et al. 2005; Paul et al. 2006) and the relevance of defence metabolites is increasingly discussed in a chemical ecological context. Moreover, recent studies on the deterrence of microbial colonization highlight that chemical antifouling defences cannot be generalized as broadly bacteriostatic or bactericidal, instead the effects are quite selective and targeted against particular microbial species (Maximilien et al. 1998; Egan et al. 2000; Kubanek et al. 2003).

Besides antimicrobial effects of secondary metabolites emanating from the host, recent studies have increasingly demonstrated that epibiotic bacteria associated with the host deter growth and attachment of co-occurring bacterial species or new epibiotic colonizers competing for the same niche (Armstrong et al. 2001; Harder et al. 2004a). A well-investigated bacterium in this context is *Pseudoalteromonas tunicata*, which has been isolated from a tunicate and a green macroalga. *P. tunicata* has been found to produce at least five extracellular compounds that inhibit other organisms from establishing themselves in a epibiotic community by inhibiting settlement of invertebrate larvae and algal spores, growth of bacteria and fungi, and surface colonization by diatoms (Holmström and Kjelleberg 1999; Holmström et al. 1996). Thus, in terms of the chemical ecology of host–epibiont associations, it seems evident that there is a significant protective role of symbiotic microbial epibionts, which in turn release antifouling compounds. However, after more than 20 years of research there is no experimental evidence demonstrating if and how host organisms selectively attract such epibionts.

With the advancement of molecular biological tools to analyse the diversity and abundance of bacteria in biofilms (Dahllöf 2002), several studies have demonstrated that quantitative and qualitative bacterial occurrence on host organisms differs significantly from inanimate reference surfaces (Harder et al. 2003, 2004b; Lee and Qian 2004; Dobretsov et al. 2006; Rao et al. 2005). These findings firstly suggest strong host defence mechanisms against non-culturable epibiotic bacteria but also support the notion of potent effects of non-culturable epibiotic bacteria against subsequent colonizers of host organisms. It will be interesting to see follow-up studies that utilize advanced molecular biological tools, such as cloning techniques, to directly test the metabolites of non-culturable symbiotic prokaryotes on epibiotic eukaryotes.

References

Armstrong E, Yan L, Boyd KG, Wright PC, Burgess JG (2001) The symbiotic role of marine microbes on living surfaces. *Hydrobiologia* 461:37–40

Berntsson KM, Jonsson PR, Lejhall M, Gatenholm P (2000) Analysis of behavioral rejection of micro-textured surfaces and implications for recruitment by the barnacle *Balanus improvisus*. *J Exp Mar Biol Ecol* 251:59–83

Bouarab K, Potin P, Weinberger F, Correa J, Kloareg B (2001) The *Chondrus crispus Acrochaete operculata* host-pathogen association, a novel model in glycobiology and applied phycopathology. *J Appl Phycol* 13:185–193

Brouns JJWM, Heijs FML (1986) Production and biomass of the seagrass *Anhalus acovoides* (Lf) Royle and its epiphytes. *Aquat Bot* 25:21–45

Butman CA (1987) Larval settlement of soft-sediment invertebrates: the spatial scales of pattern explained by active habitat selection and the emerging role of hydrodynamical processes. *Oceanogr Mar Biol A Rev* 25:113–165

Clare AS (1996) Marine natural product antifoulants. *Biofouling* 9:211–229

Clare AS, Matsumura K (2000) Nature and perception of barnacle settlement pheromones. *Biofouling* 15:57–71

Clare AS, Rittschof D, Gerhart DJ, Maki JS (1992) Molecular approaches to non-toxic antifouling. *Invert Reprod Dev* 22:67–76

Coon SL, Fitt WK, Bonar DB (1990) Competence and delay of metamorphosis in the Pacific oyster *Crassostrea gigas*. *Mar Biol* 106:379–387

Cooney RP, Pantos O, Le Tissier MDA, Barer MR, O'Donnell AG, Bythell JC (2002) Characterization of the bacterial consortium associated with black band disease in coral using molecular microbiological techniques. *Environ Microbiol* 4:401–413

Costerton JW, Lewandowski Z, Caldwell DE, Korber DR, Lappin-Scott HM (1995) Microbial biofilms. *Annu Rev Microbiol* 49:711–745

Dahllöf I (2002) Molecular community analysis of microbial diversity. *Curr Opin Biotechnol* 13:213–217

Davis AR, Targett NM, McConnell OJ, Young CM (1989) Epibiosis of marine algae and benthic invertebrates: natural products chemistry and other mechanisms inhibiting settlement and overgrowth. In: ScheuerPJ (ed.) Bioorganic marine chemistry, vol 3. Springer, Heidelberg Berlin New York, pp. 85–114

Dixon J, Schroeter SC, Kastendick J (1981) Effects of encrusting bryozoan, *Membranipora membranacea* on the loss of blades and fronds by the giant kelp, *Macrocystis pyrifera* (Laminariales). *J Phycol* 17:341–345

Dobretsov S, Dahms HU, Harder T, Qian PY (2006) Allelochemical defense of the macroalga *Caulerpa racemosa* against epibiosis: evidence of field and laboratory assays. *Mar Ecol Prog Ser* 318:165–175

Dreanno C, Matsumura K, Dohmae N, Takio K, Hirota H, Kirby R, Clare AS (2006) An α-2-macroglobulin-like protein is the cue to gregarious settlement of the barnacle *Balanus amphitrite*. *Proc Natl Acad Sci U S A* 103:14396–14401

Egan S, Thomas T, Holmström C, Kjelleberg S (2000) Phylogenetic relationship and antifouling activity of bacterial epiphytes from the marine alga *Ulva lactuca*. *Environ Microbiol* 2:343–347

Faulkner DJ (2000) Marine pharmacology. *Antonie Van Leeuwenhoek* 77:135–145

Fletcher RL, Callow ME (1992) The settlement, attachment and establishment of marine algal spores. *Br Phycol J* 27:303–329

Gil-Turness MS, Fenical W (1992) Embryos of *Homarus americanus* are protected by epibiotic bacteria. *Biol Bull* 182:105–108

Grossart HP, Kiørboe T, Tang K, Ploug H (2003) Bacterial colonization of particles: growth and interactions. *Appl Environ Microbiol* 69:3500–3509

Hadfield MG, Scheuer D (1985) Evidence for a soluble metamorphic inducer in *Phestilla sibogae:* ecological, chemical and biological data. *Bull Mar Sci* 37:556–566

Hadfield MG, Paul VJ (2001). Natural chemical cues for settlement and metamorphosis of marine-invertebrate larvae: 431–461. In: McClintock JB, Baker BJ (eds.) Marine chemical ecology. CRC, Boca Raton, pp. 1–610

Hahn KO (1989) Induction of settlement in competent abalone larvae. In HahnKO (ed.) Handbook of culture of abalone and other marine gastropods. CRC, Boca Raton, pp. 101–112

Harder T, Lam C, Qian PY (2002) Induction of larval settlement of the polychaete *Hydroides elegans* (Haswell) by marine biofilms: an investigation of monospecific diatom films as settlement cues. *Mar Ecol Prog Ser* 229:105–112

Harder T, Lau SCK, Dobretsov S, Fang TK, Qian PY (2003) A distinctive epibiotic bacterial community on the soft coral *Dendronephthya* sp. and antibacterial activity of coral tissue extracts suggest a chemical mechanism against bacterial epibiosis. *FEMS Microbiol Ecol* 43:337–347

Harder T, Dobretsov S, Qian PY (2004a) Waterborne, polar macromolecules act as algal antifoulants in the seeweed *Ulva reticulata*. *Mar Ecol Prog Ser* 274:131–141

Harder T, Lau SCK, Tam WY, Qian PY (2004b) A bacterial culture-independent method to investigate chemically mediated control of bacterial epibiosis in marine invertebrates by using TRFLP analysis and natural bacterial populations. *FEMS Microbiol Ecol* 47:93–99

Harlin MM (1973) Transfer of products between epiphytic marine algae and host plants. *J Phycol* 9:243–248

Hay ME (1986) Associational plant defenses and the maintenance of species diversity: turning competitors into accomplices. *Am Nat* 128:617–641

Hellio C, Bremer G, Pons AM, Le Gal Y (2000) Inhibition of the development of microorganisms (bacteria and fungi) by extracts of marine algae from Brittany, France. *Appl Microbiol Biotechnol* 54:543–549

Holmström C, Kjelleberg S (1999) Marine *Pseudoalteromonas* species are associated with higher organisms and produce biologically active extracellular agents. *FEMS Microbiol Ecol* 30:285–293

Holmström C, James S, Egan S, Kjelleberg S (1996) Inhibition of common fouling organisms by pigmented marine bacterial isolates. *Biofouling* 10:251–259

Huggett MJ, Williamson JE, De Nys R, Kjelleberg S, Steinberg PD (2006) Larval settlement of the common Australian sea urchin *Heliocidaris erythrogramma* in response to bacteria from the surface of coralline algae. *Oecologia* 149:604–619

Jagels R (1973) Studies of a marine grass, *Thalassia testudinum* I. Ultrastructure of the osmoregulatory leaf cells. *Am J Bot* 60:1003–1009

Jensen PR, Harvell CD, Wirtz K, Fenical W (1996) Antimicrobial activity of extracts of Caribbean gorgonian corals. *Mar Biol* 125:411–419

Johnson CR, Sutton DC (1994) Bacteria on the surface of crustose coralline algae induce metamorphosis of the crown-of-thorns starfish *Acanthaster planci*. *Mar Biol* 120:305–310

Johnson CR, Muir DG, Reysenbach AL (1991a) Characteristic bacteria associated with surfaces of coralline algae: a hypothesis for bacterial induction of marine invertebrate larvae. *Mar Ecol Prog Ser* 74:281–294

Johnson CR, Sutton DC, Olson RR, Giddins R (1991b) Settlement of crown-of-thorns starfish: role of bacteria on surfaces of coralline algae and a hypothesis for deepwater recruitment. *Mar Ecol Prog Ser* 71:143–162

Joint I, Tait K, Callow ME, Callow JE, Milton D, Williams P (2002) Cell-to-cell communication across the procaryote – eucaryote boundary. *Science* 298:1207

Kato T, Kumanireng AS, Ichinose I, Kitahara Y, Kakinuma Y, Nishihira M, Kato M (1975) Active components of *Sargassum tortile* effecting the settlement of swimming larvae of *Coryne uchidai*. *Experientia* 31:433–434

Keough MJ (1986) The distribution of a bryozoan on seagrass blades: settlement, growth and mortality. *Ecology* 67:846–857

Kolattukudy PE, Rogers LM, Li D, Hwang CS, Flaishman MA (1995) Surface signaling in pathogenesis. *Proc Natl Acad Sci U S A* 92:4080–4087

Krug PJ, Manzi AE (1999) Waterborne and surface-associated carbohydrates as settlement cues for larvae of the specialist marine herbivore *Alderia modesta*. *Biol Bull* 197:94–103

Kubanek J, Whalen KE, Engel S, Kelly SR, Henkel TP, Fenical W, Pawlik JR (2002) Multiple defensive roles for triterpene glycosides from two Caribbean sponges. *Oecologia* 131:125–136

Kubanek J, Jensen PR, Keifer PA, Sullards MC, Collins DO, Fenical W (2003) Seaweed resistance to microbial attack: a targeted chemical defense against marine fungi. *Proc Natl Acad Sci U S A* 100:6916–6921

Küpper FC, Kloareg B, Guern J, Potin P (2001) Oligoguluronates elicit an oxidative burst in brown algal kelp, *Laminaria digitata*. *Plant Physiol* 12:278–291

Laihonen P, Furman ER (1986) The site of settlement indicates commensalism between blue mussel and its epibiont. *Oecologia* 71:38–40

Lambert WJ, Todd CD, Hardege JD (1997) Partial characterization and biological activity of a metamorphic inducer of the dorid nudibranch *Adalaria proxima* (Gastropoda: Nudibranchia). *Invertebr Biol* 116:71–81

Lau SCK, Mak KKW, Chen F, Qian PY (2002) Bioactivity of bacterial strains from marine biofilms in Hong Kong waters for the induction of larval settlement in the marine polychaete *Hydroides elegans*. *Mar Ecol Prog Ser* 226:301–310

Lee O, Qian PY (2004) Potential control of bacterial epibiosis on the surface of the sponge *Mycale adhaerens*. *Aquat Microb Ecol* 34:11–21

Lemos ML, Toranzo AE, Barja JL (1985) Antibiotic activity of epiphytic bacteria isolated from intertidal seaweeds. *Microbiol Ecol* 11:149–163

Littler MM, Littler DS (1999) Blade abandonment/proliferation: a novel mechanisms for rapid epiphyte control in marine macrophytes. *Ecology* 80:1736–1746

Maida M, Coll JC, Samarco PW (1994) Shedding new light on scleractinian coral recruitment. *J Exp Mar Biol Ecol* 180:189–202

Maki JS, Mitchell R (2002) Biofouling in the marine environment. In: Bitton G (ed.) Encyclopedia of environmental microbiology. Wiley, New York, pp. 610–619

Manefield M, de Nys R, Kumar N, Read R, Givskov M, Steinberg PD, Kjelleberg S (1999) Inhibition of LuxR-based AHL regulation by halogenated furanones from *Delisea pulchra*. *Microbiology* 145:283–291

Matsumura K, Nagano M, Fusetani N (1998) Purification of a larval settlement-inducing protein complex (SIPC) of the barnacle, *Balanus amphitrite*. *J Exp Zool* 281:12–20
Maximilien R, de Nys R, Holmström C, Gram L, Givskov M, Crass K, Kjelleberg S, Steinberg PD (1998) Chemical mediation of bacterial surface colonization by secondary metabolites from the red alga *Delisea pulchra*. *Aquat Microb Ecol* 15:233–246
Michalek K, Bowden BF (1997) A natural algacide from soft coral *Sinularia flexibilis* (Coelenterata, Octocorallia, Alcyonacea). *J Chem Ecol* 23:259–273
Mitchell R, and Chet I (1975) Bacterial attack of corals in polluted seawater. *Microb Ecol* 2:227–233
Morse DE (1990) Recent progress in larval settlement and metamorphosis: closing the gaps molecular biology and ecology. *Bull Mar Sci* 46:465–483
Morse ANC, Morse DE (1984) Recruitment and metamorphosis of *Haliotis* larvae induced by molecules uniquely available at the surface of crustose red alga. *J Exp Mar Biol Ecol* 75:191–215
Mullineaux LS, Butman CA (1991) Initial contact, exploration and attachment of barnacle (*Balanus amphitrite*) cyprids settling in flow. *Mar Biol* 110:93–103
Nakanishi K, Nishijima M, Nomoto AM, Yamazali A, Saga N (1999) Requisite morphologic interaction for attachment between *Ulva pertusa* (Chlorophyta) and symbiotic bacteria. *Mar Biotech* 1:107–111
Nylund G, Cervin G, Hermansson, Pavia H (2005) Chemical inhibition of bacterial colonization by the red alga *Bonnemaisonia hamifera*. *Mar Ecol Prog Ser* 302:27–36
Nylund G, Pavia H (2005) Chemical versus mechanical inhibition of fouling in the red alga *Dilsea carnosa*. *Mar Ecol Prog Ser* 299:111–121
Paul NA, de Nys R, Steinberg PD (2006) Chemical defence against bacteria in the red alga *Asparagopsis armata*: linking structure with function. *Mar Ecol Prog Ser* 306:87–101
Qian PY, Pechenik JA (1998) Effects of larval starvation and delayed metamorphosis on juvenile survival and growth of the tube-dwelling polychaete *Hydroides elegans* (Haswell). *J Exp Mar Biol Ecol* 227:169–185
Qian PY, Rittschof D, Sreedhar B (2000) Macrofouling in unidirectional flow: miniature pipes as experimental models for studying the interaction of flow and surface characteristics on the attachment of barnacle, bryozoan and polychaete larvae. *Mar Ecol Prog Ser* 207:109–121
Rao D, Webb S, Kjelleberg S (2005) competitive interactions in mixed-species biofilms containing the marine bacterium *Pseudoalteromonas tunicata*. *Appl Environ Microbiol* 71:1729–1736
Rodriguez SR, Ojeda FP, Inestrosa NC (1993) Settlement of benthic marine invertebrates. *Mar Ecol Prog Ser* 97:193–207
Salomon CE, Deerinck T, Elissman MH, Faulkner DJ (2001) The cellular localization of dercitamide in the Palauan sponge *Oceanapia sagittaria*. *Mar Biol* 139:313–319
Sand-Jensen K (1977) Effect of epiphytes on eelgrass photosynthesis. *Aquat Bot* 3:55–63
Sieburth JM, Tootle JL (1981) Seasonality of microbial fouling on *Ascophyllum nodosum* (L.) Lejol., *Fucus vesiculosus* L., *Polysiphonia lanosa* (L.) Tandy and *Chondrus crispus* Stackh. *J Phycol* 17:57–64
Silberstein K, Chiffings AW, McComb AJ (1986) The loss of seagrass in Cockburn Sound, Western Australia. 3. The effect of epiphytes on productivity of *Posidonia australis. Hook F Aquat Bot* 24:355–371
Slattery M, McClintock JB, Heine JN (1995) Chemical defenses in Antarctic soft corals: evidence for antifouling compounds. *J Exp Mar Biol Ecol* 190:61–77
Steinberg PD, De Nys R, Kjelleberg S (2001) Chemical mediation of surface colonization, In: McClintock JB, Baker JB (eds.) Marine chemical ecology. CRC, Boca Raton, pp. 355–387
Steinberg PD, de Nys R, Kjelleberg S (2002) Chemical cues for surface colonization. *J Chem Ecol* 28:1935–1951
Swanson RL, Williamson JE, De Nys R, Kumar N, Bucknall MP, Steinberg PD (2004) Induction of settlement of larvae of the sea urchin *Holopneustes purpurascens* by histamine from a host alga. *Biol Bull* 206:161–172

Swanson RL, de Nys R, Huggett MJ, Green JK, Steinberg PD (2006) In situ quantification of a natural settlement cue and recruitment of the Australian sea urchin *Holopneustes purpurascens*. *Mar Ecol Prog Ser* 314:1–14

Tamburri MN, Zimmer-Faust RK, Tamplin ML (1992) Natural sources and properties of chemical inducers mediating settlement of oyster larvae: a re-examination. *Biol Bull* 183:327–338

Tatewaki M, Provasoli L, Pintner IJ (1983) Morphogenesis of *Monostroma oxyspermum* (Kuetz.) Doty (Chlorophyceae) in axenic culture, especially in bialgal culture. *J Phycol* 19:409–416

Terry LA, Edyvean RGJ (1981) Microalgae and corrosion. *Bot Mar* 24:177–183

Thevanathan R, Nirmala M, Manoharan A, Gangadharan A, Rajarajan R, Dhamotharan R, Selvaraj S (2000) On the occurrence of nitrogen fixing bacteria as epibacterial flora of some marine green algae. *Seaweed Res Utiln* 22:189–197

Todd CD, Keough MJ (1994) Larval settlement in hard substratum epifaunal assemblages: a manipulative field study of the effects of substratum filming and the presence of incumbents. *J Exp Mar Biol Ecol* 181:159–187

Tsukamoto S, Kato H, Hirota H, Fusetani N (1994) Narains: *N,N*-dimethylguanidinium styryl sulfates, metamorphosis inducers of ascidian larvae from a marine sponge *Jaspis* sp. *Tetrahedron Lett* 35:5873–5874

Tsukamoto S, Kato H, Hirota H, Fusetani N (1995) Pipecolate derivatives, anthosamines A and B, inducers of larval metamorphosis in ascidians, from a marine sponge *Anthosigmella aff. raromicrosclera*. *Tetrahedron* 51:6687–6694

Tsukamoto S, Kato H, Hirota H, Fusetani N (1999) Lumichrome – a larval metamorphosis-inducing substance in the ascidian *Halocynthia roretzi*. *Eur J Biochem* 264:785–789

Turner EJ, Zimmer-Faust RK, Palmer MA, Luckenback (1994) Settlement of oyster *Crassostrea virginica* larvae: effects of water flow and a water-soluble chemical cue. *Limnol Oceanogr* 39:1579–1593

Wahl M (1989) Marine epibiosis. I. Fouling and antifouling: some basic aspects. *Mar Ecol Prog Ser* 58:175–189

Wahl M (1997) Living attached: aufwuchs, fouling, epibiosis. In: Nagabushanam R, Thompson MF (eds.) Fouling organisms of the Indian Ocean: biology and control technology. Oxford and IBH, New Delhi, pp. 31–83

Wahl M, Hay ME (1995) Associational resistance and shared doom: effects of epibiosis on herbivory. *Oecologia* 102:329–340

Wahl M, Mark O (1999) The predominantly facultative nature of epibiosis: experimental and observational evidence. *Mar Ecol Prog Ser* 187:59–66

Wahl M, Hay ME, Enderlein P (1997) Effects of epibiosis on consumer-prey interactions. *Hydrobiologia* 355:49–59

Weinberger F, Friedlander M (2000) Response of *Gracilaria conferta* (Rhodophyta) to oligoagars results in defense against agar-degrading epiphytes. *J Phycol* 36:1079–1086

Wieczorek SK, Todd CD (1998) Inhibition and facilitation of the settlement of epifaunal marine invertebrate larvae by microbial biofilm cues. *Biofouling* 12:81–93

Williamson JE, De Nys R, Kumar N, Carson DG, Steinberg PD (2000) Induction of metamorphosis in the sea urchin *Holopneustes purpurascens* by a metabolite complex from the algal host *Delisea pulchra*. *Biol Bull* 198332–345

Wilsanand V, Wagh AB, Bapuji M (1999) Effect of alcohol extracts of demospongiae on growth of periphytic diatoms. *Ind J Mar Sci* 28:274–279

Yvin JC, Chevolet L, Chevolet-Maguer AM, Cochard JC (1985) First isolation of jacarone from an alga, *Delesseria sanguinea:* a metamorphosis inducer of *Pecten* larvae. *J Nat Prod* 48:814–816

Larval Settlement and Surfaces: Implications in Development of Antifouling Strategies

P. Sriyutha Murthy (✉), V. P. Venugopalan, K. V. K. Nair, and T. Subramoniam

Abstract Marine biofouling is a natural process that imposes technical operational problems and economic losses on marine-related activities. Marine biofouling communities are complex, diverse, highly dynamic ecosystems consisting of a range of organisms. Larvae of these organisms spend a part of their lives in the planktonic stage before settling on a surface. Passive transport and deposition of larvae were considered responsible for the observed spatial variation in settlement pattern, whereas breeding season and larval survival have been associated with temporal fluctuations. Hydrodynamic conditions influence the transport and deposition of larvae near the surface boundary layer, while dissolved environmental stimuli have been associated with the induction of settlement and metamorphic behaviour. Over the last two decades, chemical cues and physiological processing of the cue has been the subject of study. Sufficient information has been obtained on the settlement mechanisms, the nature of chemical substances, involvement of chemosensory receptors and signal transduction pathways downstream. Knowledge on the settlement mechanism is imperative for developing a suitable control strategy. At present, more is known about chemosensory reception and downstream processing of the sensory cue than the location of these receptors. The need to control biofouling on underwater surfaces has given rise to many different technologies. Conventional antifouling strategy employs the use of biocidal surface coatings. The rationale behind these coatings is to kill everything. Historically, different solutions for control of fouling have been employed. It was not until the development of cold-plastic antifouling paints (copper oxide and tributyltin oxide or fluoride) in the later part of the twentieth century that a truly long-lasting protection was achieved. Unfortunately, the accumulation of slow-degrading organotin moieties in the water column has resulted in sub-lethal effects on non-target organisms, which led to its progressive abandonment. Insights into the larval sensory recognition of physical cues and adhesion resulted in the development of foul release coatings based on low surface energy phenomenon. Another alternative approach for control of biofouling

P.S. Murthy
Biofouling and Biofilm Processes Section, Water and Steam Chemistry Division, BARC Facilities, Indira Gandhi Center for Atomic Research Campus, Kalpakkam 603 102, India
e-mail: psm_ murthy@yahoo.co.in, psmurthy@igcar.gov.in

Springer Series on Biofilms, doi: 10.1007/7142_2008_17

and inhibition of larval settlement lies in inhibiting the neurophysiological processes involved in larval settlement. This has been experimented upon using natural bioactive molecules and synthetic analogues, which bind to specific receptors inhibiting larval settlement. Several pharmacological compounds, natural products and synthetic analogues that inhibit the metabolic processes underlying settlement have been identified through laboratory bioassays. However, realization of these compounds into commercial coatings is yet to happen. The reasons may be attributed to reproducibility of laboratory results in ecologically realistic field experiments. The need for suitable bioassays and knowledge of the broad-spectrum activity of these compounds is obvious.

1 Introduction to Biofouling and Antifouling

The problem of biofouling presents a serious operational problem. In heat exchangers it leads to reduced heat transfer efficiency, increased fluid frictional resistance, additional maintenance and operational costs. On ship hulls the problem leads to increase in hydrodynamic drag and fuel consumption and decreases the manoeuvrability of vessels. On submerged structures the problem leads to hydrodynamic loading, material deterioration and failure of moored instruments and the optical windows of sensors. The problem of biofouling is a surface-associated phenomenon and control measures should be focussed on this aspect. Conventionally, fouling mitigation in heat transfer equipment is carried out by the addition of biocides (see Nymer et al. 2008). For protection of submerged structures and ship hulls most antifouling systems take the form of protective coatings. Understanding the mechanisms of larval settlement and metamorphosis in benthic invertebrates is important for developing suitable methods for interfering with the settlement process and inhibiting settlement. In this chapter, molecular cues and pathways involved in settlement of marine benthic organisms and the antagonistic activity of synthetic and natural compounds in settlement prevention are elucidated. In addition, an overview on the status of different antifouling strategies being researched is also presented. Benthic marine communities are dominated by diverse invertebrate fauna representing over 4,000 different species (Crisp 1974). This chapter is limited to the dominant fouling organisms such as barnacles, hydrozoans, tube-building polychaete worms and molluscs.

2 Physical Cues and Antifouling; Where do We Stand?

The possibility of larvae responding to environmental cues is debatable until they are in close proximity with the near-surface boundary layer (Butman et al. 1988). Interaction of physical processes and hydrodynamics (Pawlik et al.

1991; Walters 1992; Millineaux and Garland 1993) assist in bringing the larvae to the near-surface boundary layer, within the perception of physical or chemical cues. Sensory recognition of cues results in settlement, and the transformation of signals into the larval neuronal systems results in initiation of differentiation and metamorphosis. Physical characteristics of the substratum seem to be of secondary importance when compared to chemical and biological characteristics (Mihm et al. 1981; LeTourneux and Bourget 1988) in settlement induction. Physical factors such as light or shade, substratum type, orientation depth, gravity, hydrostatic pressure and temperature are found to influence larval behaviour and settlement in many invertebrate species (Thorson 1964; Crisp 1974; Ryland 1974; Sulkin 1984; Young and Chia 1987; Boudreau et al. 1990; Kaye and Reiswig 1991; Anderson and Underwood 1994; Holiday 1996; Connell 1999; Galsby 1999, 2000; Forde and Raimondi 2004). In comparison to the above mentioned factors, physical characteristics of the substratum like surface energy (wettability), roughness and microtopography have been shown to have a measurable effect on larval settlementand can be probably mimicked for producing antifouling surfaces. The mechanism of sensing physical cues like surface energy is thought to be initiated when the larvae abandon swimming behaviour and commence surface exploration. Alternatively, it is believed to be by the action of physiochemical forces between the surface and the larval body or antennules. Laboratory and field studies conducted by Rittschof and Costlow (1989a, b) and Roberts et al. (1991) (also, see Cooksey et al. 2008) showed similar trends in relation to surface energy by different larval groups. Larvae of barnacles preferred high-energy surfaces, bryozoans and ascidians preferred low-energy surfaces and larvae of hydroids settled equally on surfaces of all energies. A species-specific difference was observed with respect to *Balanus amphitrite* and *Balanus improvisus* cyprids. *B. amphitrite* cyprids preferred unfouled high-energy surfaces (Rittschof et al. 1984) whereas *B. improvisus* cyprids preferred hydrophilic surfaces (Dahlstrom et al. 2004). An intermediate change in surface wettability inhibited settlement by 38% for this barnacle.

An interesting aspect of these studies is that no surface energy escapes settlement by at least some group of macrofoulants. The surface energy principle has been used successfully in the development of foul release silicone coatings where additional lubricity has been provided by the addition of silicone oil. Decreasing the surface elastic modulus to about 0.01–0.1 KPa significantly decreased larval settlement in field trials as well as in laboratory experiments conducted with bryozoan larvae (Natasha et al. 2002). This response to the elastic modulus may be a result of mechanical deformation of membranes of the sensory cells used to explore surfaces. Lower elastic modulussurfaces are also known to cause alteration of ionic traffic across membranes and present a weaker signal for settlement (Natasha et al. 2002). The use of surface energy and elastic modulus principles for antifouling is still in its infancy. Species-specific data need to be generated and correlated to obtain a value for these parameters to be incorporated into antifouling coatings.

3 Induction Pathways and Physiological Events on Perception of Cues

In each of the invertebrate phyla there are representations that have shown a unique response to settlement and metamorphic cues and pathways. At the molecular and cellular level, the receptors and signal transducers that control settlement and metamorphosis are proving to be structurally, functionally and evolutionarily closely related to receptors and transducers of higher organisms. Most of the invertebrate larvae are passive to the large-scale advective processes that bring the larvae near settlement sites, whereas the irreversible phase of metamorphosis is dependent on the sensory information processed by the larvae. Broadly, the exogenous and endogenous events leading to settlement and metamorphosis (Fig. 1) can be classified into:

1. Induction by morphogenetic pathway
2. Induction by regulatory pathway
3. Induction by second messenger or diacylglycerol pathway
4. Induction by amino acid derivatives
5. Induction by catecholamine pathway
6. Induction by choline derivatives
7. Induction by ion-gated channels
8. Induction by nitric oxide synthase pathway
9. Induction by waterborne substances
10. Inducers associated with microbial biofilms
11. Inducers of biological or synthetic origin

3.1 Induction by Morphogenetic Pathway

The morphogenetic pathway proposed by Morse (1992) operates via larval sensory recognition of exogenous inducing molecules by specialized chemosensory receptors, resulting in the sequential activation of a larval membrane receptor (γ-aminobutyric acid and dihydroxyphenylalanine receptors) and membrane-associated adenyl cyclase cascade. This involves synthesis and activation of cyclic adenosine monophosphate (cAMP), calcium-regulated protein kinase (PKA), protein phosphorylation and opening of chloride or other anion channels in the chemosensory cell membrane and an efflux of chloride or other anions, resulting in excitatory depolarization or firing of the chemosensory cell (Fig. 1). This morphogenetic environmental stimulus is transduced to an electrochemical signal that is propagated in the larval nervous system.

3.2 Induction by Regulatory or Amplifier Pathway and the Second Messenger Diacylglycerol Pathway

Another signal transduction pathway involved in recognition of the chemical signal is the amplifier or regulatory pathway (Morse 1992), which amplifies the

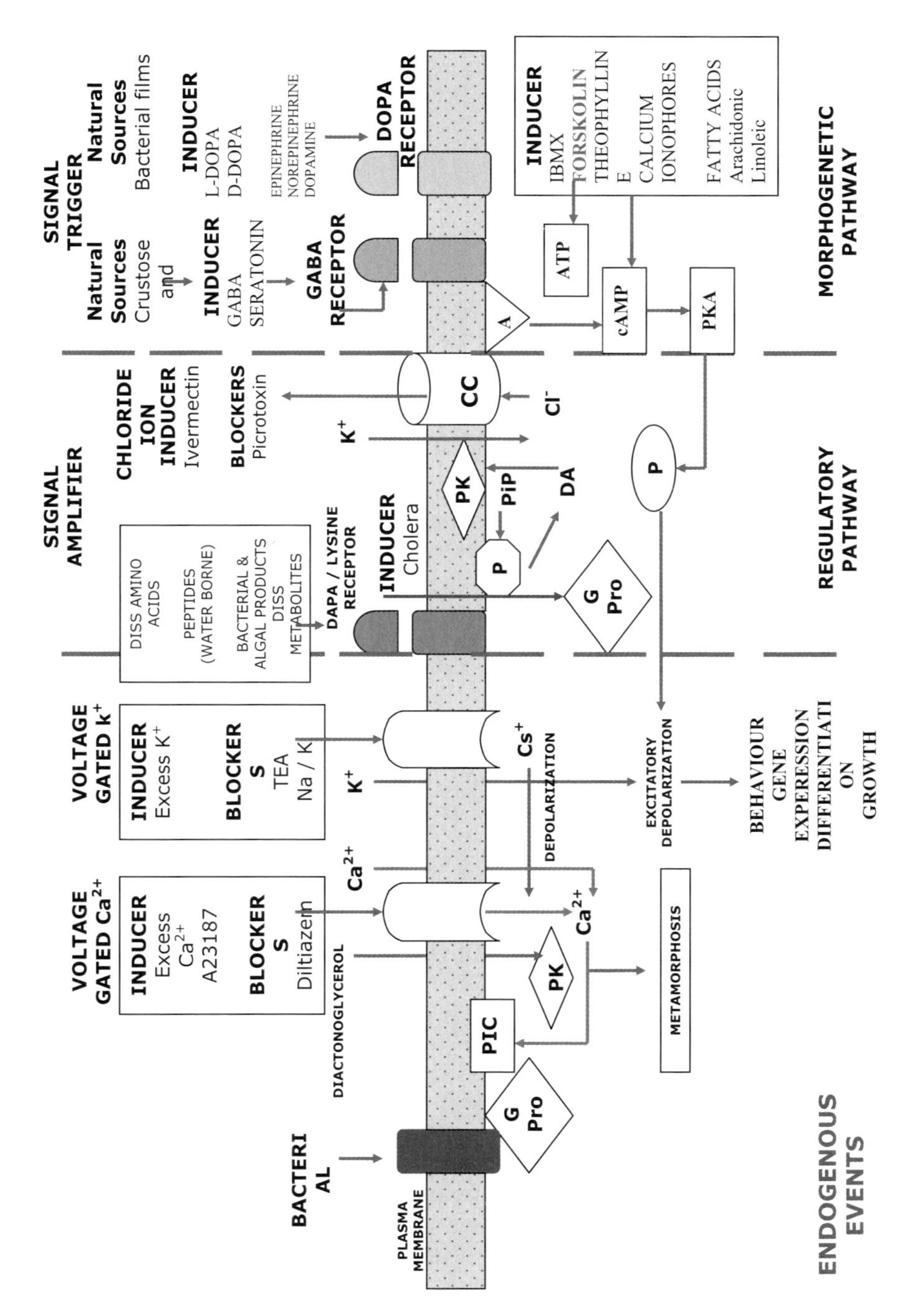

Fig. 1 Schematic overview integrating different postulated pathways inducing settlement and metamorphosis in various invertebrate larval models. *AC* adenyl cyclase; *A23187* calcium ionophore; *ATP* adenosine tri phosphate; Ca^{2+} calcium ions; *CC* chloride ion channel; *cAMP* cyclic adenosine monophosphate; *db-cAMP* dibuteryl cyclic adenosine monophosphate; *DG* diacylglycerol; *DAPA* diamino propionic acid; *GABA* γ-aminobutyric acid; *G Pro* G protein; *IBMX* isobutylmethylxanthine; K^+ potassium ions; PiP_2 phosphatidylinositol bisphosphate; *PK* protein kinase; *P* protein phosphorylation; *PL* phospholipase; *PDE* phosphodiesterase; *SITS* sulfonyl isothiocyanostilbene; *TEA* tetraethylammonium chloride

sensitivity of the larvae to the required morphogenetic signal by as much as 100-fold. This pathway, rather than inducing irreversible changes, may act as amplifier of the signal generated in the environment. The environmental signals include lysine and related diamino acids dissolved in seawater, which are found to induce settlement and metamorphosis in the abalone *Haliotis rufescens*. This lysine receptor binding signal is transduced by a receptor-associated G protein, which in turn leads to the activation of a diacylglycerol-stimulated calcium-stimulated protein kinase C (PKC) that phosphorylates a specific target protein (Fig. 1).

3.3 Induction by Amino Acid Derivatives

Gamma-aminobutyric acid (GABA), a product of glutamic acid decarboxylation, is known to hyperpolarize postsynaptic membranes by increasing membrane permeability to Cl^- ions (Kuffler et al. 1984).

3.4 Induction by Catecholamine Pathway

Artificial inducers or neurotransmitters that mimic the action of natural compounds have also been investigated in many marine invertebrate species. The ability of the neurotransmitter to mimic chemical cues emanating from the substratum implies the operation of neuronal receptors and the involvement of the nervous system in the initial processes that trigger settlement (Baloun and Morse 1984a, b; Yool et al. 1986). Experimentation on neuroactive molecules commenced after natural inducer substances eluded isolation and identification.

3.5 Induction by Choline Derivatives

Choline, a constituent of cell membranes and precursor of the neurotransmitter acetylcholine has been found to induce settlement of larvae (1) by acting directly on cholinergic receptors, (2) acting as a precursor of acetylcholine and (3) by stimulating synthesis and release of catecholamines (Hirata and Hadfield 1986; Pennington and Hadfield 1988).

3.6 Induction by Ion-Gated Channels

Another pathway by which larvae have been shown to settle and metamorphose is through ion-gated channels, such as those involving K^+, Ca^{2+}, Mg^{2+} and Na^+. In general, marine invertebrate larval settlement and metamorphosis have been understood to be under possible nervous control (Hadfield 1978; Burke 1983; Morse et al. 1984; Rittschof et al. 1986; Bonar et al. 1990). In invertebrate chemoreceptors, a stimulus-dependent increase in the cell membrane permeability can transduce chemical stimuli into electrical impulses along the nervous system (Morita 1972; Thurm and Wessel 1979; Kaissling and Thorson 1980). Perception of signals by neuronal systems has been found to involve the depolarization of specialized cells in response to an appropriate stimulus (Aidley 1978). Conduction of electrical impulses in nervous tissues depends on the maintenance of an electrical potential across the cell membrane, which is a function of differential permeability of the membrane to Na^+, K^+ and Cl^- ions (Kuffler et al. 1984). The functioning of excitable cells involves the selective movement of ions across specialized membranes. Perception of inductive cues by larvae is found to be dependent on stimulus-mediated depolarization of cells in a sensory inductive pathway (Baloun and Morse 1984a, b). Earlier studies on the nervous control of larval metamorphosis have been carried out by exposing the larvae to altered ionic composition and in the presence of neuropharmacological probes (Morse et al. 1979; Baloun and Morse 1984a; Rittschof et al. 1986).

3.7 Induction by Nitric Oxide Pathway

Nitric oxide (NO) a neurotransmitter present in neural circuits of vertebrate and arthropods has been demonstrated to play a role in larval settlement. NO (a signalling molecule) production is effected by nitric oxide synthase (NOS), which catalyses the conversion of l-arginine to l-citrulline. The pathway was first proposed for the mollusc, where inhibition of NOS activity resulted in the induction of metamorphosis in the marine snail *Ilyanassa obsoleta* (Froggett and Leise 1999). This pathway has also been demonstrated in the sea urchin *Lytechinus pictus* (Bishop and Brandhorst 2001).

4 Pharmacological Control of Larval Settlement

Several neurotransmitters and neuromodulators play inductive or inhibitory roles in the pathways that govern larval settlement and metamorphosis. Environmental cues influencing larval settlement have been studied but specific molecular details and downstream neuroendocrine actions that control the events of larval metamorphosis have often been bypassed. Pharmacological approaches are helpful in understanding the internal biochemical pathways and this section deals with agonists and antagonists that effectively block the neurotransmitter pathway, resulting in settlement inhibition of fouling organisms.

4.1 Barnacles

B. amphitrite has been the most widely studied organism with respect to its settlement induction and inhibition. Larval sensory recognition of environmental cues in barnacles is by physical surface contact of antennules, wherein structures proposed to be situated on the fourth antennular segment are employed for chemical recognition (Nott and Foster 1969; Hoeg et al. 1988; Clare and Nott 1994). Multiple settlement induction pathways have been demonstrated for this species of barnacle. Settlement-inducing protein complex(SIPC) is a glycoprotein of high molecular mass consisting of three major subunits of 76, 88 and 98 kDa with a lentil lectin (LCA)-binding sugar chain. It was found to be synthesized during larval development and induced settlement in *B. amphitrite* cyprids (Matsumara et al. 1998).

An increase in acetylcholine by partial inhibition of acetyl cholinesterase activity increased settlement (Faimali et al. 2003) and induced muscular contraction and cement gland exocytosis in *B. amphitrite* cyprids. Inhibition of acetyl cholinesterase (AchE), the lytic enzyme of acetylcholine, by inhibitors such as eserine, methomyl and mercaptodimetur (carbamates) was found to promote settlement. In contrast, settlement was inhibited by the cholinomimetic drugs atropine, an alkaloid and natural antagonist of muscarinic receptors (mAChRs), and nicotine, an agonist of nicotinic receptors (nAChRs) affecting cholinergic neurotransmission through binding of cholinergic receptors. Incidentally, a substance that inhibits AcHE activity has been isolated from the sponge *Reniera sarai* (Sepcic et al. 1998; Faimali et al. 2003) and can be used as a potential lead in settlement inhibition of these barnacles. Apart from acetylcholine, cAMP was also found to induce settlement in *B. amphitrite* through the morphogenetic pathway (Clare et al. 1995).

Activation of adenylate cyclase with forskolin and the inhibition of phosphodiesterase with 3-isobutyl-1-methylxanthine, caffeine and theophylline significantly increased the settlement of *B. amphitrite* cyprids. No significant increase in settlement was observed with the cAMP analogues dibutyryl cAMP (db-cAMP), dibutyryl cyclic GMP, 8-(4-chlorophenylthio) (CPT) cAMP or papaverine(phosphodiesterase inhibitors), but db-cAMP induced metamorphosis in *B. amphitrite* (Rittschof et al. 1986). Induction of metamorphosis by db-cAMP by triggering the second messenger pathway was also observed in *B. amphitrite* cyprids; however, the response was not significantly greater than with cAMP. Treatment with forskolin significantly increased the cAMP titre of cyprids (Clare et al. 1995). Miconazole nitrate, an adenylate cyclase inhibitor that inhibited settlement by blocking cAMP production, could be used as an antifouling agent. In addition, PKC has been shown to be involved in barnacle settlement and metamorphosis (Yamamoto et al. 1995); specific inhibitors of these enzymes may be effective antifouling agents.

Involvement of another group of receptors known as tyrosine kinase-linked receptors and their transduction pathways in barnacle settlement induction was demonstrated by Okazaki and Shizuri (2000). AG-879, a nerve growth factor receptor-tyrosine kinase (NGFR-TK) inhibitor; tyrphostin 9, a platelet-derived growth factor receptor-tyrosine kinase (PDGRF-TK) inhibitor; tyrphostin 25, an epidermal growth factor receptor-tyrosine kinase (EGFR-TK) inhibitor; insulin receptor-tyrosine kinase (InsR-TK) inhibitor; phospholipase C (PLC) inhibitor; Wortmanin, a phoshatidylinositol-3 kinase (PI3K) inhibitor; and PD-98059, a mitogen-activated protein kinase (MAPK) inhibitor all suppressed cyprid settlement in a dose-dependent manner (Okazaki and Shizuri 2000). These compounds and their synthetic analogues may be tried out in antifouling paint matrices to test their efficacy in the field.

Metamorphosis is dependent on the settlement success of larvae in invertebrates. Competent barnacle cyprids engaged in active probing of substrata secrete temporary cement, which enables the larvae to explore the substratum and also acts as a settlement pheromone for interactions among larvae (Clare et al. 1994). Both temporary and permanent adhesives are required for successful settlement of larvae. Secretion of these cements from the cement glands to the antennular sacs is under neuronal control (Okano et al. 1996). The catecholamines dopamine and noradrenaline were found to initiate settlement in the barnacle *Megabalanus rosa* (Okano et al. 1996). Secretion of dopamine and noradrenaline suggests the involvement of adrenergic receptors in the settlement process. In comparison with *B. amphitrite*, larvae showed that α-adrenergic receptor antagonist like phentolamine inhibited settlement but induced metamorphosis without prior attachment at concentrations of 0.1–10 mM (Yamamoto et al. 1998). These findings resulted in the screening of various adregenergic antagonists like phentolamine, prazosin, atipamezole, medetomidine, clonidine and idazoxanfor settlement inhibition acti-vity. Medetomidine and clonidine repeatedly inhibited settlement of *B. amphitrite* cyprids (Dahlstrom et al. 2000).

Medetomidine revealed a strong tendency to accumulate in solid/liquid phase boundaries and seems to be an attractive candidate for incorporation into antifouling coatings. Surface affinity studies with medetomidine showed that the compound adsorbed strongly to hydrophilic polystyrene compared to hydrophobic polystyrene without change in chemical structure, which further provides an attractive feature for incorporation into existing coatings as boosters (Dahlstrom et al. 2004).

α-Adrenergic antagonists idozaxon and phentolamine inhibited settlement of *B. amphitrite* larvae (Dahms and Qian 2004). An important feature of these vertebrate α-adrenergic antagonists is their hydrophobic nature and their tendency to accumulate at surfaces, which contributes to their successful inhibition of larval settlement. Structural analogues of medetomidine and clonidine inhibited settlement in barnacles. Among the seven analogues tested guanbenz [α(2)-agonist, I(2) ligand], moxonidine [α(2)-agonist, I(1) ligand], BU 224 [I(2) ligand], metrazoline [I(2) ligand], cirazoline [α(1)-agonist, I(2) ligand] and tetrahydrozoline [α-agonist I(2) ligand] inhibited settlement of *B. improvisus* cyprids (Dahlstrom et al. 2005).

Another antifouling strategy is based on settlement inhibition of larvae by blocking ion-gated channels. Excess potassium ion, magnesium ion and calcium ion inhibited settlement in *B. amphitrite* (Rittschof et al. 1986). This effect may be due to changes in the membrane potential caused by a potassium electrochemical gradient (Hodgkin and Horowicz 1959). Potassium ion affected young cyprids while other cations had more pronounced effects on older cyprids. Results indicate that firing of external receptors becomes less important in aged cyprids and settlement inhibition may be related to physiological stress on the increasingly fragile larvae. Delaying the settlement of cyprids may result in natural inhibition and appears to be an easy method for fouling control. Compounds that inhibit settlement and metamorphosis in invertebrate larvae increase the probability of death, as the larvae have to spend more time in the planktonic stage, which increases mortality. Settlement of *B. amphitrite* larvae was inhibited by sulfonyl isothiocyanostilbene (SITS), a calcium channel blocker and TEA (tetraethylammonium chloride), a potassium channel blocker, in the presence of a settlement induction factor. Calmodulin, a major intracellular calcium-binding protein present in adults and cyprids, was also found to be involved in the settlement process (Yamamoto et al. 1998). Picrotoxin, an effector of chloride channels, was found to induce settlement in barnacle cyprids (Rittschof et al. 1986). Antagonists of these compounds could be used for inhibiting settlement of barnacle cyprids.

4.2 Hydrozoans

As for barnacles, multiple settlement induction pathways were observed in hydrozoans. Activation of *Hydractinia echinata* larvaetakes place in cells located at the anterior end as a result of activation of a kinase C-like enzyme, which directly leads to the closure of K^+ channels. Closure of these channels causes depolarization and thus release of an internal signal (Leitz and Klingmann 1990). Another induction pathway is through the phosphotidylinositol/diacylglycerol/protein kinase C (PI/DAG/PKC) system, where the metamorphic signal produced by bacterial cues is transduced in membranes (Leitz and Muller 1987). Induction by bacterial films causes an increase in endogenous diacylglycerol, the physiological activator of PKC, suggesting that the bacterial inducer acts by activating receptor-regulated phospholipid metabolism. Exogenous diacylglycerol leads to membrane translocation of PKC, indicative of activation (Schneider and Leitz 1994). Subsequent labelling studies by Leitz et al. (1994) showed that [^{14}C]-arachidonic acid release was also involved in the induction process.

Diacylglycerols such as 1,2,-*sn*-dioctanoylglycerol induced metamorphosis. The inductive activity was suppressed by compounds like sphingosine and K-252a, which inhibit mammalian PKC (Leitz and Klingmann 1990). An investigation with compounds known to block transmembrane ion transport has enabled researchers to understand the role of these ions in the induction process. Oubainblocks Na^+/K^+-ATPase, which maintains electrical potential across cell membranes (Kuffler et al. 1984), and has been found to inhibit metamorphosis in

H. echinata larvae, which are induced by bacterial films (Muller and Buchal 1973). Other modulators of intracellular Ca^{2+} (inositol 1,4,5-triphosphate) or compounds that regulate the calcium-dependent PKC have been implicated in the cascade of events leading to metamorphosis of *H. echinata* (Leitz and Muller 1987; Leitz and Klingmann 1990).

4.3 Tubeworms

Among the tube-building polychaete worms, *Hydroides elegans*and *Phragmatopoma lapidosa californica*are the most extensively studied organisms (see Cooksey et al. 2008). Biofilms, adult extracts and dissolved free amino acids have all been shown to induce settlement and metamorphosis in *H. elegans* (Beckmann et al. 1999; Harder and Qian 1999). Structural and functional relationships have often been implied between the neuroactive substances and the natural inducers. *H. elegans* failed to respond to G-protein activator Gpp[NH]p or the inhibitor GDP-β-S, negating the involvement of G protein-coupled receptors in the induction process (Holm et al. 1998). PKC and the adenyl cylase pathway were not involved in the induction process.

The response of *H. elegans* larvae to inhibitors of mRNA and protein synthesis revealed that larvae were induced to metamorphose by exposing them to bacterial film or to a 3-h pulse of 10 mM CsCl in the presence of the gene transcription inhibitor DRB (5,6-dichloro-1-β-d-ribofuranosylbenzimidazole) or the translation inhibitor emetine. DRB and emetine inhibited the incorporation of radiolabelled uridine into RNA and radiolabelled methionine into peptides, respectively, indicating that they were effective in blocking the appropriate synthesis. Results of this study demonstrated that induction of metamorphosis in *H. elegans* does not require de novo transcription or translation (Carpizo-Ituarte and Hadfield 2003).

Abnormal metamorphosiswas observed with the neurotransmitters GABA, choline chloride, 3,4-dihydroxyphenyl l-alanine(l-DOPA) and the ionic channel blockers and potassium ions (Bryan et al. 1997). Neurotransmitter blockers idazoxan and phentolamine inhibited settlement of *H. elegans* (Dahms and Qian 2004; see Harder 2008). Several investigations revealed that density and specific species of biofilm bacteria were preferred by these larvae. Incorporation of succinic acid, a fungal metabolite isolated from sponge surface-associated fungus, was effective in inhibiting settlement of these larvae (Yang et al. 2007). Abnormal metamorphosis was induced by choline succinyl choline, serotonin and Oubain in *P. lapidosa californica* larvae (Pawlik 1990). Larvae did not respond to GABA, dopamine, epinephrine and norepinephrine. Both l- and d-DOPA induced normal metamorphosis in these larvae. Tetraethylammonium chloride (TEA), sulfonyl isothiocyanostilbene (SITS) and picrotoxinhad no apparent effect on the larvae. The cAMP pathway was also not active in this polychaete group, as evident from the failure to respond to db-cAMP, cholera toxin and isobutylmethylxanthine (IBMX). The divalent cation Ca^{2+} induced metamorphosis of the polycheate *Phragmatopoma californica* larvae. Studies by Ilan et al. (1993) showed that the larvae were induced to settle by excess calcium ions in the external medium (Ilan et al. 1993). This effect was found to be specific for calcium ions and was not simply the result of osmotic changes, as an excess of Mg^{2+} ions did not elicit this effect. The calcium ionophore A23187 and the aromatic calcium channel blockers diltiazem, verapamil, D600 and nifidipine, known to block Ca^{2+} channels in other systems (see Lewandowski and Beyenal 2008), also induced metamorphosis. These results indicate direct control of the morphogenetic pathway by calcium ion and complexities of the calcium regulation of this process (Ilan et al. 1993).

4.4 Molluscs

Chemosensory and molecular mechanisms governing the perception of cues and internal transduction of the signal has been extensively studied in molluscs due to their commercial importance. Among the bivalves, *M. edulis*and the oysters *Crassostrea virginica*, *C. gigas*and *C. madrasensis*constitute important aquaculture and fouling species, and have been extensively studied with respect to their settlement behaviour. The blue mussel is distributed in Europe and North America. l-DOPA was found to induce settlement in the mussel, *M. edulis* (Cooper 1982; Dobretsov and Qian 2003). Isobutylmethylxanthine also induced metamorphosis whereas the neurotransmitter GABA had no effect on this species (Dobretsov and Qian 2003).

Triacylglycerols (TAG) from the diatom *Chaetoceros muelleri*were found to improve the settlement rate of larvae of the blue mussel *Mytilus* sp. Free fatty acid content was found to increase with TAG levels (Pernet et al. 2003). The ability of the neuroactive compounds to influence metamorphosis in oysters has been extensively studied. l-DOPA has been shown to induce larval metamorphosis in *C. gigas* (Coon et al. 1985). l-DOPA and the catecholamines(namely dopamine, adrenaline and noradrenaline) are derivatives of tyrosine and function as hormones, neurotransmitters and pigments. These tyrosine derivatives were found to induce settlement of many mollusc larvae. l-DOPA was found to induce both settlement and metamorphosis in the oyster *C. gigas*, whereas epinephrine and norepinephrine were found to induce metamorphosis without settlement (Coon et al. 1985). Dopamine was found to induce low levels of settlement of this species. Among the neurotransmitters tested (l-DOPA, d-DOPA, dopamine, GABA, acetylcholine chloride, acetylcholine iodide and choline chloride) in the Indian oyster *C. madrasensis*, l-DOPA alone induced settlement at a concentration of 10^{-5} M.

The cAMP pathway has not been found to be active in oyster species studied viz: *C. virginica* (Bonar 1976), and *C. madrasensis* (Murthy 1999; Murthy et al. 1999). The other compounds tested were cAMP and db-cAMP, isobutylmethylxanthine, forskolin, 12-*O*-tetradecanoyl phorbol-13-acetate (TPA) and phorbol-12–13-dibutyrate (PuDB), which did not induce settlement and metamorphosis of *C. madrasensis* larvae (Murthy 1999; Murthy et al. 1999). Choline derivatives tested against the larvae of the Pacific oyster *C. gigas* (Coon et al. 1985) showed low and inconsistent levels of metamorphosis. Seratonin (5-hydroxytryptamine), a derivative of tryptophan and a modulator of vertebrate and invertebrate nervous systems (Kuffler et al. 1984), was found to induce metamorphosis of the mud snail *Ilyanassa obsoleta* (Levantine and Bonar 1986; Couper and Leise 1996). In contrast, the compound did not induce settlement in the larvae of the molluscs *Haliotis rufescens* (Morse et al. 1979), *Phestilla sibogae* (Hadfield 1984) and *C. gigas* (Coon and et al. 1985). The larvae of the Indian oyster *C. madrasensis* responded to changes in external ionic concentration (Murthy 1999; Murthy et al. 1999). An increase in settlement and metamorphosis was observed with an increase in calcium ion concentration above that found in standard seawater (12.4 mM) up to 21 mM. Settlement and metamorphosis decreased with further increase in concentration of the ion above 21 mM. Similarly, concentrations of the ion (1–6 mM) below those found in standard seawater showed low levels of settlement compared to standard seawater. The larvae failed to survive when placed in calcium-free seawater. Results suggest that the induction process may be due to an influx of Ca^{2+} ion across the larval cell membranes.

The carboxylic calcium ionophore A23187 induced settlement and metamorphtosis in the larvae in a concentration-dependent manner. Inhibition of settlement and metamorphosis was observed with respect to calcium channel blockers(diltiazem, verapamil and nifidipine). These results suggest the functional role of Ca^{2+} channels and the presence of active sites on the larval surface, which mediate in the induction process (Murthy 1999; Murthy et al. 1999).

Alteration in concentrations of potassium ion induced settlement in the Sydney rock oyster *Saccostrea commercialis* (Neil and Holiday 1986). Baloun and Morse (1984b) have demonstrated that induction by excess K^+ ions could be attributed to depolarization of certain "externally accessible excitable cells" by this ion. Larvae of the prosobranch gastropod *Crepidula fornicata*are known to metamorphose in response to dopamine. In contrast, blocking of dopamine receptors might prevent excess potassium ions from stimulating metamorphosis as these receptors follow the excitatory depolarization pathways, which could be used in an antifouling strategy. Response of these larvae to putative dopamine antagonists varied with age and time of exposure. Chlorpromazineblocked the inductive action by excess potassium in the initial stages and D_1 antagonist R(+)-Sch-23309 and D_2 antagonist spiperone(SPIP) had similar effects.

The latent induction by chlorpromazine (an inhibitor of nitric oxide synthase) suggests that endogenous nitric oxide may play a natural role in inhibiting metamorphosis. In short, exposing larvae of *C. fornicata* to excess K^+ leads to a shutdown of NO synthesis via a dopaminergic pathway. Alternatively, the inductive action by chlorpromazine may be due to the activation of the endogenous cAMP concentrations, again acting downstream from the steps acted on directly by excess K^+ (Pechenik et al. 2002). Similarly, the mud snail *L. obsoleta* larvae metamorphose by the NO pathway. NO donors block pharmacologically induced metamorphosis in the snail. In contrast, nitric oxide synthase (NOS) allows competent larvae to become juveniles (Froggett and Leise 1999). Neuroanatomical studies revealed endogenous NO levels to increase throughout the planktonic stage and decrease with the onset of metamorphosis, suggesting the involvement of the nitroergic signalling pathway in mollusc embryos (Leise et al. 2001).

Comparative analysis of induction pathways in various invertebrate larvae indicates the involvement of biogenic amines (l-DOPA, norepinephrine, epinephrine, serotonin, lisuride, isoproterenol), which mediated neurotransmission to play a major role in settlement of many of the invertebrate species. Although these molecules many not have a commercial importance they do indicate that chemicals or synthetic analogues that specifically alter biogenic amine-mediated pathways may be promising antifouling compounds. A study on larval settlement inhibition by pharmaceutical compounds and toxins with known modes of action seems to be a viable solution. Laboratory bioassays using larval models, and field assays using these molecules individually or in combination, would yield substantial results.

5 Inhibition of Larval Settlement by Natural Products and Synthetic Analogues

Many marine organisms are able to resist fouling by producing metabolites that deter settlement of fouling organisms. This concept has been addressed repeatedly by academic researchers for incorporating active natural compounds into antifouling coatings. Experimental approaches have demonstrated two types of natural compounds: (1) compounds extracted from marine organisms and (2) compounds released into the surrounding water bathing these organisms. Research on natural product antifoulants aims to demonstrate the potency of an organic extract from organisms to inhibit settlement of a target species. This is followed by a bioassay-guided fractionation and identification of molecules. From the perspective of potency, the US Navy program on natural antifouling compounds has prescribed that extracts should be active at $<25\ \mu g\ mL^{-1}$ in static bioassays (Rittschof 2001). The therapeutic ratio of the effective median concentration to the 50% lethal concentration (EC_{50}/LC_{50}) should be larger than ten (Rittschof 2001).

Further, secondary metabolites are complex molecules and the active component involved in inhibition has often been difficult to identify. This poses a problem in creating structural analogues for incorporation into antifouling coatings. Experimental approaches to natural product-based antifoulants have rallied around sponges and octocorals, which are a rich source of secondary metabolites. Terpenoids, steroids, saponins, brominated tyrosine derivatives, sterols and fatty acid-related compounds have all been shown to possess antifouling activity. However, most of the studies have failed to elucidate the mode of action or mechanism of inhibition by a particular natural compound.

It is worthwhile confronting the problem by conducting assays that involve structure–activity relationships. Priorities should be to screen for natural compounds possessing the ability to interfere with identified neuronal signalling pathways in larvae, the synthesis and screening of synthetic analogues to these molecules and the enhancement of the activity of analogues through modification of their chemical structure. Among hundreds of antifouling compounds isolated, some examples of natural compounds belonging to different classes that show promising activity and whose structure–activity relationships have been worked out are outlined below.

5.1 Terpenoids

Examples of compounds whose structure–activity relationships have been studied are the isocyanoterpenoids, belonging to the terpenoid class identified by Kitano et al. (2004); Kalihiene (Okino et al. 1995); 10β-formamidokalihinol; 10β-formamidokalihino-A (Hirota et al. 1998); Kalihinol A (Yang et al. 2006); 15-formamidokalihinene (Okino et al. 1996); 10-isocyano-4-cadinene (Okino et al. 1996) and 10-formamide-4-cadinene (Nogata et al. 2003). Isocyanoterpenoids, like some of the natural terpenoids, are known to inhibit settlement of *B. amphitrite* larvae at concentrations of 0.1 μg mL^{-1}. A structure–activity analysis has shown that an isocyano group and a hydrophobic site at a suitable position are important in the expression of potent antifouling activity. Isocyanocylohexane synthesized with an ester function showed potent antifouling activity against barnacles.

Synthetic derivatives of a sesquterpene hydroquinone avarol from the sponge *Dysidea avara*and the corresponding quinine avarone obtained by oxidation were found to decrease settlement of *B. amphitrite* cyprids. Among these, avarol was the most toxic and caused mortality of larvae (Tsoukatou et al. 2007). Sesquiterpeneslike elatol, isolated from the red algae of the genus *Laurencia*, showed potent activity against *B. amphitrite* cyprids (Steinberg et al. 1998). Diterpenoids like dictyol, pachydictyol A and dictyodial isolated from brown algae *Dictyota* spp. are antifouling against bryozoan *Bugula neritinina* larvae (Schmitt et al. 1995). Higher concentrations (5 μg mL^{-1}) were toxic to the larvae. Pukalide, another diterepenoid isolated from the gorgonian *Leptogorgia virgulata*, and renillafoulin A, isolated from the sea pen *Renilla reniformis* (Rittschof et al. 1986), inhibited barnacle settlement.

11β,12β-Epoxypukalide, a cembranolide diterpene isolated from the sea fan *Phyllogorgia dilatata*,was able to prevent mussel byssus attachment (Epifanio et al. 2006). Epoxypukalide isolated from a different species of gorgonian, namely *Leptogorgia virgulata,* also inhibited the settlement of barnacle *B. amphitrite* (Rittschof et al. 1985). Even though the source of these molecules differed (isolated from different gorgonian species) the molecule (epoxypukalide) exhibited antifouling activity. However, both studies were carried out independently and on two different fouling species. Compounds that posses such multifunctional roles should be taken up for subsequent field trials for commercialization. Labdane, a diterpene isolated from the pulmonate *Trimusculus reticulates*, inhibited the settlement of the tubeworm *Phragmatopoma californica*.

5.2 *Alkaloids*

Alkaloids have been shown to influence the central and peripheral nervous system; nicotinic receptors and some pyridyl alkaloids also activate certain chemoreceptor neurons. Anabaseine [2-(3-pyridyl)-3,4,5,6-tetrahydropyridyl] and 2,3′-bipyridyl (2,3-BP) are two nemertine alkaloids that are known to influence crustaceans. Structural isomers of bipyridyl (2,3′-BP) tested showed varying degrees of antifouling activity against *B. amphitrite* larvae (Kem et al. 2003).

5.2.1 3-Alkylpyridinium Compounds

Another group of molecules that have been successfully demonstrated to have antifouling activity are the polymeric 3-alkylpyridinium salts (poly-APS) isolated from the Mediterranean sponge *Reniera sarai*. The structure of the large poly-APS isolated constituted of *N*-butyl(3-butylpyridinium) repeating subunits. This molecule acts as an aetylcholinesterase inhibitor (Sepcic et al. 1998). The compounds were toxic to the *D*-shaped veliger larvae of the mussel *Mytilus galloprovincialis* (Faimali et al. 2003). The structure–activity relationship of poly-APS has revealed a dual mechanism of action, viz: a detergent-like property exhibited by poly-APS that means it could serve as a surfactant for cell membranes, and AChE inhibitory activity that might be involved in the antifouling mechanism. Interestingly the compound was not found to be toxic to cyprids, but inhibited settlement (Sepcic and Turk 2006). The antifouling activity of natural poly-APS was far more effective than the synthetic analogues of polymeric 3-alkylpyridinium salts developed (Faimali et al. 2005). The low toxicity and high solubility and stability of poly-APS make these compounds promising candidates for incorporation into antifouling paints.

5.3 *Peptide Analogues*

Tegtmeyer and Rittschof (1989) demonstrated the involvement of synthetic analogues of barnacle pheromone in settlement and metamorphosis of *B. amphitrite* larvae as early as 1989. Compared to organic compounds, cyclopeptides isolated from the sponge *G. barrette*, have also shown inhibition of settlement of the barnacle *B. improvisus* and the blue mussel *Mytilus edulis* in the field. Barettin, cyclo [(6-bromo-8-en-tryptophan) arginine], and 8,9-dihydrobarettin constituted the class of cyclopeptidestested. A significant aspect of this study by Sjogren et al. (2004) is that the activities of these compounds are not lost on incorporation into self-polishing marine paints. Among various synthetic analogues of barettin tested, benzo(g)dipodazines inhibited settlement but were not toxic to cyprids (Sjogren et al. 2006). Synthetic analogues of l-arginine, l-tryptophan, 5-bromo-d,l-tryptophan,

6-bromo-d,l-tryptophan and 6-fluoro-d,l-tryptophan were tested for structure–activity relationships and did not shown any effect against *B. improvisus* cyprid larvae (Sjogren and Bohlin 2004). As in the case of poly-APS, settlement inhibition was reversible when incubated in normal seawater.

5.4 Sterols, Fatty Acids and Tyrosine Derivatives

Lactones, furansand furanones are some examples of compounds that have shown multiple antifouling activities. Lactones prevented barnacle settlement through the signal transduction pathway. Furanones exhibited maximum activity among the polyketides isolated from the red algae *Delisea pulchra* (de Nys et al. 1995; Steinberg et al. 2001). Among the steroids, epidioxysteroids isolated from the marine sponge *Acanthella cavernosa* inhibited settlement and prolonged the swimming time of cyprids (Tomono et al. 1998). 1-*O*-palmityl-*sn*-glycero-3-phosphocholine (lyso-PAF) extracted from the marine sponge *Crella incrustans* inhibited *B. neritina* larvae (Butler et al. 1996). Ceratinamides A and B and psammaplysin A derived from the sponge *Psamplysia purpurea* were effective against barnacle cyprids (Tsukamoto et al. 1996). Moloka'iamine and its analogues also exhibited settlement inhibitory activity in barnacles, with less toxicity to the cyprids (Schoenfeld and Ganem 1998; Schoenfeld et al. 2002).

Oxime substituents, a form of bromotyrosine derivative (bastadin-3, hemibastadin-1, aplysamine-2 and psammaplin A), inhibited settlement in the barnacle *B. improvisus*. The synthetic analogue of hemibastadin-1, debromohemibastadin-1, inhibited cyprid settlement more effectively than the natural compound hemibastadin-1. The mechanism of action of the compound inhibiting settlement is being studied (Ortlepp et al. 2007). Heterocyclic compounds like pseudoceratidine from *P. purpurea* and mauritiamine, an oroidin dimer from the sponge *Agelas mauritiana,* showed antifouling activity (Tsukamoto et al. 1996). Among all these, 2,5,6-tribromo-1-methylgramine isolated from the bryozoan *Z. pellucidum* showed a tenfold increase in antifouling activity compared to tributiyl tin oxide (Kon-ya et al. 1994).

Over 500 analogues have been synthesized and tested for antifouling activity. A structure–activity relationship study with 155 indole derivatives led to the discovery of the non-toxic antifoulants 5–6-dichlorogramine, 5-chloro-2-methylgramineand 5,6-dichloro-1-methylgramine(DCMG). These were incorporated into silicone paint matrices, which remained barnacle-free for a period of 1.5 years, which is comparable to the commercial counterpart BIOX (Kawamata et al. 2006). An advantage of this compound is that leaching of DCMG could be controlled by the addition of an acrylic acid–styrene copolymer(ASP) to improve the performance of the silicone coating(Kawamata et al. 2006).

Studies in search of non-toxic antifoulants over the last decade have yielded more than 100 compounds and over 500 synthetic analogues being tested and qualified in laboratory bioassays against target species. Another paradigm in this area is that most of the compounds were identified using barnacle larval assays. However, their activity on other invertebrate larval groups is still unknown. Only five of these compounds have been tested in the field. These compounds need to be evaluated under dynamic field conditions before ruling out their potential.

A problem encountered by researchers in this effort is incorporation of these compounds into commercial coating formulations. In some cases simple matrices like rosin, hydrogels and phytagels have been used to qualify extracts in the field. But these are far away from commercial coatings, which come as a package containing base, tie and topcoat systems, where the structural integrity of natural compounds may be lost on incorporation into these coatings. These coatings further vary in their physical and chemical properties. Compatibility of the compounds with the coating matrix and interaction with additives during polymerization further compound the problem and is difficult to predict the outcome of such procedures. Academic–industry–research collaborations suggested by Rittschof (2001), (see Lewandowski and Beyenal 2008) seem to be a practical way to solve the problem.

A few interesting leads have been obtained with compounds like isocyanoterpenoids and 3-alkylpyridinum salts (poly-APS), which inhibit acetyl cholinesterase activity, that need to be pursued in field assays. Similarly, serotonin uptake blockers and a-adrenergic antagonists, like medetomidine, also need to be qualified in field assays. Partial success has been achieved with the compound furanone (halogenated furanones from the Australian seaweed *Delisea pulchra*) incorporated into polymer matrices and qualified in field trials (deNys et al. 1993; de Nys and Steinberg 2002). However, commercialization of this product involves issues like registration and environmental clearances. To date, more than 100 compounds possessing antifouling activity have been isolated and characterized, and their effective inhibitory concentrations have been listed by Rittschof (2001) and Fusetani (2004). However, these compounds have not yet been realized into antifouling coatings.

Several compounds have shown promising activity against one species of foulant or another but the successful realization of these compounds into antifouling coatings is yet to happen. This is because the identified compounds are specific in action, inhibiting settlement of a particular larval form (narrow spectrum) whereas in the field a multitude of organisms are found to settle (broad spectrum). A classical example of this is the study by Pereira et al. (2002), who adopted an alternative approach to screening and qualifying crude extracts through direct field assays. Extracts of the gorgonian *Phyllogorgia dilatata* and sponges *Aplysina fluva*and *Mycale microsigmatosa* were tested. *A. fluva* extracts failed whereas *M. microsigmatosa* and *P. dilatata* extracts inhibited only barnacle settlement. These results once again point to the narrow spectrum of activity exhibited by natural products. A solution to this lies in conducting bioassays with larvae of the major fouling organism to qualify and rank natural compounds. Alternatively, compounds that act downstream from the initial sensory transduction, i.e., Ca^{2+} and K^{+} channel blockers, perhaps are broad spectrum and more likely to have less organismic specificity.

Other problems concerned with natural products are the effective concentration for incorporation into a coating matrix and the leaching rate of the compounds in the environment. For realization of natural product antifoulants, research should aim at determining the concentration of secondary metabolites experienced by settlers in the field (Dobretsov et al. 2006; see Nedved and Hadfield 2008) to arrive at the leaching rate of compounds from paint matrices. A more practical solution to the problem would be to first identify potential candidate compounds (compounds effective against different larval forms) and to incorporate them in a single coating matrix in the field for testing natural products as antifoulants.

6 Current Status of Antifouling Methods and the Need for Alternative Strategies

As a consequence of the 2001 International Maritime Organization restrictions on the use of tributyltin (TBT), replacement coatings have to be environmentally acceptable as well as maintain a certain lifetime. Tin-free self-polishing copolymer (SPC) and foul release coatings are currently in use commercially. For a detailed chronological description of developments in the coatings industry refer to the review by Yebra et al. (2004). In the aftermath of restrictions on TBT, metallic species like copper and zinc have again found increasing use and are delivered through a self-polishing mechanism. The SPCs work on the principle of hydrolysis and erosion. However, recently copper has also been under scrutiny. A copper concentration as low as 5 $\mu g\ L^{-1}$ can be lethal to invertebrates and the USEPA has prescribed a stipulated limit of 1,000 $\mu g\ L^{-1}$ in drinking water (Chambers et al. 2006). With copper being reviewed and with the increased tolerance exhibited by some macrophytes like *Enteromorpha* sp. to this metal, alternatives are being sought. Booster biocides (pesticides and herbicides) have been incorporated into coatings to increase the life and functionality of copper-based antifouling systems. Among the various booster biocides used, Irgarol 1,051 and Diuron are facing regulation by the UK Health and Safety Executive (Chesworth et al. 2004; Lambert et al. 2006). The effectiveness of copper-based coatings depends on the ability of the coatings to continuously leach booster biocides. Accumulation and persistence of these biocides in sediments is being monitored and a worldwide environmental effect (Konstantinou and Albanis 2004) of key booster biocides used in antifouling coatings is also under scrutiny. Currently, booster biocides are offering an interim solution to the antifouling problems, but they face the threat of being phased out in the near future due to their toxic effects on the environment.

On the other hand, foul release coatings have offered a promising alternative for the industry. Two types of foul releasecoatings are in use, namely those based on fluoropolymersand those based on silicones. These coatings are in use currently and have been listed by Swain (1988) and Yebra et al. (2004). The coatings are effective for fast-moving vessels and at high speeds (10–20 knots) but under static conditions they do become fouled. The fouling layer is sloughed off once the vessel is in motion. This is because of the poor adhesion strengthof organisms on these coatings. These non-stick coatings aid removal of fouling through shear and tensile stresses as well as their own weight by lowering the thermodynamic work of adhesion. A combination of the critical surface energy (19–24 mN m^{-1}) and low elastic modulus allows the interfacial joint between the animal's adhesive and the coating to fracture and fail (see Thomas et al. 2008). These coatings work on the principle of cohesive failure of the bioadhesive to bind to the substratum. Comparison of the adhesive strength of common foulants on RTV silicones showed that oysters (*Crassostrea virginica*) and tubeworms (*Hydroides dianthus*) had higher adhesion strength than barnacles (*Balanus eburneus*), (Kavanagh et al. 2001).

Continuous improvements in the surface properties of silicone coatings have revealed critical surface tensions of 20–30 mN m^{-1} to easily release more biofouling than other materials at the same critical surface tension. Oils added to these coatings selectively further diminish the attachment strength of organisms without affecting the critical surface tension (Meyer et al. 2006). Cyprids showed an inverted behaviour, preventing adhesion on coatings containing silicone oil at concentrations $\geq$ 5% (Afsar et al. 2003). A comparison of vessel speeds on foul release properties showed that barnacles were removed at 7 knots and that 18 knots was required to remove weeds, whereas a speed of 30 knots was ineffective in removing slime films (Ryle 1999). Decreasing the surface energy of materials renders them brittle, which is another problem to be overcome. These disadvantages plus high cost and poor mechanical properties are yet to be solved. Efforts are underway to improve these parameters but it remains to be seen whether these coatings will move into the future or will be replaced by more efficient competing technologies.

7 Status of Alternative Antifouling Methods

Challenges related to the development of alternative antifouling coatings depend on the chemical nature of the compounds, developing polymer systems compatible with the additives as well as with anticorrosive undercoats with appropriate mechanical and application properties. Several alternatives are being actively researched. These involve natural surface microtopographies, nanostructured surfaces, microbial metabolites, synthetic polymers and biomimetics. Increased attention is being directed to understanding the induction process and downstream processing of the signal by larvae. Attention is also drawn towards natural antifouling strategies employed by living marine organisms and mimicking their effects.

Since bioadhesion and surface wettability are influenced by microscale topography, investigators have mimicked the surfaces of marine organisms (see Thomas et al. 2008). One such proposition for study is the skin of pilot whales, which is free of fouling. Studies have revealed a hydrated jelly nanorough surface characterized by patterns of nanoridges with low pore sizes (Baum et al. 2003). Similarly, the skins of porpoise and killer whales posses a surface rich in glycoproteinaceous material with a low surface energy. The antifouling property of these surfaces may be due to the high shear flow of water across the surface. Alternatively, the whales may constantly secrete hydrolytic enzymes that may also help maintain the surfaces clean. The effects of each of these individual components are being investigated.

Biomimics is another area of emerging interest in antifouling research. Novel antifouling coatings were designed to mimic the placoid scales on shark skin surface (Sharklet AF) and were effective in reducing fouling by the algae *Ulva* (Carman et al. 2006). Engineered surface topographies of polydimethyl siloxanes (feature height/feature width) showed that cyprid settlement was inhibited at high aspect ratios. A barnacle-specific coating, Sharklet AF, which has a defined topography (40 μm feature height with an aspect ratio of two), reduced cyprid settlement by 97% (Schumacher et al. 2007). The presence of silicone oilsdid not reduce the settlement of zoospores of the green fouling alga *Ulva*. However, the presence of oils with altered topographic features reduced settlement of algae. The alga settled more in channels than pillars (Hoipkemeier-Wilson et al. 2004).

The low surface energy of the polymer polyethylene glycol (PEG) has been the target of recent investigations. These polymers are being manipulated for antifouling properties. Maintaining the mechanical properties and altering the surface characteristics by using side group-modified polystyrene-based surface-active block co-polymers (SABC) has been shown to inhibit zoospore settlement of the green alga *Enteromorpha* (Youngblood et al. 2003). Modification of biomaterial (l-DOPA) with poly(ethylene glycol) and subsequent coating on surfaces resulted in surfaces that resisted cell attachment, exhibiting antifouling properties (Dalsin and Phillip 2005). Conjugation of the methoxy-terminated poly(ethylene glycol) polymer to the adhesive amino acid l-DOPA and coating on titanium surfaces gave surfaces that performed well compared to glass, uncoated titanium and a polydimethylsilicone elastomer. Reduced settlement of diatoms *N. perminuta* and zoospores of the algae *U. linza* has shown that the biogenic amine is an effective antifouling agent (Statz et al. 2006). Both of these studies have given potential leads that need investigation in the field.

Surface properties (both physical and chemical) of mollusc shells and invertebrates are now being studied (Scardino et al. 2003; Scardino and de Nys 2004). Surface microtopography is found to influence settlement. A homogenously ridged surface (ridges with uniform distance of 1–2 μm with a mean depth of 1.5 μm) of *Mytilus galloprovincialis* shell prevented settlement compared to the heterogeneous surface of *Pinctada imbricate* with repeating structural patterns (Scardino et al. 2003). Scardino and de Nys (2004) demonstrated that the mussel *M. galloprovincialis* was fouling-free, whereas high-resolution biomimics of shells and sanded moulds fouled after 6–8 weeks. Results point to the fact that apart from surface microtopography, other factors may be involved that contribute to fouling deterrence. Species-specific fouling patterns with differential fouling between shell regions were observed with the pearl oyster species *Pinctada fucata*, *Pteria penguin* and *Pteria chinensis*. High resolution resin replicates of the surface of the blue mussel *Mytilus edulis* and the crab *Cancer pagurus*with a surface microtopography of <500 μm were effective in inhibiting settlement (Bers and Wahl 2004). Further, larval settlement of the barnacle *Semibalanus balanoides* was reduced on isotropic resin surfaces prepared from the molluscs *M. edulis* and *Perna perna* compared to anisotropic surfaces (Bers et al. 2006). Periostracum of the mussels *M. edulis* and *Perna perna* were found to possess a generic anti-settlement property against cyprids of *B. balanoides* (Bers et al. 2006).

Involvement of surface-bound compounds in the antifouling property of shells was demonstrated by Bers and Wahl (2004), where moderately polar and non-polar fractions showed antifouling activity against *B. amphitrite* cyprids and the marine bacteria *Cobetia marina* and *Marinobacter hydrocarbonoclasticus*. Moderate wettability of surface structures of the sea stars *Linckia laevigata*, *Fronia indica*, *Cryptasterina pentagona* and *Archaster typicus* are not fouled in nature (Guenther and De Nys 2006). Mixed responses to fouling inhibition by shells were observed in these studies. These studies have demonstrated that the micro-/nanotopography of surfaces play a role in larval settlement. Micro- or nanopatterning of surfaces through nanocoatings may provide interesting leads for antifouling.

A novel concept being researched is the use of enzymes as antifoulants. Two approaches are being followed: (1) direct antifouling where enzymes themselves are used as antifoulants and (2) indirect antifouling where certain enzymes are used to release active components from paint matrixes. An example is the hydrolyzing of adhesive polymers using commercial enzymes for biofouling prevention. This showed that alcalase, a group of serine-proteases, exhibited antifouling activity against the spores of the green alga *Ulva linza* and the barnacle *B. amphitrite*. Results showed that effective hydrolysis could not be achieved once curing of the juvenile barnacle cement commenced (Pettitt et al. 2004). However, problems that plague natural product antifoulants seem to hold good for these compounds as no efficient broad-spectrum antifouling coating based on a single or multiple enzymes has yet been achieved. Research inputs have warranted the creation of new legislative issues, such as which part of the enzyme system should be considered as a biocide for product registration purposes (Olsen et al. 2007).

Marine bacteria could provide a source of biologically active metabolites for the antifouling industry (Steinberg et al. 2001). The major advantage of bacterial metabolites is the production of large water-borne (>100 kDa) polar compounds (Dobretsov and Qian 2002; Dobretsov et al. 2004; Harder et al. 2004), compared to secondary metabolites produced by macroorganisms with low solubility and activity (Steinberg et al. 2002). Many studies have aimed at characterizing antifouling compounds produced by bacteria isolated from epibionts like algae (Dobretsov and Qian 2002; Dobretsov et al. 2004; Harder et al. 2004; Lee and Qian 2004), sponges (Holmstrom et al. 1992; Kon-ya et al. 1995), ascidians (Olguin-Uribe et al. 1997; Zapata et al. 2007) and seaweeds (Armstrong et al. 2000). Ubiquinoneis one example of a compound produced by a species of *Alteromonas*, which colonizes the surface of the sponge *Halichondria okadali* (Kon-ya et al. 1995). Bacterial extracts were tested with acrylic base paint resin. These formulations were found to inhibit marine fouling bacterial adhesion (Armstrong et al. 2000). As a further step, Holmstrom et al. (2002) produced inhibitory "living paints"by incorporating live cells of *Pseudoalteromonas tunicata* into hydrogels. Partial success was achieved, as the paints were inhibitory for a period of 15 days. Investigations on the use of microbial natural products are in their infancy compared to the large potential of microbes to produce diverse compounds. The problems relate to compound supply; simple molecules that allow easy structural modifications are advantageous with microorganisms whose potential has not been fully utilized.

Another innovative approach for fouling control is creation of physical analogues of natural processes (Cowling et al. 2000). The logic behind this approach is that active substances that alter the characteristics, properties and behaviour of materials may have potential in antifouling surfaces. Compounds like benzalkonium chloride (BCI), propyl 4-hydroxygenzoate and 2,4-hexadienoic acid were incorporated into the hydrophilic carrier 2-hydroxyethylmethacrylate gels. Mixed results were obtained where polymers containing propyl 4-hydroxygenzoate and 2,4-hexadienoic acid were found to accumulate microfouling within a period of 1 month, whereas benzylalkonium chloride-incorporated polymers were fouling-free for up to 5 months.

8 Concluding Remarks

From the studies reviewed here, two coating systems: self polishing copoly-mers (SPC) containing a biocide (copper alone or with boosters) and foul release coatings (polydimethyl siloxanes) working on the principle of low surface energy are currently in use, even though they are not as effective as tributyltin paints. Among these, the boosters containing SPC coatings face a threat of fresh regulations by regional pesticide authorities, restricting their use due to the toxic nature of their ingredients. Foul release coatings have been efficient on vessels operating at high speeds. Versions of these coatings for small craft are currently being released into the market. The disadvantage of these coatings is that they do not resist fouling when the vessel is alongside and serve as a medium for translocation of alien species. The long-term fouling behaviour of these coatings under different environmental conditions needs to be documented for assessing the success of these coatings.

The biomimetic approach adopts the natural defence mechanisms of marine organisms to solve the problem of biofouling. However, no single natural compound or a particular microtextured or micropatterned surface has been shown to exhibit a broad spectrum of activity as the settling preferences of marine invertebrates are diverse. Studies replicating natural antifouling mechanisms, wherein both the physical and chemical defences exhibited by organisms are mimicked together, would offer valuable inputs to antifouling research. Currently, polydimethyl siloxanes have been the focus of such investigations for incorporation of natural products, synthetic analogues, boosters, microtexturing, nanopatterning and incorporation of nanoparticles to improve their efficiency. Fluoropolymers are being researched as alternatives, and additives involving incorporation of natural product analogues and physical modification of the surfaces using nanomaterials are also being pursued. It is to be seen whether these coatings remain or are replaced by more efficient technologies.

References

Afsar A, De Nys R, Steinberg P (2003) The effects of foul-release coatings on the settlement and behaviour of cyprid larvae of the barnacle *Balanus amphitrite,* Darwin. *Biofouling* 19:105–110

Aidley DJ (1978) The physiology of excitable cells. Cambridge University Press, Cambridge, UK

Anderson MJ, Underwood AJ (1994) Effects of substratum on the recruitment and development of an intertidal estuarine fouling assemblage. *J Exp Mar Biol Ecol* 184:217–236

Armstrong E, Boyd KG, Burgess JG (2000) Prevention of marine biofouling using natural compounds from marine organisms. *Biotechnol Annu Rev* 6:221–241

Baloun AJ, Morse DE (1984a) Modulation by unsaturated fatty acids of norepinephrine and adenosine induced formation of cyclic AMP in brain slices. *J Neurochem* 42:192–197

Baloun AJ, Morse DE (1984b) Ionic control of settlement and metamorphosis in larval *Haliotis rufescens* (Gastropoda). *Biol Bull* 167:124–138

Baum C, Simon F, Meyer W, Fleischer I, Siebers D, Kacza J, Seeger J (2003) Surface properties of the skin of the pilot Whale *Globicephala melas*. *Biofouling* 19:181–186

Beckmann M, Harder T, Qian PY (1999) Induction of larval attachment and metamorphosis in the serpulid polychaete Hydroides elegans by dissolved free amino acids: mode of action in laboratory bioassays. *Mar Ecol Prog Ser* 190:167–178

Bers A, Wahl M (2004) The influence of natural surface microtopographies on fouling. *Biofouling* 20(1):43–51

Bers AV, Prendergast GS, Zurn CM, Hansson L, Head RM, Thomason JC (2006) A comparative study of the anti-settlement properties of *Mytillid* shells. *Biol Lett* 2:88–91

Bishop CD, Brandhorst BP (2001) NO/cGMP signaling and HSP90 activity represses metamorphosis in the sea urchin *Lytechinus pictus*. *Biol Bull* 201:394–404

Bonar DB (1976) Molluscan metamorphosis: a study in tissue transformation. *Am Zool* 16:573–591

Bonar DB, Coon SL, Walch M, Weiner RM, Fitt W (1990) Control of oyster settlement and metamorphosis by endogenous and exogenous chemical cues. *Bull Mar Sci* 46(2):484–498

Boudreau B, Bourget E, Simard Y (1990) Benthic invertebrate larval response to substrate characteristics at settlement: shelter preferences of the American lobster *Homarus americanus*. *Mar Biol* 106:191–198

Bryan PJ, Qian PY, Kreider JL, Chia FS (1997) Induction of larval settlement and metamorphosis by pharmacological and conspecifics associated compounds in the serpulid polychaete *Hydroides elegans*. *Mar Ecol Prog Ser* 146:81–90

Burke RD (1983) The induction of marine invertebrate larvae: stimulus and response. *Can J Zool* 61:1701–1719

Butler AJ, van Altena IA, Dunne SJ (1996) Antifouling activity of lyso-platelet-activating factor extracted from Australian sponge *Crella incrustans*. *J Chem Ecol* 22:2041

Butman CA, Grassle JP, Webb CM (1988) Substrate choices made by marine larvae settling in still water and in a flume flow. *Nature* 333:771–773

Carman ML, Estes TG, Feinberg AW, Schumacher JF, Wilkerson W, Wilson LH, Callow ME, Callow JA, Brennan AB (2006) Engineered antifouling microtopographies-correlating wettability with cell attachment. *Biofouling* 22(1–2):11–21

Carpizo-Ituarte E, Hadfield MG (2003) Transcription and translation inhibitors permit metamorphosis up to radiole formation in the Serpulid polychaete *Hydroides elegans*. *Haswell Biol Bull* 204:114–125

Chambers LD, Stokes KR, Walsh FC, Wood RJK (2006) Modern approaches to marine antifouling coatings. *Surf Coat Technol* 201:3642–3652

Chesworth JC, Donkin ME, Brown T (2004) The interactive effects of the antifouling herbicides Irgarol 1051 and Diuron on the seagrass *Zostera marina* (L). *Aquat Toxicol* 66:293–305

Clare AS, Nott JA (1994) Scanning electron microscopy of the fourth antennular segment of *Balanus amphitrite amphitrite*. *J Mar Biol Assoc UK* 74:967–970

Clare AS, Freet RK, McClary Jr M (1994) On the antennular secretion of the cyprid of *Balanus amphitrite amphitrite* and its role as a settlement pheromone. *J Mar Biol Assoc UK* 74:243–250

Clare AS, Thomas RF, Rittschof D (1995) Evidence for the involvement of cyclic AMP in the pheromonal modulation of barnacle settlement. *J Exp Biol* 198:655–664

Connell SD (1999) Effects of surface orientation on the cover of Epibiota. *Biofouling* 14(3):219–226

Cooksey KE, Wigglesworth-Cooksey B, Long RA (2008) A strategy to pursue in selecting a natural antifoulant: a perspective. Springer Ser Biofilms. doi: 10.1007/7142_2008_11

Coon SL, Bonar DB, Weiner RM (1985) Induction of settlement and metamorphosis of the pacific oyster, *Crassostrea gigas* (Thunberg) by l-DOPA and catecholamines. *J Exp Mar Biol Ecol* 94:211–221

Cooper K (1982) A model to explain the induction of settlement and metamorphosis of eyed pediveligers of the blue mussel *Mytilus edulis* L by chemical and tactile cues. *J Shellfish Res* 2:117

Couper JM, Leise EM (1996) Serotonin injections induce metamorphosis in larvae of the Gastropod mollusk *Ilyanassa obsolete*. *Biol Bull* 191:178–186

Cowling MJ, Hodgkiess T, Parr ACS, Smith MJ, Marrs SJ (2000) An alternative approach to antifouling based on analogues of natural processes. *Sci Total Environ* 258:129–137

Crisp DJ (1974) Factors influencing the settlement of marine invertebrate larvae. In: Chemoreception in marine organisms. Academic, London, pp. 177–277

Dahlstrom M, Martensson LGE, Jonsson PR, Arnebrant T, Elwing H (2000) Surface active adrenoceptor compounds prevent the settlement of cyprid larvae of *Balanus improvisus*. *Biofouling* 16(2–4):191–203

Dahlstrom M, Jonsson H, Jonsson PR, Elwing H (2004) Surface wettability as a determinant in the settlement of the barnacle *Balanus improvisus* (Darwin). *J Exp Mar Biol Ecol* 305:223–232

Dahlstrom M, Lindgren F, Berntsson K, Sjogren M, Martensson LG, Jonsson PR, Elwing H (2005) Evidence for different pharmacological targets fro imidazoline compounds inhibiting settlement of the barnacle *Balanus improvisus*. *J Exp Zool* 303(7):551–562

Dahms HU, Jin T, Qian PY (2004) Adrenoreceptor compounds prevent the settlement of marine invertebrate larvae: *Balanus amphitrite* (Cirripedia), *Bugula neritina* (Bryozoa) and *Hydroides elegans* (Polychaeta). *Biofouling* 20(6):313–323

Dalsin JL, Messersmith PB (2005) Bioinspired antifouling polymers. *Mater Today* 8:38–46

de Nys R, Steinberg PD (2002) Linking marine biology and biotechnology. *Curr Opin Biotechnol* 13(3):244–248

de Nys R, Wright AD, Koning GM, Sticher O (1993) New halogenated furanones from the marine alga *Delisea pulchara* (fimbriata). *Tetrahedron* 49:11213–11220

de Nys R, Steinberg PD, Willemsen P, Dworjanyn SA, Gabelish CB, King RJ (1995) Broad spectrum effects of secondary metabolites from the red alga Delisea pulchra in antifouling assays. *Biofouling* 8:259

Dobretsov S, Qian PY (2002) Effect of bacteria associated with the green alga *Ulva reticulate* on marine micro- and macrofouling. *Biofouling* 18:217–228

Dobretsov S, Qian PY (2003) Pharmacological Induction of larval settlement and Metamorphosis in the blue mussel *Mytilus edulis* L. *Biofouling* 19(1):57–63

Dobretsov S, Dahms H, Qian PY (2004) Antilarval and antimicrobial activity of waterborne metabolites of the sponge Callyspongia (*Euplacella pulvinata*): evidence of allelopathy. *Mar Ecol Prog Series* 271:133–146

Dobretsov S, Dahms HU, Qian PY (2006) Inhibition of biofouling by marine microorganisms and their metabolites. *Biofouling* 22(1–2):43–54

Epifanio RA, da Gama BAP, Pereira RC (2006) 11b-12b-Epoxypukalide as the antifouling agent from the Brazilian endemic sea fan *Phyllogorgia dilatata* Esper (Octocorallia, Gorgoniidae). *Biochem Syst Ecol* 34:446–448

Faimali M, Falugi G, Gallus I, Piazza V, Tagliafierro G (2003a) Involvement of acetyl choline in settlement of *Balanus amphitrite*. *Biofouling* 19:213–220

Faimali M, Sepcic K, Turk T, Geraci S (2003b) Non-toxic antifouling activity of polymeric 3-alkylpyridinium salts from the mediterranean sponge *Reniera sarai* (Pulitzer-Finali). *Biofouling* 19 (1):47–56

Faimali M, Garaventa F, Mancini I, Sicurelli A, Guella G, Piazza V, Greco G (2005) Antisettlement activity of synthetic analogues of polymeric 3-alkylpyridinium salts isolated from the sponge *Reniera sarai. Biofouling* 21(1):49–57

Forde SE, Raimondi PT (2004) An experimental test of the effects of variation in recruitment intensity on intertidal community composition. *J Exp Mar Biol Ecol* 301:1–14

Froggett SJ, Leise EM (1999) Metamorphosis in the marine snail *Ilyanassa obsoleta,* yes or no? *Biol Bull* 196:57–62

Fusetani N (2004) Biofouling and antifouling. *Nat Prod Rep* 21:94–104

Galsby TM (1999) Effects of shading on subtidal epibiotic assemblages. *J Exp Mar Biol Ecol* 234:275–290

Galsby TM (2000) Surface composition and orientation interact to affect subtidal epibiota. *J Exp Mar Biol Ecol* 248:177–190

Guenther J, DeNys R (2006) Differential community development of fouling species on the pearl oysters *Pinctada fucata*, Pteria penguin and *Pteria chinensis* (Bivalvia, Pteriidae). *Biofouling* 22(3–4):163–171

Hadfield MG (1978) Metamorphosis in marine Molluscan larvae: an analysis of stimulus and response. In: ChiaFS, Rice ME (eds.) Settlement and metamorphosis of marine invertebrate larvae. Elsevier, New York, pp. 165–175

Hadfield MG (1984) Settlement requirements of Molluscan larvae new data on chemical and genetic roles. *Aquaculture* 39:283–298

Harder T (2008) Marine epibiosis – concepts, ecological consequences and host defense. Springer Ser Biofilms. doi: 10.1007/7142_2008_16

Harder T, Qian PY (1999) Induction of larval attachment and metamorphosis in the serpulid polychaete *Hydroides elegans* by dissolved free amino acids: isolation and identification. *Mar Ecol Prog Ser* 179:259–271

Harder T, Dobretsov S, Qian PY (2004) Waterborne polar macromolecules act as algal antifoulants in the seaweed *Ulva reticulata. Mar Ecol Prog Ser* 274:133–141

Hirata KY, Hadfield MG (1986) The role of choline in metamorphic induction of *Phestilla* (Gastropoda: Nudibranchia). *Comp Biochem Physiol* 84C:15–21

Hirota H, Okino T, Yoshimura E, Fusetani N (1998) Five new antifouling sesquiterpenes from two marine sponges of the genus *Axinyssa* and the Nudibranch *Phyllidia pustulosa. Tetrahedron* 54:13971–13980

Hodgkin AL, Horowicz P (1959) The influence of potassium and chloride ions on the membrane potential of single muscle fibres. *J Physiol* 148:127–160

Hoeg J, Hosfeld B, Gram-Jensen P (1988) TEM studies on the lattice organs of cirripede cypris larvae (Crustacea Thecostraca, Cirripedia). *Zoomorphology* 118:195–205

Hoipkemeier-Wilson L, Schumacher JF, Carman ML, Gibson AL, Feinberg AW, Callow ME, Finlay JA, Callow JA, Brennan AB (2004) Antifouling potential of lubricious, micro-engineered, PDMS elastomers against zoospores of the green fouling alga *Ulva* (Enteromorpha). *Biofouling* 20(1):53–63

Holiday JE (1996) Effects of surface orientation and slurry coating on settlement of Sydney rock *Saccostrea commercialis* oysters on PVC slats in hatchery. *Aquaculture Eng* 15(3):159–168

Holm ER, Nedved BT, Carpizo-Ituarte E, Hadfield MG (1998) Metamorphic signal transduction in *Hydroides elegans* (Polychaeta: Serpulidae) is not mediated by a G protein. *Biol Bull* 195:21–29

Holmstrom C, Rittschof D, Kjelleberg S (1992) Inhibition of settlement by larvae of *Balanus amphitrite* and *Ciona intestinalis* by a surface colonizing marine bacterium. Appl Env Microbiol 2111–2115

Holmstrom C, Egan S, Franks A, McCloy S, Kjelleberg S (2002) Antifouling activities expressed by marine surface associated *Pseudoalteromonas* species. *FEMS Microbiol Ecol* 41:47–58

Ilan M, Jensen RA, Morse DE (1993) Calcium control of metamorphosis in polycheate larvae. *J Exp Zool* 267:423–430

Kaissling KE, Thorson J (1980) Insect olfactory sensilla: structural, chemical and electrical aspects of the functional organizations. In: Satelle DB, Hall LM, Hildebrand JG (eds.) Receptors for neurotransmitters, hormones and pheromones in insects. Elsevier/North Holland Biomedical, New York, pp. 261–282

Kavanagh CJ, Schultz MP, Swain GW, Stein J, Truby K, Darkangelo-Wood C (2001) Variation in adhesion strength of *Balanus eburneus*, *Crassostrea virginica* and *Hydroides dianthus* to fouling-release coatings. *Biofouling* 17:155–167

Kawamata M, Kon-ya K, Miki W (2006) 5,6-Dichloro-1-methylgramine, a non-toxic antifoulant derived from a marine natural product. *Prog Mol Subcell Biol* 42:125–139

Kaye HR, Reiswig HM (1991) Sexual reproduction in four Caribbean commercial sponges. III. Larval behavior, settlement and metamorphosis. *J Invert Reprod* 19:25–35

Kem WR, Soti F, Rittschof D (2003) Inhibition of barnacle larval settlement and crustacean toxicity of some hoplonemertine pyridyl alkaloids. *Biomol Eng* 20:355–361

Kitano Y, Nogata Y, Shinshima K, Yoshimura E, Chiba K, Tada M, Sakaguchi I (2004) Synthesis and anti-barnacle activities of novel isocyanocylohexane compounds containing an ester or ether functional group. *Biofouling* 20(2):93–100

Konstantinou IK, Albanis TA (2004) Worldwide occurrence and effects of antifouling paint booster biocides in the aquatic environment review. *Environ Int* 30:235–248

Kon-ya K, Shimidzu N, Adachi K, Miki W (1994) 2,5,6-Tribromo-1-methylgramine, an antifouling substance from the marine bryozoan *Zoobotryon pellucidum*. *Fish Sci* 60(6):773–775

Kon-ya K, Wataru M, Endo M (1995) l-Tryptophan and related compounds induce settlement of the barnacle *Balanus amphitrite* Darwin. *Fish Sci* 61(5):800–803

Kuffler SW, Nicholls JG, Martin AR (1984) From neuron to brain. Sinauer Associatea, Sunderland, MA, pp. 651

Lambert SJ, Thomas KV, Davy AJ (2006) Assessment of the risk posed by the antifouling booster biocides Irgarol 1051 and Diuron to freshwater. *Chemosphere* 63:734–743

Lee OO, Qian PY (2004) Potential control of bacterial epibiosis on the surface of the sponge *Mycale adherens*. *Aquat Microb Ecol* 34:11–21

Leise EM, Thavradhara K, Durhan NR, Turner BE (2001) Serotonin and nitric oxide regulate metamorphosis in the marine snail *Ilyanassa obsoleta*. *Am Zool* 41:258–267

Leitz T, Klingmann G (1990) Metamorphosis in *Hydractinia*: studies with activators and inhibitors aiming at protein kinase C and potassium channels. *Rouxs Arch Dev Biol* 199:107–113

Leitz T, Muller WA (1987) Evidence for the involvement of PI-signalling and diacylglycerol second messengers in the initiation of metamorphosis in the hydroid *Hydractinia echinata*. *Fleming Dev Biol* 121:82–89

Leitz T, Beck H, Stephan M, Lehmann WD, Petrocellis LDE, Marzo VDI (1994) Possible involvement of arachidonic acid and eicosanoids in metamorphic events in *Hydractina echinata* (Coelenterata; Hydrozoa). *J Exp Zool* 269:422–431

LeTourneux F, Bourget E (1988) Importance of physical and biological settlement cues used at different spatial scales by the larvae of *Semibalanus balanoides*. *Mar Biol* 97:57–66

Levantine PL, Bonar DB (1986) Metamorphosis of *Ilyanassa obsoleta* natural and artificial inducers. *Am Zool* 26:14A

Lewandowski Z, Beyenal H (2008) Mechanisms of microbially influenced corrosion. Springer Ser Biofilms. doi: 10.1007/7142_2008_8

Matsumara K, Mori S, Nagano M, Fusetani N (1998) Lentil lectin inhibits adult extract-induced settlement of the barnacle *Balanus amphitrite*. *J Exp Zool* 280:213

Meyer A, Baier R, Wood CD, Stein J, Truby K, Holm E, Montemarano J, Kavanagh C, Nedved B, Smith C, Swain G, Wiebe D (2006) Contact angle anomalies indicate that surface-active eluates from silicone coatings inhibit the adhesive mechanisms of fouling organisms. *Biofouling* 22(5–6):411–423

Mihm JW, Banta WC, Loeb GI (1981) Effects of adsorbed organic and primary fouling films on bryozoan settlement. *J Exp Mar Biol Ecol* 54:167–179

Millineaux LS, Garland ED (1993) Larval recruitment in response to manipulated field flows. *Mar Biol* 116:667–683

Morita H (1972) Primary process of insect chemoreception. *Adv Biophys* 3:161–198
Morse DE, Hooker N, Duncan H, Jensen L (1979) Gamma-amino butyric acid – a neurotransmitter induces planktonic abalone larvae to settle and begin metamorphose. *Science* 204:407–410
Morse ANC, Froyd CA, Morse DE (1984) Molecules from cyanobacteria and red algae that induce larval settlement and metamorphosis in the mollusc *Haliotis rufescens*. *Mar Biol* 81:293–298
Morse DE (1992) Molecular mechanisms controlling metamorphosis and recruitment in abalone larvae. In: Shepherd SA, Tegner MJ, Guzman del Proo SA (eds.) Abalone of the world. Blackwell, Oxford, pp. 107–119
Muller WA, Buchal G (1973) Metamorphose-Induktion bei Planulalarven. II Induktion durch monovalente Kationen: Die Bedeutung des Gibbs-Donnan-Verhältnisses und der Na^{+}/K^{+}-ARPase. *Wilhelm Roux Arch* 173:122–135
Murthy PS (1999) Studies on the factors influencing larval settlement and metamorphosis in *Crassostrea madrasensis* (Preston) and its control using bioactive compounds. PhD thesis, University of Madras, pp. 105
Murthy PS, Venugopalan VP, Nair KVK, Subramoniam T (1999) Chemical cues inducing settlement and metamorphosis in the fouling oyster *Crassostrea madrasensis*. *J Ind Inst Sci* 79:513–526
Natasha IG, William B, Loeb GI (2002) Aquatic biofouling larvae respond to differences in the mechanical properties of the surface on which they settle. *Biofouling* 18 (4):269–273
Nedved BT, Hadfield MG (2008) *Hydroides elegans* (Annelida: Polychaeta): a model for biofouling research. Springer Ser Biofilms. doi: 10.1007/7142_2008_15
Neil JA, Holliday JE (1986) Effects of potassium and copper on the settling rate of Sydney Rock Oyster (*Saccostrea commercialis*) larvae. *Aquaculture* 58:263–267
Nogata V, Yoshimura V, Shinshima K, Kitano Y, Sakaguchi I (2003) Antifouling substances against larvae of the barnacle *Balanus amphitrite* from the marine sponge, *Acanthella cavernosa*. *Biofouling* 19:193–196
Nott JA, Foster BA (1969) On the structure of the antennular attachment organ of the cypris larva of *Balanus balanoides*. *Phil Trans R Soc* 256:115–133
Nymer M, Cope E, Brady R, Shirtliff ME, Leid JG (2008) Immune responses to indwelling medical devices. Springer Ser Biofilms. doi: 10.1007/7142_2008_4
Okano K, Shimizu K, Satuito CG, Fusetani N (1996) Visualization of cement exocytosis in the cypris cement gland of the barnacle *Megabalanus rosa*. *J Exp Biol* 199:2131–2137
Okazaki Y, Shizuri Y (2000) Effect of inducers and inhibitors on the expression of bcs genes involved in cypris larval attachment and metamorphosis of the barnacles *Balanus amphitrite*. *Int J Dev Biol* 44:451–456
Okino T, Yoshimura E, Hirota H, Fusetani N (1995) Antifouling kalihinenes from the marine sponge *Acanthella cavernosa*. *Tetrahedron Lett* 36(47):8637–8640
Okino T, Yoshimura E, Hirota H, Fusetani N (1996) New antifouling sesquiterpenes from four nudibranchs of the family Phyllidiidae. *Tetrahedron* 52(28):9447–9454
Olguin-Uribe G, Abou-Mansour E, Boulander A, Debard H, Francisco C, Combaur G (1997) 6-Bromoindole-3-carbaldehyde, form an *Acinetobacter* sp. bacterium associated with the ascidian *Stomoza murrayi*. *J Chem Ecol* 23:2507–2521
Olsen SM, Pedersen LT, Laursen MH, Kill S, Dam-Hohansen K (2007) Enzyme-based antifouling coatings: a review. *Biofouling* 23(5):369–383
Ortlepp S, Sjogren M, Dahlstrom M, Weber H, Ebel R, Edrada R, Thomas C, Schup P, Bohlin L, Proksch P (2007) Antifouling activity of bromotyrosine-derived sponge metabolites and synthetic analogues. Mar Biotechnol 23
Pawlik JR (1990) Natural and artificial induction of metamorphosis of *Phragmatopoma lapidosa californica* (Polychaeta: Sabellariidae) with a critical look at the effects of bioactive compounds on marine invertebrate larvae. *Bull Mar Sci* 46(2):512–535
Pawlik JR, Butman CA, Starczak VR (1991) Hydrodynamic facilitation of gregarious settlement of a reef-building tube worm. *Science* 251:421–424
Pechenik JA, Li W, Cochrane DE (2002) Timing is everything: the effects of putative dopamine antagonists on metamorphosis vary with larval age and experimental duration in the prosobranch gastropod *Crepidula fornicate*. *Biol Bull* 202:137–147

Pennington KT, Hadfield MG (1988) Larvae of a Nudibranch mollusc (*Phestilla sibogae*) metamorphose when exposed to common organic solvents. *Biol Bull* 177:350–355

Pernet F, Tremblay R, Bourget E (2003) Settlement success, spatial pattern and behavior of mussel larvae *Mytilus* sp. in experimental down-welling systems of varying velocity and turbulence. *Mar Ecol Prog Ser* 260:125–140

Pereira RC, Carvalho AGV, Gama BAP, Countinho R (2002) Field experimental evaluation of secondary metabolites from marine invertebrates as antifoulants. *Braz J Biol* 62(2):311–320

Pettitt ME, Henry SI, Callow ME, Callow JA, Clare AS (2004) Activity of commercial enzymes on settlement and adhesion of cypris larvae of the barnacle *Balanus amphitrite*, spores of the green alga *Ulva linza*, and the diatom *Navicula perminuta Biofouling* 20(6):299–311

Rittschof D (2001) Natural product antifoulants and coating development. In: McClintock JB, Baker BJ (eds.) Marine chemical ecology. CRC, Boca Raton, pp. 543–566

Rittschof D, Costlow JD (1989a) Surface determination of macro invertebrate larvae settlement. In: Klekowsky RZ, Styczynska-Jureqwicz E, Falkowsky J (eds.) Proceedings of the 21st European marine biology symposium, Gdansk, 14–19 Sept 1996. Ossolineum, Gdansk, pp. 155–163

Rittschof D, Costlow JD (1989b) Bryozoan and barnacle settlement in relation to initial surface wettability: a comparison of laboratory and filed studies. In: Ros JD (ed.) Topics in marine biology: proceedings of 22nd European marine biology symposium, Barcelona, 17–22 August 1987: Scientia Marina, Barcelona, pp. 411–416

Rittschof D, Branscomb ES, Costlow JD (1984) Settlement and behavior in relation to flow and surface in larval barnacle, *Balanus amphitrite* Darwin. *J Exp Mar Biol Ecol* 82:131–146

Rittschof D, Hooper IR, Branscomb ES, Costlow JD (1985) Inhibition of barnacle settlement and behavior by natural products from whip corals, *Leptogorgia virgulata* (Lamarck, 1815). *J Chem Ecol* 11:551–563

Rittschof D, Maki J, Mitchel R, Costlow JD (1986) Ion and neuropharmacological studies of barnacle settlement. *Neth J Sea Res* 20:269–275

Roberts D, Rittschof D, Holm E, Schmidt AR (1991) Factors influencing initial larval settlement: temporal, spatial and surface molecular components. *J Exp Mar Biol Ecol* 150:203–211

Ryland RS (1974) Behaviour, settlement and metamorphosis of bryozoan larvae: a review. *Thalassia Jugoslavica* 10:239–262

Ryle M (1999) Are TBT alternatives as good? *The Motor Ship*. pp. 34–39

Scardino A, Nys RD, Ison O, O'Connor W, Steinberg P (2003) Microtopography and antifouling properties of the shell surface of the bivalve molluscs *Mytilus galloprovincialis* and *Pinctada imbricate*. *Biofouling* 19:221–230

Scardino AJ, de Nys R (2004) Fouling deterrence on the bivalve shell *Mytilus galloprovincialis*: a physical phenomenon. *Biofouling* 20(4–5):249–257

Schmitt TM, Hay ME, Lindquist N (1995) Constraints on chemically mediated coevolution: multiple functions for seaweed secondary metabolites. *Ecology* 76:107–123

Schneider T, Leitz T (1994) Protein kinase C in hydrozoans: involvement in metamorphosis of Hydractinia and in pattern formation of Hydra. *Roux's Arch Dev Biol* 203:422–428

Schoenfeld RC, Ganem B (1998) Synthesis of ceratinamine and moloka'iamine: antifouling agents from the marine sponge *Pseudoceratina purpurea*. *Tetrahedron Lett* 39:4147–4150

Schoenfeld RC, Conova S, Rittschof D, Ganem B (2002) Cytotoxic, antifouling bromotyramines: a synthetic study on simple marine natural products and their analogues. *Bioorg Med Chem Lett* 12:823–825

Schumacher JF, Aldred N, Callow ME, Finlay JA, Callow JA, Clare AS, Brennan AB (2007) Species-specific engineered antifouling topographies: correlations between the settlement of algal zoospores and barnacle cyprids. *Biofouling* 23(5):307–317

Sepcic K, Turk T (2006) 3-Alkylpyridinium compounds as potential non-toxic antifouling agents. In: Fusetani N, Clare AS (eds.) Antifouling compounds. Progress in molecular and subcellular biology, vol 42. Springer Berlin Heidelberg New York, pp. 105–124

Sepcic K, Marcel V, Klaebe A, Turk T, Síuput D, Fournier D (1998) Inhibition of acetylchloinesterase by an alkylpyridinium polymer from the marine sponge, *Reniera sarai*. *Biochim Biophys Acta* 1387:217–225

Sjogren M, Dahlstrom M, Goransson U, Jonsson P, Bohlin L (2004) Recruitment in the field of *Balanus improvisus* and *Mytilus edulis* in response to the antifouling cyclopeptides barettin and 8,9-dihydrobarettin from the marine sponge *Geodia barrette*. *Biofouling* 20(6):291–297

Sjogren M, Johnson AL, Hedner E, Dahlstrom M, Goransson U, Shirani H, Bergman J, Jonsson PR, Bohlin L (2006) Antifouling activity of synthesized peptide analogs of the sponge metabolite. *Barettin Peptides* 27:2058–2064

Statz A, Finaly J, Dalsin J, Callow M, Callow JA, Messersmith PB (2006) Algal antifouling and fouling-release properties of metal surfaces coated with a polymer inspired by marine mussels. *Biofouling* 22(5–6):391–399

Steinberg PD, De Nys TD, Kjelleberg S (1998) Chemical inhibition of epibiota by Australian seaweeds. *Biofouling* 12(1–3):227–244

Steinberg PD, De Nys R, Kjelleberg S (2001) Chemical mediation of surface colonization. In: McClintock J, Baker B(eds.) Marine chemical ecology. CRC, Boca Raton, pp. 325–353

Steinberg PD, De Nys R, Kjelleberg S (2002) Chemical cues for surface colonization. *J Chem Ecol* 28:1935–1951

Sulkin SD (1984) Behavioral basis of depth regulation in the larvae of brachyuran crabs. *Mar Ecol Prog Ser* 15:181–205

Swain G (1988) Proceedings of the international symposium on sea water drag reduction. Naval Undersea Warfare Center, Newport, pp. 155–161

Tegtmeyer K, Rittschof D (1989) Synthetic peptide analogs to barnacle settlement pheromone. *Peptides* 9:1403–1406

Thomas JG, Corum L, Miller K (2008) Biofilms and ventilation. Springer Ser Biofilms. doi: 10.1007/7142_2008_7

Thorson G (1964) Light as an ecological factor in the dispersal and settlement of larvae of marine bottom invertebrates. *Ophelia* 1:187–208

Thurm U, Wessel G (1979) Metabolism-dependant transepithelial potential differences at epidermal receptors of arthropods. *J Comp Physiol* 134A:119–130

Tomono Y, Hirota H, Fusetani N (1998) Antifouling compounds against barnacle (*Balanus amphitrite*) larvae from the marine sponge *Acanthella cavernosa*. In: Watanabe Y, Fusetani N (eds.) Sponge sciences – multidisciplinary perspectives. Springer Berlin Heidelberg New York, pp. 413–424

Tsoukatou M, Marechal JP, Hellio C, Novakovic I, Tufegdzic S, Sladic D, Gasic MJ, Clare AS, Vagias C, Roussis V (2007) Evaluation of the activity of the sponge metabolites Avarol and Avarone and their synthetic derivatives against fouling micro- and macroorganisms. *Molecules* 12:1022–1034

Tsukamoto S, Kato S, Hirota H, Fusetani N (1996) Pseudoceratidine: a new antifouling spermidine derivative from the marine sponge *Pseudoceratina purpurea*. *Tetrahedron Lett* 37(9):1439–1440

Walters LJ (1992) Field settlement locations on subtidal marine hard substrata: is active larval exploration involved? *Limnol Oceanogr* 37:1101–1107

Yamamoto H, Satuito CG, Yamazaki M, Natoyama K, Tachibana A, Fusetani N (1998) Neurotransmitter blockers as antifoulants against planktonic larvae of the barnacle *Balanus amphitrite* and the mussel *Mytilus galloprovincialis*. *Biofouling* 13:69–82

Yamamoto H, Tachilbana A, Matsumura K, Fusetani N (1995) Protein kinase C (PKC) singal transduction system involved in larval metamorphosis of the barnacle, *Balanus amphitrite*. *Zool Sci* 12:391–396

Yang LH, Lee O, Jin T, Li XC, Qian PY (2006) Antifouling properties of 10b-formamidokalihinol-A and kalihinol A isolated from the marine sponge *Acanthella cavernosa*. *Biofouling* 22(1–2):23–32

Yang LH, Lau SCK, Lee OO, Tsoi MMY, Qian PY (2007) Potential roles of succinic acid against colonization by a tubeworm. *J Exp Mar Biol Ecol* 349:1–11

Yebra DM, Kiil S, Dam–Johansen K (2004) Antifouling technology – past, present and future steps towards efficient and environmentally friendly antifouling coatings. *Prog Org Coat* 50:75–104

Yool AJ, Grau SM, Hadfield MG, Jensen RA, Markell DA, Morse DE (1986) Excess potassium induces larval metamorphosis in four marine invertebrate species. *Biol Bull* 170:255–266
Young CM, Chia FS (1987) Abundance and distribution of pelagic larvae as influenced by predation behavior and hydrographic factors. In: Giese AC, Pearse JS (eds.) Reproduction of marine invertebrates, vol 9. Blackwell, Palo Alto, pp. 385–453
Youngblood JP, Andruzzi L, Ober CK, Hexemer A, Kramer EJ, Callow JA, Finaly JA, Callow ME (2003) Coatings based on side-chain ether-linked poly(ethylene glycol) and fluorocarbon polymers for the control of marine biofouling. *Biofouling* 19:91–98
Zapata M, Silva F, Luza Y, Wilkens M, Riquelme C (2007) The inhibitory effect of biofilms produced by wild bacterial isolates to the larval settlement of the fouling ascidia *Ciona intestinalis* and *Pyura praeputialis*. *Electron J Biotechnol* 10(1):149–159

Macrofouling Control in Power Plants

R. Venkatesan and P. Sriyutha Murthy (✉)

Abstract Macrofouling organisms readily colonize artificial man-made structures, cooling water intake tunnels, culverts, pump chambers and heat exchangers. Cooling water systems if not properly treated invariably become susceptible to biofouling. The problem is particularly severe in the tropics and is site-, season-, and substratum-specific. Further, cooling systems serve as a source of macrofouling organisms and breeding grounds wherein invertebrate larvae are produced and colonize equipment downstream like pipelines, valves and heat exchangers. Once-through seawater or freshwater systems encounter severe macrofouling-associated problems like flow reduction, increased pressure drop across heat exchangers and equipment breakdown. Biocidal dose and regime for cooling water systems and heat-exchangers have to be tailor-made for a power plant and should be effective in controlling microbial biofouling as well as hard foulants (barnacles, mussels, tubeworms and oysters). With regard to macrofouling control in condenser-cooling systems of power plants, chlorination has been the method of choice for fouling control over the years due to its low cost, easy availability and handling, and known degradation pathways. Increasing awareness on the toxic effects of chlorination by-products and better understanding of the biocidal action, environmental issues and higher dosages required for sanitization of surfaces has resulted in replacement of chlorine by stronger oxidizing biocides like chlorine dioxide. Experimental studies using coastal seawater in plate heat exchangers, has revealed a chlorine residual of 1.0 ppm to prevent settlement of invertebrate larvae. However, an intermittent chlorination dose of 1.2 ppm residuals at a frequency of 0.5–2 h was sufficient in controlling slime formation. Side-stream monitoring of these heat exchangers in a nuclear power plant cooling circuit revealed barnacle fouling in spite of continuous chlorination of 0.2–0.3 ppm residuals and shock doses of 0.4–0.6 ppm twice a week for 8 h. In an operational plant, continuous monitoring of the fouling situation using side-stream monitoring devices is to be practised and the biocidal dose and regime

P.S. Murthy
Biofouling and Biofilm Processes Section, Water and Steam Chemistry Division, BARC Facilities, Indira Gandhi Center for Atomic Research Campus, Kalpakkam, 603 102, India
e-mails: psm_murthy@yahoo.co.in, psmurthy@igcar.gov.in

Springer Series on Biofilms, doi: 10.1007/7142_2008_14

altered to overcome any spikes in settlement. This is essentially because biocidal doses required to kill established fouling communities are far higher than those for inhibiting settlement. Even if killing is achieved, accumulation of dead shell biomass (barnacles and tubeworms) often results in loading on equipment surfaces and increases surface roughness, facilitating settlement of other fouling organisms.

1 Introduction

Industrial fouling involves inorganic, organic, particulate and biological fouling (see Smeltzer 2008). Biofouling in industrial water systems is a recalcitrant problem not easily controlled and even then at a significant cost. Operationally, the problem due to biofouling manifests when "biofilm development exceeds a given threshold of interference" (Flemming and Griebe 2000). Microbial biofouling can alternatively be considered as a biofilm reactor in the wrong place, as cooling systems offer large surface areas for colonization and nutrients for growth (Flemming 2002). Biofouling in recirculating freshwater systems is less pronounced than in once-through seawater systems and generally manifests in the form of condenser slime (biofilms) in power-plant cooling circuits.

About 150 species of macrofoulants have been listed for seawater-cooled systems whereas Asiatic clams and Zebra mussels are the main macrofoulants in freshwater-cooled systems (Claudi and Mackie 1994). Macrofouling in the cooling water systems of power plants results in reduction of flow to the condenser tubes, blockage of intake pipes, condenser tube blockage, mechanical damage to pumps and condenser tubes, promotion of microfouling and enhanced corrosion of condenser tubes. Further settlement of hard-shelled fouling organisms causes damage to material integrity and often results in failure of equipments. Compared to shell and tube heat exchangers, plate heat exchangers (PHE) are finding increased applications recently in nuclear, thermal and desalination plants. Long-term data on the use of these heat exchangers in power plant cooling circuits revealed accumulation of corrosion products and particulate fouling on the plates. Initiation time for biological fouling in industrial systems ranges from a few hours to about 400 h (Bott 1993). For a successful antifouling strategy an integrated approach of monitoring the surfaces, including analysis of fouling situations, is important (Flemming 2002). In seawater-cooled systems macrofouling is a predominant problem and control measures should aim at a biocidal dose and regime to prevent settlement of organisms and development of biofilms, i.e. restricting biofilm development at the given threshold of interference. Environmental parameters such as fluid velocity, temperature, pH value, nutrient levels, cell concentration and surface roughness have all been demonstrated to have a measurable effect on the development of both biofilms and settlement of macrofouling organisms. A biocide dose should be effective from zero hours to keep the surface clean. The old approach of increasing biocide dosage to remediate a biofouling problem frequently fails in practice. Since biofouling is a surface-associated phenomenon it should be treated as such, by targeting treatments for controlling surface-associated or sessile organisms (Donlan 2000; Flemming 2002).

One of the major factors affecting power plant operation is the performance of the cooling water system (Brankevich et al. 1990; Bell 1977). This has led to the development of specific antifouling measures for power plants. The most common countermeasure as practised in a majority of power plants comprises filtration of debris and fouling organisms through intake and travelling water screens, biocide dosing (chlorine) and thermal shock treatment. Compared to chlorine dioxide, the use of a much stronger oxidizing agent such as ozone is not popular. Due to the high production cost and considering the volume of water to be treated in once-through systems, this does not seem to be a viable alternative. The success of these methods is dependent on the nature of the fouling organisms in a given geographical location (Strauss 1989; Sasikumar et al. 1992), quality of cooling water (Corpe 1977), thermal and biocidal tolerance ranges of different size classes of macrofouling organisms existing in a cooling circuit (Jenner 1980; Sasikumar et al. 1992; Rajagopal et al. 1995) and siltation (Jenner et al. 1998). The development of a macrofouling community in the cooling water system occurs as a result of passage of planktonic larvae of invertebrates followed by a settlement phase (see Murthy et al. 2008), during which the organisms metamorphose into adults by producing an outer shell. The high flow velocity ensures a continuous supply of oxygen, food for the growth of macrofoulants, avoids the accumulation of waste and results in increased colonization of macrofoulants (Jenner 1980). Over a period of time if the situation is left untreated a uniform fouling layer accumulates on the surfaces with a thickness reaching 30 cm from the wall (Kovalak et al. 1993). These cause problems in the condenser section as the fouling layer grows in thickness they are sloughed off due to high velocities, and such clusters are deposited in the heat exchangers blocking the flow (Kovalak et al. 1993).

The problem of macrofouling in heat exchangers leads to flow blockage in shell and tube heat exchangers (Fig. 1a) and deposition in plate heat exchangers (Fig. 1b). Fouling causes irreversible mechanical damage to the equipment surface due to the hard calcareous shells of fouling organisms. A 5 mm Hg reduction in condenser backpressure is equal to 0.5% improvement in turbine heat rate, which is approximately equal to an additional 3 MW(e) of generating capacity (Drake 1977). On the other hand, fouling of heat exchanger surfaces results in reduced heat transfer efficiency and increased fluid frictional resistance,resulting in additional maintenance and operating costs (Bott and Tianqing 2004). Apart from macrofouling, even a 250 micron-thick layer of slime may result in up to a 50% reduction in heat transfer in heat exchangers (Goodman 1987). In the case of heat exchangers, a decrease of the overall heat transfer coefficient due to fouling deposits leads to overdesign, and increased energy and cleaning costs, which are substantial (Bott 1995). As far as overdesign is concerned, fouling in plate exchangers leads to higher operational costs compared to shell and tube exchangers because of their higher efficiency (Muller-Steinhagen 1993; Hesselgreaves 2002; Kukulka and Devgun 2007). In seawater flow systems the fouling layer on a heat exchanger surface comprises chiefly of inorganic and biological fouling. If biofilm formation precedes that of inorganic film, the inorganic film develops in the channels existing in the biofilm

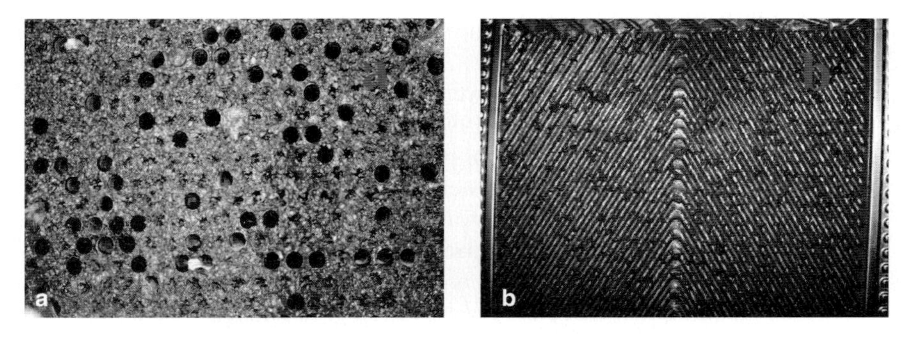

Fig. 1 **a** Flow blockage of tubular heat exchangers (PSWHX) of Madras Atomic Power station after one year of operation. **b** Fouling by particulate material and corrosion product on titanium plate heat exchanger surface after 69 days of operation at 0.5 m s^{-1} velocity in a side-stream study at the Madras Atomic Power Station, Kalpakkam

matrix. If inorganic fouling precedes organic fouling then it would attract adhesion of bacteria (Sheikholeslami 2000). However, these mechanisms are yet to be substantiated through experimental data. Some of the factors influencing fouling of heat exchangers are the material of construction of a heat exchanger, and its surface roughness will influence the development of biofilms (Mott and Bott 1991; Mott et al. 1994). Vieira et al. (1992) demonstrated that initial attachment of *Pseudomonas fluorescens* was more pronounced on aluminium plates than on brass or copper, which may be attributed to the release of toxic ions by these surfaces. Rabas et al. (1993) demonstrated that fouling was higher on spirally indented tubes than plain tubes. On the other hand, rough surfaces were more hospitable to microorganisms than smooth surfaces (Reid et al. 1992). On a rough surface, valleys provide shelter against removal by shear stress and hills act as nucleation sites; therefore the extent of fouling is high on rough surfaces. Surface properties like adsorption, surface charge and corrosiveness were also found to affect fouling (Epstein 1983).

2 Types and Features of Industrial Cooling Water Systems

Choice on the type of cooling water system is influenced by plant location and availability of water suitable for cooling purposes. Once-through cooling systems are used in plants sited beside large water bodies (sea, large flowing rivers and estuaries) that have the ability to dissipate waste heat from the steam cycle. For detailed system analysis and design of structures refer to Jenner et al. (1998), Neitzel and Dalling (1984). Design characteristics of once-through systems may allow or even increase the rate of fouling by promoting conditions that are conducive to sedimentation, macrofoulants and corrosion. Intake structures of once-through systems vary from plant to plant depending on environmental considerations and flow requirements. Most of the flow-through systems comprise of an offshore intake system (bored tunnel or a buried culvert), which conveys the water to a pump

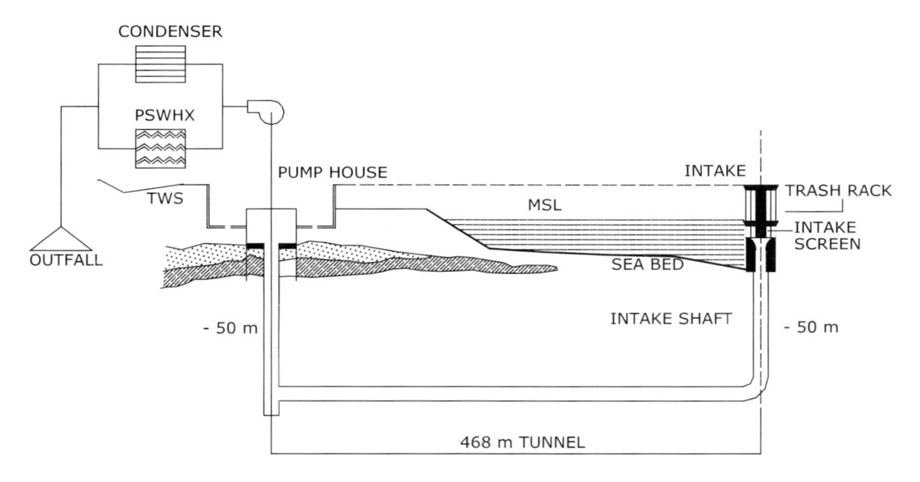

Fig. 2 Schematic view of a cooling water system of a power station (Madras Atomic Power Station, Kalpakkam, located on the east coast of India). *TWS* travelling water screens, *PSWHX* process seawater heat exchangers, *MSL* mean sea level

house located onshore from where the water is pumped through the condensers, and returns via a shore-based outfall (Fig. 2). Near shore intakes are not preferred as they may result in a severe siltation problem. Some of the design features incorporate physical water treatment methods like water velocities in the cooling water tunnel designed around 1.5–3.0 m s^{-1} to prevent sedimentation, and provision of trash screens at the offshore intake point and travelling water screens before the pumps. In addition, once-through systems typically have high flow velocities and mass flow rates for minimizing temperature effects on receiving waters. A typical 500 MW(e) unit would have a flow of 30 m^3 s^{-1} at an average velocity of 3 m s^{-1} in the cooling water circuits. Design factors influencing macrofouling of once-through systems are (1) flow velocity, (2) flow pattern, (3) frequency of use, (4) valve leaks, (5) unreliable and ineffective biocidal systems, (6) compartment size, (7) system configuration and (8) water temperature (Neitzel et al. 1984).

3 Economic Losses Due to Biofouling in Power Plants

An estimate of economic losses due to biofouling problems is large and emphasizes the importance of biofouling control measures in power plants. Most of the literature on losses due to biofouling dates to the late 1990s. Costs for one day of unplanned outage of a 235 MW(e) power plant can be around 0.3% of the earning. Hence control measures adopted to maintain cooling water cleanliness reflects on indirect earnings to the industry. Losses as a result of shutdown of a 235 MW(e) power station due to biofouling were estimated to be about Rs. 40 lakhs a day (about US$100,000) (Venugopalan and Nair 1990). The cost of removing macrofouling organisms from screening houses alone for two European power plants cost around US$ 25–30,000 every 2 years (Kovalak et al. 1993). In the USA, approximately 4% of failures in power plants >600 MW(e) is due to biofouling of condensers (Meesters et al. 2003). Fouling by Asiatic clams *Corbicula asiatica* in condenser tubes of power plants alone costs the USA over US$1 billion annually (Chow 1987; Strauss 1989). The problem is more pronounced in heat exchangers, where an increase in condenser backpressure due to fouling in a 250 MW(e) plant costs about US$250,000 annually (Chow 1987).

Cleaning of biofouling organisms from cooling water-circuits of power plants is a very expensive option for plant operations. Studies by Coughlan and Whitehouse (1977) showed that between 1957 and 1964, some 4,000 condenser tubes failed due to mussel fouling, leading to leakage of cooling water into the boiler. Apart from the loss of power generation, these leaks contaminated the feed-water system and accelerated the boiler waterside corrosion, resulting in boiler tube failures. This has necessitated the inlet culverts to be drained for manual cleaning once a year. The average quantity of mussels removed was estimated at 40 tons per year and the maximum was 130 tons per year. Similarly, 300 tons of mussels were removed from the Pools power station (Dorset). About 300 tons of mussels were removed following shock chlorination treatment from the atomic power station intake tunnel in a single occasion (Rajagopal et al. 1996). Similarly, 4,000 man-hours were used to clean the circuits and remove mussels (360 m^3) at the power station at Dunkerque (Whitehouse 1985). Another example of intense fouling is the Carmarthen Bay power station, where within a year of commissioning the problem became so severe that the plant was shut down periodically (James 1985). The underwater cooling conduits of the Tanagwa power station in Japan showed a fouling thickness of 70 cm (Kawabe and Treplin 1986).

4 Biofouling Control Methods in Once-Through Seawater-Cooled Systems

An important consideration in the operation of equipment subject to biofouling is its mitigation. Mitigation techniques broadly fall into (1) physical (Bott and Tianqing 2004; Melo and Bott 1997) and (2) chemical methods (Sohn et al. 2004; Meyer 2003; Ludensky 2003; Walsh et al. 2003; Rajagopal et al. 2003; Butterfield et al. 2002a, b; Prince et al. 2002; Ormerod and Lund 1995).

4.1 Physical Control Methods

4.1.1 Flow

Water flow is a major factor influencing settlement of marine invertebrate larvae (Table 1). Velocity is a design factor for cooling water systems (Strauss and Puckorius 1984; Tuthill 1985; Johnson et al. 1986). Flow velocities in cooling water conduits must be measured very close to the wall (1 mm) surface rather than

Table 1 Velocity as an antifouling measure

Velocity (m s^{-1})	Effects on organisms	References
Less than 0.9	Allows sedimentation and microfouling to occur	
>0 but <0.3	Allows Asiatic clam larvae to settle	Jenner et al. (1998)
>0.1 but <1.2	Allows blue mussel and oyster larval settlement	
>0.1 up to 1.5	Allows mussel settlement in intake tunnel *Branchidontes variablis*, *B. striatulus* and *Modiolus philippinarum*	Rajagopal et al. (2006)
>3.0	Does not detach mussels	Syrett and Coit (1983)
>0.01 up to 1.0	Allows Zebra mussel larvae to settle	Tuthill (1985)
0.15	Allows Zebra mussel larvae to settle. Data from ten European power plants	Kovalak et al. (1993)
Above 2.0	Inhibits Zebra mussel larval settlement	Leglize and Ollivier (1981)
<0.2; >0.6; >0.9	Settlement; growth inhibition; detachment of colonies of bryozoans	Aprosi (1988)
1.8–2.2	Allows settlement of mussels, barnacles, hydroids in circular conduits	Jenner and Khalanski (1998)
Up to 1.4	Allows mussel, barnacle, hydroid settlement in large rectangular conduits of (5–11 m^2)	Jenner et al. (1998)
>4	Required for preventing erosion corrosion of metal structures	Jenner et al. (1998)

the mean water velocity, as settling larvae are subjected to the velocity at the near surface boundary rather than in the bulk water. At high flow rates, the shear stress of the water often exceeds the shear strength of many organisms, hence they do not settle (Collins 1964). However this is not the situation in operational power plants, where barnacles were found to colonize even at velocities of 3.0 m s^{-1}, making the surface rough and creating sites for further settlement of mussels (Syrett and Coit 1983). Hence it is imperative to adopt a chemical control strategy to control the settlement and growth of macrofoulants. Conventionally a 1,000 MW(e) capacity power plant requires cooling water at the rate of 30 m^3 s^{-1} (Whitehouse 1985), which is drawn at a velocity of 2.0–3.0 m s^{-1} through inlet pipelines. In general, flow across the heat exchanger tubes is maintained around 1.4–2.0 m s^{-1}. Analysis of operational and experimental data from different power plants shows that velocities in the range of 3.5–4.0 m s^{-1} are required to prevent settlement of marine organisms. However, most power stations operate at velocities of 1.4–1.8 m s^{-1} across the heat exchangers and at around 2.0–3.0 m s^{-1} in pipe sections and cooling water circuits, which does not prevent the settlement of macrofoulants. The successional pattern of macrofouling organisms in cooling water circuits are also known to be driven by flow velocity. Larval forms capable of settling at high velocities are the primary colonizers and these established organisms were found to baffle water currents, which enabled attachment of larvae that preferred low velocities to settle (Corfield et al. 2004). An example of this effect was reported by Jenner (1980), whose study showed that barnacles were found to settle before mussels as they can attach at much higher velocities and their shells provide the roughness required for mussels to settle. In practice, there are low flow regions associated with the geometry of the cooling water circuit that may favour increased settlement, and it appears impossible to maintain a constant velocity throughout the cooling water circuit. Increasing the velocity requires additional pumping costs and does not seem to be a viable method. A possible antifouling method postulated by Jenner (1983) for controlling biofouling using velocity is to decrease the flow rate instead of increasing the flow rate, taking into consideration the sinking rate of different organisms at low flow velocities. However, this proposition is yet to be substantiated through experimental studies to be adopted in power stations. Alternatively the use of high velocities (1.5–2.5 m s^{-1}) with a smooth finish to the surface of the culverts could theoretically prevent settlement of mussel spat (Jenner 1982). With the advent of foul release coatings this seems to be a viable option, along with the flow and biocidal regimes in force. In general, the use of flow as an antifouling method is unlikely to be effective in power stations.

4.1.2 Travelling Screens

Another problem encountered in once-through offshore/near shore intake systems of power plants is impingement by large fish, driftwood, seaweed, jellyfish etc. The problem is overcome by the provision of single or double trash racks for the offshore intake systems, which serve as the first line of defense (Brankevich et al. 1990). Usually, travelling water screens are provided ahead of the heat exchangers to filter out the floating debris and adult macroorganisms. Screens of different mesh sizes (10 mm UK and Japan; 4 mm France and Italy; 4–10 mm Netherlands; 10–25 mm India) are in general use. Downstream of service water pumps, the water passes through basket strainers for removing particles and the water is taken to the condensers for cooling. Seawater-cooled plants are designed to minimize the number of components that interface directly with seawater because of the corrosive nature of seawater. Intermediate closed-cycle loops filled with demineralized water are used for cooling the auxiliary systems (process water heat exchangers). The use of travelling water screens is indispensable for plant operation.

4.1.3 Mechanical Cleaning Techniques

In spite of the presence of physical control methods like travelling screens and biocidal programmes, heat exchangers are often subjected to sedimentation and biofilm accumulation. Mechanical online and offline methods are available for cleaning of shell and tube heat exchangers. In the case of plate heat exchangers, online mechanical cleaning has been found to be economically non-viable and technically unfeasible warranting the use of chemical treatments. Ceramic, glass or sponge rubber balls have been used for online cleaning. Brush type online cleaning devices are also available. Two types of automatic cleaning systems are employed in power stations during normal plant operations: the Amertap system and the American M.A.N. brushes system. The Amertrap system can be operated on an intermittent or continuous basis depending on the severity of problem in the shell and tube heat exchangers. The Amertrap system comprises sponge rubber balls, slightly bigger in size than the tubes, that circulate along the length of the tubes (Bell 1977; Brankevich et al. 1990; Fritsch et al. 1977). The constant rubbing action keeps the surfaces clean and removes biofilms. The balls are collected in an outlet chamber and are again pumped into the heat exchanger. The American M.A.N system uses flow-driven brushes that are passed through the condenser tubes intermittently by reversing the flow. The brushes abrasively remove fouling and corrosion products. Automatic online mechanical cleaning methods are the most economical and are practised invariably in most of the power stations around the world. Even though these are a crude methods, no alternative technology is available at hand and the sponge rubber ball cleaning is again an indispensable method for microbial biofouling control in heat exchangers.

Offline cleaning is done by the hydrolazing method(specialized high pressure water jet cleaning) with a pressure of 10,000–20,000 psi for cleaning heat exchanger tubes. Tubes with scales showed that at a low pressure of 10,000 psi cleaning was relatively poor compared to 20,000 psi. Cleaning of dry tubes was more efficient with brass and spin grit brushes compared to wet tubes, which may be attributed to the lubricating effect of water between brush tips and tube surface. Hydrolazing was effective in cleaning wet tubes (Young et al. 2000). Other types of mechanical cleaning techniques involve moulded plastic cleaners (pigs) that are useful for cleaning light silt deposits. Spirally formed, indented or finned brushes are used for cleaning tubes. Hard calcite depositsare difficult to clean even by acids; rotary cutters similar to the ones used for cutting glass with a Teflon body are used for cleaning. Compressed airdriven devices are also available for cleaning of heat exchanger tubes. Offline cleaning methods are practised in power stations when the thermal resistance values in heat exchangers drop beyond an acceptable level. This results in shutdown of the equipment and affects production costs.

4.1.4 Thermal Treatment

Cooling water requirement of power plants are sized according to the upper thermal limits of discharge, i.e. 7–10°C above the ambient prevailing at a given location. Thermal treatment of cooling water circuits and inlet conduits is an effective environmentally friendly method for control of biofouling in power plants, wherein the cooling water temperature is raised above the thermal tolerance level (Table 2) of fouling organisms (Brankevich et al. 1990). The exact temperature and time required for mortality of fouling organisms is dependent on many factors; the main factor being the acclimation temperature, i.e. the difference between the ambient and treated temperature. A second factor is the rate of acclimation: if the temperature increase is slow, the mussels are found to acclimatize to the rate. Another factor

Table 2 Thermal tolerancelevels of common fouling organisms

Temperatures (°C)	Effect on organisms	References
35–37	Kills most macrofouling organisms	Jenner (1980); Gunasingh et al. (2002)
37 for 30 min; 38 for 15 min; 39 for 5 min	Causes 100% mortality in the mussel *Mytilus edulis*	Jenner (1980)
43 for 30 min	Causes 100% mortality in the green mussel *Perna viridis*	Rajagopal et al. (1995)
39 for 30 h; 43 for 30 min; >45	Tolerates; 100% mortality within 2.15 h; 100% mortality immediately for *Branchidontes striatulus*	Rajagopal et al. (1995)
35–47	Causes 100% mortality for barnacle *Megabalanus tintinabulum*	Sasikumar et al. (1992)

is genetic variations in local populations (Claudi and Mackie 1994). Before implementing a thermal treatment programme, the thermal tolerance of the major foulant species at a particular facility needs to be determined. This can be derived through simple experimentation and following the multiple regression formulae developed by McMahon et al. (1993) for 50% LT_{50} and 100% LT_{100} mortality of fouling organisms:

$$LT_{50}=34.57-0.035(\text{min}/1°\text{c}) + 0.149(°\text{C acclimation temperature})$$

$$LT_{100}=36.10-0.040(\text{min}/1°\text{c}) + 0.147(°\text{C acclimation temperature})$$

Thermal treatment procedures are very effective against macrofoulants compared to microfoulants as they are known to exist in condenser tubes (condenser slime) experiencing elevated temperatures (50–70°C). Thermal treatment methods have been successfully implemented at the Commonwealth Edison Heat plant, where 100% mortality of mussels was achieved by raising the water temperature from 31.6 to 37.2°C and maintaining this temperature for a 6 h period (Claudi and Mackie 1994). The method is also used in some power stations that have the option of dual intake pipelines and facilities for recirculation of water from the heat exchangers. One of the pipelines is used as an intake and the heated effluent from the heat exchanger is circulated through the other. After a certain period the direction of flow is reversed in these pipelines (Jenner 1982). The thermal method has been effective at Marsden B power station where the cooling circuits were treated with elevated temperatures (51.7°C). However, the periodicity of such operations depends on the intensity of fouling at a location. The effectiveness of thermal treatment is also dependent on the appropriate choice of water temperature, duration of exposure and frequency of exposure. For thermal treatment to be efficient, exposure periods no longer than 3 h should be adopted if it is to be economically and environmentally sustainable (Jenner 1982) for operational power plants. However, cooling water systems have a variety of macrofouling organisms, with different tolerance levels. Knowledge of the thermal tolerance of fouling species and information about the biological community that exists in a cooling water system is important for designing treatment strategies. Major disadvantages of thermal shock treatments are in regard to meeting the environmental regulations governing discharge of heated water and the availability of the option for thermal backwashing and losses involved in shutdown of the plant during the backwashing period.

4.2 Chemical Methods

4.2.1 Advantages and Disadvantages of Oxidizing Biocides

Oxidizing biocides are in use for treating cooling waters. In the oxidizing biocide category, chlorine has been the most extensively used and cost-effective biocide. The order of volatility is ozone > chlorine > chlorine dioxide > chloramines > hypochlorous acid > hypobromous acid.

Efficient chlorination treatments suitable both for biofilm and macrofouling control in condensers must be worked out for a given location. The addition of chlorine to water can be viewed as an instantaneous reaction resulting in an equilibrium mixture of hypochlorous acid (HOCl) and hypochlorite ions. Hypochlorous acid is the active biocide and its stability is dependent on the pH of the solution. At a low pH value of 6.0–7.0 relatively more concentration of hypochlorous acid is present than at a seawater pH value of 8.2. In addition, hypochlorous acid (HOCl) reacts with organic matter/nitrogenous compounds and is consumed readily (chlorine demand). This necessitates increased dosing to overcome the demand. In chlorination of natural waters, the chlorine demand has to be ascertained before administering the biocide. Chlorine demand is also found to vary seasonally and the demand of tropical coastal seawater usually varies between 0.7 and 1.0 mg L^{-1} (Murthy et al. 2005). To reduce biofouling, chlorination of seawater is usually practised, with typical applied doses of 0.5–1.0 mg L^{-1} (expressed as Cl_2) and a resultant residual oxidant level of 0.1 ± 0.3 mg L^{-1} in the cooling water. Chlorination can be an effective control technique for both bivalves and microbial slime. Different chlorine doses and regimes have been tested for fouling control (Jenner et al. 1998; Rajagopal et al. 1994, 2003; Rajagopal 1997; Gunasingh et al. 2002).

Some of the common chlorination practices adopted in power stations are:

- *Low level continuous chlorination*: Continuous application of chlorine at residuals of 0.1–0.2 mg L^{-1} is used to deter larval forms from settling. Mussel larvae close their shells in the presence of chlorine and the velocity in the system will flush them out without allowing the larvae to colonize the substratum (Claudi and Mackie 1994).
- *Intermittent treatment*: This method came into practice to reduce the cost of the treatment programme and also to meet the biocide discharge criteria. In addition, mussels are constrained to close their shell valves in response to continuous chlorination. However, studies by Rajagopal et al. (2003) have shown the method to

be ineffective as mussels were able to tide over periods of chlorine dosages by closing their shell valves. Alternatively, the more intelligent version of intermittent chlorination, namely pulse chlorination, developed at KEMA (Poleman and Jenner 2002) has been effective in achieving killing of mussels as well as reducing the biocide inventory and environmental burden.

- *End of the season chlorination*: This has been practised in some of the European power stations (Jenner and Janssen-Mommen 1993) where chlorine levels of 0.5 mg L^{-1} were maintained for 2 weeks at the end of the breeding season to cause 95% mortality of newly settled mussels.

Growing concerns over the harmful effects of chlorination by-products, i.e. trihalomethanes (THMs; volatile), haloacetonitriles (HANs; semi-volatile), halophenols (HPhs) and haloacetic acids (HAAs), resulted in chlorination being disallowed in several of the US, UK, Canadian and European power stations. Use of chlorine is subjected to increasing environmental regulations (such as the new Biocidal Product Directive, 98/8/CE, in European countries). The USEPA chronic and acute marine water quality guidelines for chlorine are 0.0075 and 0.013 g m^{-3}, respectively (USEPA 2002). CORMIX modelling done by the National Institute of Water and Atmospheric Research (NIWA) show that an eightfold dilution of the cooling water plume occurs in the mixing zone (Oldman et al. 2004). If residual concentrations in the cooing water outfall are in the range of 0.1 g m^{-3} after reasonable mixing, the maximum chlorine concentration would be 0.013 g m^{-3}, equivalent to the USEPA guideline (Corfield et al. 2004). The Safe Drinking Water Act (1979) enacted by the USA prescribes the maximum contaminated levels of total THM (TTHM) to 0.10 mg L^{-1} (100 ppb) and the disinfectants and disinfectant by-product rule has fixed the limits at 0.80 mg L^{-1} TTHM (USEPA 1994). Alternatives to conventional chlorination in power stations depends on the cost of the products proposed from the market, roughly these products are one to three orders of magnitude costlier than sodium hypochlorite.

An alternative biocide for controlling biofouling in power plant cooling systems is bromine. Bromine is a chemical halogen similar to chlorine and was introduced commercially in 1980. Since then, plant chemists have had the option of choosing either one or both biocides for their cooling systems. Bromination has been used for some time along with chlorine and can significantly reduce the total disinfectant and halogen application rates because bromine oxidants generated in water are more effective for controlling biofouling than their chlorine counterparts at high pH values, above the 8.0 found in seawater. Several forms of bromine are available, which include activated bromine, sodium bromide, bromine chloride and proprietary mixtures of bromine and chlorine. Commercial formulations like the Active Bromide (NALCO Chemicals), BromiCide (Great Lakes Chemical Corporation) and Starbex, a sodium hypobromite compound for microfouling control, have been adopted by some power stations along with chlorine dosing. Sodium bromide can be used to convert hypochlorous acid (HOCl) into hypobromous acid (HOBr). Literature on the toxicity of this biocide to marine organisms is limited. However, when used in combination with chlorination it is effective in reducing the total halogen load, and the bromine oxidants that are generated are more effective for controlling biofouling at pH values above 8.0 (Fisher et al. 1999).

Currently, chlorine dioxide is being adopted in several European power stations because of its effectiveness in killing macrofoulants as well as against microbial biofouling and because of the lesser formation of organo-halogenated by-products. Typical doses of ClO_2 for seawater cooling systems range from 0.05 to 0.1 mg L^{-1} (Petrucci and Rosellini 2005). Another oxidizing biocide being used for fouling control is ozone. The use of ozone as a biocide is still a very expensive method, estimated around 3.8 times that of the cost of sodium hypochlorite (Duvivier et al. 1996).

Oxidizing biocides have a similarity in their mode of action on biological organisms. The toxicity of chlorine has been reported to be due to the destruction of the respiratory membrane by oxidation (Bass and Heath 1977), oxidation of enzymes containing a sulfhydryl moiety (Ingols et al. 1953) and ion imbalances (Vreenegoor et al. 1977). An EPRI report (Electric Power Research Institute 1980) attributed the toxic effect of chlorine on mussels to a weakening of the strength of the byssal threads. The principle effect of chlorine was to depress the activity of the foot of mussels, leading to a reduction in the number of threads formed. Chlorinated mussels, with their weaker attachment systems, were swept from the walls of the cooling system (Claudi and Mackie 1994). In comparison, the biocidal action of ozone is on the bacterial membrane glycoproteins, glycolipids and certain amino acids such as tryptophan. Ozone also acts on the sulfhydryl groups of certain enzymes, resulting in disruption of normal cellular enzymatic activity. Bacterial death is rapid and is often attributed to changes in cellular permeability followed by cell lysis. Ozone also acts on the nuclear material, modifying the purine and pyramidine bases of nucleic acids (Roy et al. 1981).

The choice of the biocide for cooling water systems is primarily governed by the cost. The dose and regime depends on the nature and intensity of fouling organisms at a given geographical location, and on environmental conditions. There is no such concept as a best dose or a best biocide. Biocide doses and regimes must be tailor-made for each of the cooling water systems. Chlorine is effective but may require very high concentrations, which are not environmentally acceptable. Hence alternative biocides like chlorine dioxide or ozone may be considered, but here cost becomes a limiting factor. Hence, power plant operators have to strike a balance between cost and the cleanliness required. A comparative account of the properties and effectiveness of different oxidizing biocides are given in Table 3. The table has been synthesized based on experience and on data published by Jenner et al. 1998; Claudi and Mackie 1994; Corfield et al. 2004; Rajagopal et al. 1997; Cristiani 2005.

4.2.2 Biocidal Requirements for Prevention of Larval Settlement in Cooling Water Systems

Usually the problem of biofouling gains attention when it interferes with the performance of the station, even in the presence of a biocidal programme in place. This is due to the inadequacy of the biocidal programme in overcoming sudden increases in macrofouling settlement. Hence continuous surveillance, detection of fouling

Table 3 Comparison of properties of different oxidizing biocides

Parameters	Chlorine (Cl_2)	Bromine (Br)	Chlorine dioxide (ClO_2)	Ozone (O_3)	Peraacetic acid
Conc. used in CWS (doses)	0.2–1.0 mg L^{-1}	0.1–0.5 mg L^{-1}	0.1–0.5 mg L^{-1}	0.01–0.3 mg L^{-1}	1.5–3.0 g m^{-3}
Activity	Narrow spectrum at low concentrations	Moderately effective	Broad spectrum at low concentrations	Broad spectrum at low concentrations	Moderately effective
Contact time	Seconds to minutes	Seconds to minutes	Seconds to minutes	Seconds	Minutes
pH	Not very effective at pH higher than 7	Effective up to pH 9.0	Very effective up to pH 11	Not effective above pH 8.5	Effective up to pH 9.0
Temperature effects	Cannot be used at higher temperatures	Cannot be used at higher temperatures	Cannot be used at higher temperatures	Cannot be used at higher temperatures	Cannot be used at higher temperatures
Corrosiveness	Corrosive to handle	Not very corrosive	Not very corrosive	Moderately corrosive	Corrosive to iron substrates
By-products	Produces toxic trihalomethanes; regulations on upper toxic levels	Used along with chlorine	Does not produce toxic by-products;chlorite ions are generated	Bromate and assimilable organic carbon	Readily biodegradable
Reaction with organics	Reacts with organics and is consumed. Reacts with NH_3	Does not react with organics	Does not react with organics and NH_3. Reacts with secondary amines	Reacts with NH_3. Removes organic matter, odour	Reacts with sulfites and sulfides
Storage	Conc. decreases slowly with time	Can be prepared fresh and dosed	Conc. decreases rapidly with time	Cannot be stored	Can be stored
Cost (arbitrary units)	Cheap	2.0 times cost of chlorine	2.5 times cost of chlorine	3.8 times cost of chlorine	10–20% more than chlorine

organisms, monitoring of the efficiency of the biocide, and fine tuning the dosages would help in reducing the biocidal requirement of power plants. Flow-through power stations use different biocidal doses and regimes for control of macrofouling organisms. In practice, a low level of continuous chlorination (0.1–0.2 mg L^{-1} residuals) coupled with shock dosing (0.5–1.0 mg L^{-1} residuals for 30 min once a week) is employed in power stations. Studies on the dosages required to prevent settlement of organisms are limited except for the available literature based on operational experiences at power stations. The gap in knowledge is due to the complexity of the cooling systems (geometry, flow, surface characteristics, diversity of organisms, cost involved in a biocidal programme and knowledge about larval settlement behaviour) encountered and to the interfacing of engineering aspects with biology and toxicology. Further complexity arises in the scaling of laboratory results to real-time cooling circuits. In general dosages required to inhibit or prevent settlement would be far less compared to those required for killing established fouling communities (Claudi and Mackie 1994).

Field observation on the effectiveness of continuous chlorination revealed a residual of 0.25 mg L^{-1}, sufficient for preventing attachment and growth of *Mytilus* species at water velocities as low as 0.4 m s^{-1} (Elecric Power Research Institute 1980). Laboratory studies showed that a total residual oxidant (TRO) level of 0.1 mg L^{-1} prevented attachment of mussels to concrete panels at water velocities as low as 0.76 m s^{-1} (Elecric Power Research Institute 1980). Alternatively, continuous chlorination at 0.2 mg L^{-1} had no effect on settlement of the blue mussel *Mytilus edulis*(L) at Maasvlakte power station, Rotterdam (Jenner 1983). Comparison of results from these two studies reveal the inadequacy of chlorine, i.e. through interaction with organics and being unavailable for killing, and the site-specific requirements of biocidal doses. Levels of 0.2–0.5 mg L^{-1} delayed settlement of 30% of mussel larvae (Khalanski and Bordet 1980). Compared to mussels, barnacles were more resistant to continuous low-level chlorination and required higher dosages. From the studies carried out at Astoria power station (NY, USA), a chlorine residual of 0.1 mg L^{-1} for 1 week during the spat settlement season reduced the density of settlement 15-fold. However, these concentrations were not effective in preventing the settlement of the barnacle *Balanus eburneus* (Sarunac et al. 1994). Continuous chlorination at concentrations above 0.8 mg L^{-1} were required to prevent settlement of coelenterate Hydroids and tubeworms on steel surfaces in flow chambers (Fig. 2a), whereas an intermittent chlorination of 1.2 mg L^{-1} with a 2 h-on/2 h-off regime was effective in bringing down the settlement of these foulants on plate heat exchanger surfaces (Murthy et al. 2005). In comparison, continuous application of chlorine dioxide at residuals of 0.1–0.2 mg L^{-1} resulted in clean surfaces (Ambrogi 1997) and elimination of the Mediterranean hydroid *Laomedea flexuosa*at residuals of 0.1–0.2 mg L^{-1} (Geraci et al. 1993). Chlorine dioxide treatments (0.22 mg L^{-1}) adopted at the Brindisi Nord power station on the Adriatic showed that test panels placed inside the condenser boxes were clean of both macrofouling and slime in comparison to periods before switching to chlorine dioxide, when 20 × 30 cm panels would accumulate a wet weight of 160 g over 3 months (Ambrogi 1997). Similarly, at the Taranto steel plant located in the south of Italy, fouling biomass on experimental panels were 60 kg m^{-2} $year^{-1}$. Continuous treatment with chlorine dioxide at a dose of 0.5 mg L^{-1} resulted in clean surfaces, as observed from fouling collectors (Belluati et al. 1997). The higher biocidal action against macrofouling organisms at low concentrations has resulted in many stations turning to chlorine dioxide. At present, the studies at the Brindisi Nord power station and the Taranto steel plant are the only literature available on dosages required to prevent settlement of larvae by chlorine dioxide. Application of ozone to cooling water systems was also found to be effective in preventing the settlement of mussel larvae by inhibition of production of byssal thread at concentrations in the range 20 –30 mg L^{-1}. Studies by Lewis et al. (1993) have indicated that a minimum contact time of 5 h was required for 100% mortality of veligers and post-veligers at concentrations of 0.5 mg L^{-1} at 15–20 C water temperatures. These biocides have to be evaluated under dynamic conditions at varying velocities to assess their efficacy and to arrive at some minimum dosages for cooling water systems.

4.2.3 Biocidal Requirements for Killing Established Fouling Communities in Cooling Water Systems

Often in an operating plant the problem of biofouling gains attention and importance when it leads to breakdown of equipment. The usual situation one encounters in an operational plant is an established fouling community as a result of lack of surveillance and monitoring and fine tuning of the biocidal dose and regime according to the requirements to keep biofouling at bay. As a result, the biofouling load exceeds the threshold limits and one is faced with the challenge of killing and removing the established fouling communities. It is all the more important to keep the cooling water systems clean from macrofouling as killing is not cleaning. An established fouling community offers surface roughness for larvae to colonize the substratum. In the case of macrofouling by hard-shelled organisms like barnacles, oysters and tubeworms irreversible damage to the surface occurs and can be cleaned only by mechanical methods like chiselling. Not all places in a cooling water systems are accessible to cleaning and may result in replacement of the equipment. In many power stations, bivalves are the most dominant of the fouling organisms. Dosages and regimes required for preventing bivalve settlement are different to those required for removal or killing of already settled mussels. Further, byssal threads of mussels dead or detached tend to remain in the system leading to under-deposit corrosion and can enhance attachment opportunities for incoming fouling larvae (Claudi and Mackie 1994). Discontinuous chlorination was not effective in killing mussels even at concentrations of 0.5–1.5 mg L^{-1} residuals. The biocidal action of chlorine in killing mussels of the species *Mytilus edulis* and *Mytilus galloprovincialis* was found to be dependent on temperature. Residuals of 0.2–1.0 mg L^{-1} required 15–135 days for mortality (Lewis 1983). Toxicity modelling showed a tenfold decrease in the required killing time for mussels, when comparing mortality rates at 10 C and 25 C. Low-level continuous chlorination was more effective against mussel spat than on adults (Travade and Khalanski 1986) Adult mussels were able to survive the continuous chlorination (residuals of 0.29 and 0.49 mg L^{-1}) practised at the Gravelines plant. Adult mussels survived up to 5 months whereas the recently settled mussels were far more sensitive to chlorine. Settlement of spat was inhibited and a large portion of existing spat were found to detach and die at continuous residuals of 0.49 mg L^{-1} within 20–30 days. At residuals of 0.29 mg L^{-1} growth inhibition and detachment of spat was observed (Travade and Khalanski 1986).

Intermittent chlorinationwas ineffective in removing mussel community lodged in the intake tunnel of the Madras atomic power station (MAPS), India. Alternatively, a continuous high-level chlorination at residuals of 1.4 mg L^{-1} followed by continuous low-level chlorination at 0.2 mg L^{-1} dislodged the mussel community and about 187 tons of fouling biomass was collected in the travelling water screens (Rajagopal et al. 1996). Low-level continuous chlorination of 0.2 mg L^{-1} led to reduction in the growth of the shell of the mussel *Mytilus edulis* as observed from the growth rates of mussels in the cooling culverts (Thompson et al. 2000). The findings were consistent with introduced mussels also exhibiting the same trend. High-level continuous chlorination has also proven to be effective in eliminating mussels due to two processes: a decrease in water filtration rate, which deprives the mussel of its food, or a progressive intoxication by oxidant compounds absorbed within small amounts of seawater in the mantle cavity (Khalanski and Bordet 1980). In comparison, low continuous chlorine residuals of 1.0 mg L^{-1} took 468 h (7 mm) and 570 h (25 mm) for 100% mortality in the brackish water mussel *Brachidontes striatulus,* whereas high chlorine residuals of 5.0 mg L^{-1} took 102 h (7 mm) and 156 h (25 mm) for 100% mortality (Rajagopal et al. 1997).

In a cooling water system different species of mussels co-exist, hence species-specific variability in tolerance of mussels to chlorination is an important aspect in framing a treatment regime. Small sized mussels are more susceptible to chlorination than larger ones (Rajagopal et al. 2003). Similarly, response of different species of tropical marine mussels, *Perna viridis*, *Perna perna*, *Brachidontes striatulus*, *Brachidontes variabilis*and *Modiolus philippinarum,* to chlorination showed that reduction in physiological activities is the lowest in *P. viridis* and the highest in *B. variabilis* (Rajagopal et al. 2003). Mussels were able to tide-over continuous low-level or intermittent chlorination by closing their shell valves to overcome the period. Consequently, the technique of pulse chlorination(Poleman and Jenner 2002; European IPPC Bureau 2000) developed by KEMA has been found to be effective in controlling bivalve fouling in European power stations as off-treatment intervals occur when the mussels have shut their valves tight in response to the biocide. Chlorine must be applied continuously at least during spawning seasons to control bivalve settlement. On the other hand, semi-continuous treatments coupled with high frequency treatments (i.e. 15 min-on/15 min-off and 15 min-on/30 min-off) has shown good results for controlling mussels at residuals of 0.5 mg L^{-1} (Wiancko and Claudi 1994). Compared to mussels, oysters (*Crassostrea madrasensis*) attach to surfaces by cementing one of the valves to the substratum, posing severe problems. A continuous residual of 1.0 mg L^{-1} took 21 days for 100% mortality in the size group 13 mm and 31 days for the size group 64 mm (Rajagopal et al. 2003). Compared to mussels, barnacles tolerated high chlorine residuals of 1.0 mg

L^{-1} for up to 15 days, where 75% of them survived up to 5 days only (Turner et al. 1948). Another species of barnacles, *Balanus improvisus,* required 2.5 mg L^{-1} for 100% mortality at short exposure times (5 min) (McLean 1973).

Further chlorine dioxide residuals of 0.2 mg L^{-1} were found to kill bivalve mussels more rapidly than chlorine at concentrations of 1.1 mg L^{-1} (Jenner et al. 1998). Comparison of the efficacy of chlorine and chlorine dioxide on killing mussels has shown chlorine dioxide to be more effective at a concentration of 1.1 mg L^{-1}. Long-term semi-continuous addition of chlorine dioxide at residuals of 0.2 mg L^{-1} with time intervals of 1 h-on and 2 h-off is as efficient as continuous treatment (Belluati et al. 1997). Chlorine dioxide has also been reported to be effective against serpulid worms at a concentration of 0.2 mg L^{-1}. Experimental runs with chlorine and chlorine dioxide conducted at the Vandellos II nuclear power station on the Mediterranean coast of Catalonia in Spain showed that macrofouling was eliminated at chlorine dioxide concentrations of 0.16–0.20 mg L^{-1} (residuals of 0.04 mg L^{-1}) and chlorine at 1.1–1.2 mg L^{-1} (residuals of 0.3–0.4 mg L^{-1}). However, the cost difference between chlorine dioxide and electro-chlorination was found to be 30% (Jenner et al. 1998). In contrast to chlorine dioxide, a far lower concentration of ozone (0.1 mg L^{-1}) was required to eliminate bryozoans (*Plumatella emarginata*) (Duvivier et al. 1996). Chlorine dioxide has shown to be effective against established fouling communities compared to chlorine and seems a promising candidate for cooling water systems in the future. The cost economics of application of chlorine dioxide needs to be worked out for the treatment to be widely adopted. In comparison, studies using ozone for treating cooling water systems in power plants are also limited. Concentrations of 0.25–0.5 mg L^{-1} were effective in eliminating the blue mussel (*Mytilus*) from European power stations (Claudi and Mackie 1994). In another study, concentrations of 0.5 mg L^{-1} ozone were required for a period of 7–12 days for 100% mortality of mussels (Lewis et al. 1993). The features of ozone that make it attractive for treating once-through cooling water systems are also its major drawbacks. One of the major disadvantages of ozone is that it dissipates more rapidly in water, which in a way minimizes the downstream environmental impact. However, the short life of ozone in water requires multiple injection points in the cooling water system to protect downstream equipment, which would be probably very expensive and is the main reasons why this biocide is not so popular for large once-through cooling systems.

4.2.4 Fouling Control in Once-Through Freshwater Cooling Systems

In freshwater systems, fouling by Asiatic clams, Zebra mussels and weeds poses a severe problem. In once through systems using freshwater clogging of heat exchangers by Asiatic clams has been related to changes in flow configuration in the service water systems. Continuous chlorination of 0.6–0.8 mg L^{-1} was required for controlling Asiatic clam settlement. Fouling by colonial hydroid *Cordylophora caspia*is a problem in several European and American power stations. Chlorine residuals of 0.2–5.0 mg L^{-1} with exposure time of 105 min and short intermittent exposure of 20 min did not kills the animals but reduced their growth (Folino-Rorem and Indelicato 2005).

Compared to marine mussels, the freshwater mussel *Dreissena polymorpha*was less tolerant to chlorine. Continuous chlorination employed at the Ontario Hydro experimental station on Lake Erie at Nanticoke showed that at residuals of 0.3 mg L^{-1}, attachment of the mussels was inhibited. Discontinuous treatments (half hour on, once in 12 h) were ineffective even at higher dosages of 0.5–1.5 mg L^{-1}. Similarly, another power station (Cleveland Electric illuminating company installation on Lake Erie) operating on a discontinuous mode failed to kill the mussels at 0.3 mg L^{-1} (Barton 1990). In contrast, semi-continuous chlorination has shown promising results at residuals of 0.5 mg L^{-1} with high frequency treatments like 15 minon/15 min-off and 15 min-on/30 min-off (Wiancko and Claudi 1994). In comparison, response of the Zebra mussels to shock chlorination showed that two successive shocks of 200 mg L^{-1} once every 24 h resulted in 100% mortality of mussels in 9 days (Khalanski 1993). The action of chlorine in killing Zebra mussels was found to be dependent on water temperature. For 95% mortality at 10°C, a time period of 42 days was required as against 7 days at a water temperature of 25°C (Van Benschoten et al. 1993). The study also demonstrated that compared to chlorine, chloramine concentrations above 1.5 mg L^{-1} were effective in controlling veligers of Zebra mussels in both static and flow-through tests. Exposure times of 1,080 h at 0.25 mg L^{-1} and 252 h at 3.0 mg L^{-1} are required for 100% mortality of these mussels (Rajagopal et al. 2003). Continuous chlorination at residuals of 0.5 mg L^{-1} was effective in killing the Asiatic clams*C. fluminea* with periods ranging from 2–3 weeks (Dohorty et al. 1986; Ramsey et al. 1988). In addition, monochloramine (NH_2Cl) was found to be effective against the Asiatic clams (Belanger et al. 1991). With respect to the freshwater Zebra mussel-*Dreissena polymorpha*brief exposure to chlorine dioxide at a concentration of 10 mg L^{-1} for 13 min or 50 mg L^{-1} for 3.2 min kills 50% of adult mussels, whereas at concentrations of 2 mg L^{-1} no mortality is observed (Montanat et al. 1980). Concentrations of 5 mg L^{-1} in closed recirculating systems of the Seraing power station on the river Meuse were found to be effective, giving 100% mortality of the bivalves (*Corbicula* sp. and *Dreissena* sp.) over a period of 18 days (Jenner et al. 1998). Synthesis of the above information reveals that biocidal requirements for fouling control in freshwater once-through systems are far less than for seawater-cooled once-through systems. In freshwater cooling systems (where Zebra mussels and Asiatic clams are the dominant foulers) chlorine has been found to be the most effective and commonly used method of mussel control in Europe, Asia and North America (Jenner et al. 1998; Claudi and Mackie 1994; Rajagopal 1997).

5 New Approaches for Fouling Control in Heat Exchangers

5.1 Electrolytically Generated Biocides

Currently, electrochemical methods are being tested for treating industrial waters with the goal of combating fouling without adversely affecting the environment. Metal ions particularly silver, copper (Cu anodes) hydrogen peroxideand potassium permanganate can be electrolytically generated (Martinez et al. 2004). In heat exchangers made of titanium, anodic polarization by a current of some tens of milliamps per square metre applied to titanium causes a low production of oxidant species (chlorine or bromine) at the metal–seawater interface. However, with heavy metal ions there is always the problem of occurrence of resistance in the organisms. The current is low but is enough to inhibit the growth of titanium on plates and tubes. Experiments to this effect at the Venetian Lagoon demonstrated the control of settlement of macroorganisms and algae with a polarization of about 100–200 mA m^{-2}. This technique is very interesting for heat exchangers considering the effect of the pH decrease in the water close to the anodic polarized surface of titanium (Cristiani 2005). Electrolytically generated biocides are particularly useful in cooling systems to combat biofouling of sensors for temperature, conductivity and pH. This technique is still in infancy and in-situ studies demonstrating this effect are lacking. Further application of this technique to large cooling systems seems to be an unviable proposition.

5.2 *Surface Modification Approach to Control Biofouling*

Surfactants or surface active agents alter the surface tension within the biofilm and at the biofilm–substratum interface, allowing enhanced penetration by biocide molecules and also more effective removal of the biofilm deposits from the surfaces. However, they only address one of the forces that provide cohesion and adhesion of fouling layers. Several studies on the positive effects of the use of surfactants have been reported from cooling water systems. Some of the most effective surfactants reported are ethylene oxide/propylene oxide block copolymer (Donlan et al 1997), dimethlyamide (DMATO) (Lutey et al. 1989), dinonylsulfosuccinate (Wright and Michalopoulos 1996), a combination of peracetic acid with ethylene oxide/propylene oxide (Meade et al. 1997), sodium dodecyl sulfate (SDS) in combination with urea (Whittaker et al. 1984), and Tween 20 (Fletcher et al. 1991). Recently, low-energy surfaces have been prepared by ion implantation (Yang et al. 1994). Low-energy surfaces can increase the induction period of fouling and facilitate detachment of foulants (Yang et al. 1994; Forster et al. 1999) during which stable nucleation takes place at localized sites and the lateral growth of individual nucleation sites results in a complete coverage of the surface. Another study to minimize particle adhesion on stainless steel plate heat exchangers used TiN sputter coatings, which decreased the surface energy and resulted in less deposition of particles (Rosmaninho et al. 2005). Ion-sputtered diamond-like carbon (Forster et al. 1999), self-assembled monolayers (SAMs) and electroless plated surfaces (Yang et al. 2000b) have been used to mitigate fouling due to the weak adhesion strength between the fouling layer and the heat transfer surface. SAMs of low surface energy can prolong the induction period of fouling. Also, thermally resistant (Yang et al. 2000a). SAM surfaces based on Si wafers exhibit no significant change after heat treatment up to 200°C (Shin et al. 1999) and SAM films of hexadecyl disulfide can withstand temperatures of up to 225°C (Nuzzo et al. 1987). SAMs can also protect metals against corrosion as they act as effective barriers against diffusion of oxygen and water. tThe cross-linking SAMs can result in more robust films with improved levels of protection (Itoh et al. 1995).Thus, SAM surfaces have great potential for use as heat transfer surfaces to reduce fouling. The technique has not been used widely in heat exchangers except for some small stations that require an improvement in the rate of heat transfer. The technique as such seems to be interesting for plate-type heat exchangers and needs to be evaluated under field conditions, provided the film lasts over an extended period of time. However, a breakthrough against fouling has not been achieved yet by the use of surfactants. Surfactants comprise only one of many more components of integrated antifouling strategies.

6 Concluding Remarks

The incidence of macrofouling in cooling water circuits of power plants varies considerably depending on the location and design of systems. The use of trash racks at the offshore intake point and travelling water screens before the pumps is a mandatory technique for removing debris and detached fouling biomass from clogging the heat exchangers. Thermal treatment is an effective option but many stations do not have the facility of recirculating effluent water in the cooling circuits. Chlorine or sodium hypochlorite is in common use internationally and requires high doses (0.5–1.0 mg L^{-1}) to overcome the demand in water and for killing macrofouling organisms. However, for effective plant operation the issue of killing established macrofoulants is secondary to preventing their settlement and colonization, right from the initial stages of commissioning the cooling water circuit. A fouled circuit provides a source of larvae, which colonize systems downstream, and dosages required for inhibiting settlement are far less than those for killing established communities. In general, power stations adopt low-level continuous chlorination (with residuals of 0.1–0.2 mg L^{-1}) at the outfall coupled with periodic shock or booster doses of the biocide depending on the intensity of fouling. The use of various techniques of chlorination, like shock chlorination and targeted chlorination of heat exchangers, has offered temporary relief to certain sections or equipment in the circuit. With the advent of the technique of pulse chlorination, up to 50% reduction in chlorine consumption can be achieved and bring down the environmental burden of toxic by-products (Poleman and Jenner 2002; European IPPC Bureau 2000). Commercial variants of bromine are in use in some of the European and Indian power stations. However, growing awareness of keeping biofouling levels within the threshold to minimize plant shutdown and the increasing regulations on effluent discharge has resulted in plant operators adopting the stronger oxidant, i.e. chlorine dioxide. The low concentrations required for killing and its environmentally safe nature have resulted in its use in power plants in spite of the higher cost of this biocide. Prior to the commissioning of a power station, effective biocidal dose and concentration need to be worked out on a site-specific basis. Dosages worked out elsewhere will not be effective for a given location.

Side-stream monitors(Biobox) (Jenner et al. 1998) or electrochemical probes need to be installed and monitored regularly. Spikes in settlement observed in side-stream monitors can be taken as a signal to alter the biocide dose or regime to achieve killing of new settlers. Continuous surveillance and monitoring of cooling water and fine tuning of the biocidal programme will ensure that biofouling levels are maintained well within the threshold limits. As biofouling is a surface-associated phenomenon, a combined approach of treating the cooling water and surface protection by the use of foul release coatings would offer long term solutions to macrofouling problems in cooling systems. Fouling release coatings have demonstrated their ability to resist macrofouling at high water velocities and are a potential option for cooling circuits (Leitch 1993; Kilgour and Mackie 1993; Claudi and Mackie 1994). With regard to heat exchanger fouling, mechanical methods like sponge rubber ball cleaning together with biocidal treatment is the only available method of control. In shell and tube heat exchangers flow blockage due to clogging of tubes by macrofoulants and biofilm is the primary problem. Contrary to the concept that high shear forces created by chevron angles in plate heat exchangers retard fouling, barnacle fouling on these heat exchangers has been observed. Sedimentation and accumulation of corrosion products on these heat exchangers is a problem to be overcome. Since online cleaning methods are not available for these heat exchangers a control strategy should take into account a biocide, cleaners and a corrosion inhibitor for optimum performance of these heat exchangers.

References

Ambrogi R (1997) Environmental impact of biocidal antifouling alternative treatments on seawater once-through cooling systems. In: Chlorine dioxide and disinfection. Proceedings of first European symposium on chlorine dioxide and disinfection, Rome, 7–8 Nov 1996. Collana Ambiente 17 pp. 119–132

Aprosi G (1988) Bryozoans in cooling water circuits of a power plant. Verh Int Verein Limnol 23:1542–1547

Barton LK (1990) Control of zebra mussels at CEI facilities. Annual meeting of the American power conference, 23–25 April 1990, Chicago, pp. 1001–1005

Bass ML, Heath AG (1977) Cardiovascular and respiratory changes in rainbow trout, *Salmo gairdneri* exposed intermittently to chlorine. Water Res 11:497–502

Belanger SE, Cherry DS, Farris JL, Sappington KG, Cairns J (1991) Sensitivity of the Asiatic clam to various biocidal control agents. J Am Water Works Assoc 83:79–87

Bell KJ (1977) The effect of fouling on heat exchanger design, construction and operation. In: Proceedings of the Ocean Thermal Energy Conversion (OTEC) biofouling and corrosion symposium, pp. 19–29

Belluati M, Bartole L, Bressan G (1997) Chlorine dioxide and disinfection. In: Proceedings of first European symposium on chlorine dioxide and disinfection, Rome, 7–8 Nov 1996. Collana Ambiente 17

Bott TR (1993) Aspects of biofilm formation and destruction. Corrosion Rev 11(1–2):1–24

Bott TR (1995) Fouling of heat exchangers. Elsevier, Amsterdam, p. 576

Bott TR, Tianqing L (2004) Ultrasound enhancement of biocide efficiency. Ultrason Sonochem 11:323–326

Brankevich GJ, De Mele MLF, Videla HA (1990) Biofouling and corrosion in coastal power plant cooling water systems. MTS J 24:18–28
Butterfield PW, Camper AK, Ellis BD, Jones WL (2002a) Chlorination of model drinking water biofilm: implications for growth and organic carbon removal. Water Res 36:4391–4405
Butterfield PW, Camper AK, Biederman A, Bargmeyer AM (2002b) Minimizing biofilm in the present of iron oxides and humic substances. Water Res 36:3898–3910
Chow W (1987) Targeted chlorination schedules and corrosion evaluations. American Power Conference Chicago
Claudi R, Mackie GL (1994) Practical manual for Zebra Mussel monitoring and control. Lewis, Boca Raton, p. 227
Collins TM (1964) A method for designing seawater culverts using fluid shear for the prevention of marine fouling. CERL report no RD/1/N93/64. Central Electricity Research Laboratories, Leatherhead, pp. 1–5
Corfield J, Spooner D, Ray D, Clearwater S, Macaskill B, Roper D (2004) Biofouling issues for the Marsden B power station cooling water system. NIWA client report HAM 2005–106 NIWA Project MRP05251. National Institute of Water and Atmospheric Research, Wellington, NZ
Corpe WA (1977) Marine microfouling and OTEC heat exchangers. Proceedings of the Ocean Thermal Energy Conversion (OTEC) biofouling and corrosion symposium, pp. 31–44
Coughlan J, Whitehouse J (1977) Aspects of chlorine utilization in the UK. Chesapeake Sci 18(1):102–111
Cristiani P (2005) Solutions to fouling in power station condensers. Appl Therm Eng 25:2630–2640
Dohorty FG, Farris JL, Cherry DS, Cairns Jr. J (1986) Control of freshwater fouling bivalve *Corbicula fluminea* by halogenation. Arch Environ Contam Toxicol 15:535–542
Donlan RM (2000) Biofilm control in industrial water systems: approaching an old problem in new ways. In: Evans LV (ed.) Biofilms: recent advances in their study and control. Harwood Academic, pp. 333–361
Donlan RM, Elliott DL, Gibbon DL (1997) Use of surfactants to control silt and biofilm deposition onto PVC fill in cooling water systems. International water conference, paper no IWC-97–73
Drake RC (1977) Increasing heat exchanger efficiency through continuous mechanical tube maintainance. In: Jensen LD (ed.) Biofouling control procedures technology and ecological effects. Marcel Dekker, New York pp. 43–53
Duvivier L, Leynen M, Ollivier F, Van Damme A (1996) Fighting zebra mussel fouling with ozone. In: Proceedings of the EPRI PSE Service water systems reliability improvement seminar 25–27 June 1996, Daytona Beach, FL, p. 14
Electric Power Research Institute (EPRI) (1980) Review of open literature on effects of chlorine on aquatic organisms. Report to the EPRI by Oak Ridge National Laboratory. EPRI, Palo Alto, CA
Engineering Sciences Data Unit (ESDU) (2000) Heat exchanger fouling in the pre-heat train of a crude oil distillation unit. ESDU item 00016, ESDU, London
Epstein N (1983) Thinking about heat transfer fouling: a 5 × 5 matrix. Heat Transfer Eng. 14(1):25–36
European IPPC Bureau (2000) Document on the application of best available techniques to industrial cooling systems, November 2000. BREF (11.00) Cooling systems. European IPPC Bureau, Sevilla. http://eippcb.jrc.es. Last accessed 14 July 2008
Fisher DJ, Burton DT, Yonkos LT, Turley SD, Ziegler GP (1999) The relative acute toxicity of continuous and intermittent exposures of chlorine and bromine to aquatic organisms in the presence and absence of ammonia. Water Res 33(3):760–768
Flemming HC (2002) Biofouling in water systems-cases, causes and countermeasures. Appl Microbiol Biotechnol 59:629–640
Flemming HC, Griebe T (2000) Control of biofilms in industrial waters and processes. In: Walker J, Surman S, Jass (eds.) Industrial biofouling. Wiley, London, pp. 125–137
Fletcher M, Lessmann JM, Loeb GI (1991) Bacterial surface adhesives and biofilm matrix polymers of marine and freshwater bacteria. Biofouling 4:129–140

Folino-Rorem NC, Indelicato J (2005) Controlling biofouling caused by the colonial hydroid *Cordylophora caspia*. Water Res 39:2731–2737

Forster M, Aufustin W, Bohnet M (1999) Influence of the adhesion force crystal/heat exchanger surface on fouling mitigation. Chem Eng Process 39:449–461

Fritsch A, Adamson W, Castelli V (1977) An evaluation of mechanical cleaning methods for removal of soft fouling from heat exchanger tubes in OTEC power plants. Proceedings of the Ocean Thermal Energy Conversion (OTEC) biofouling and corrosion symposium. pp. 159–166

Geraci S, Ambrogi R, Festa V, Piraino S (1993) Field and laboratory efficacy of chlorine dioxide as antifouling in cooling systems of power plants. Oebalia 19:383–393

Goodman PD (1987) Effect of chlorination on materials for seawater cooling system: a review of chemical reactions. Br Corros J 22(1):56–62

Gunasingh M, Jesudoss KS, Nandakumar K, Satpathy KK, Azariah J, Nair KVK (2002) Lethal and sub-lethal effects of chlorination on green mussel Perna viridis in the context of biofouling control in a power plant cooling water system. Mar Env Res 53:65–76

Hesselgreaves JE (2002) An approach to fouling allowances in the design of compact heat exchangers. App Therm Eng 22:755–762

Ingols RS, Wyckoff HA, Kethley TW, Hodgden HW, Fincher EL, Hildebrand JC, Mandel JE (1953) Bactericidal studies of chlorine. Ind Eng Chem 45:996–1000

Itoh M, Nishihara H, Aramaki K (1995) Preparation and evaluation of two dimensional polymer films by chemical modification of an alkanethiol self assembled monolayer for protection of copper against corrosion. J Electrochem Soc 142:3696–3704

James WG (1985) Mussel fouling and use of exomotive chlorination. Chem Ind 994–996.

Jenner HA (1980) The biology of the mussel *Mytilus edulis* in relation to fouling problems in industrial cooling water systems. Janvier 434:13–19

Jenner HA (1982) Physical methods in the control of mussel fouling in seawater cooling systems. Trib Cebedeau 35:287–291

Jenner HA (1983) Control of mussel fouling in The Netherlands: experimental and existing methods. In: Tous IAD, Miller MJ, Mussalli YG (eds.) Condenser macrofouling control technologies: the state of the art. Electric Power Research Institute, Hyannis. MA pp. 12–24

Jenner HA, Davis MH (1998) A new design in monitoring fouling: KEMA Biofouling Monitor Fawley Biobox. In: Abstracts from the eighth international zebra mussel and other nuisance species conference, Sacramento CA, 16–19 March 1998

Jenner HA, Janssen-Mommen JPM (1993) Monitoring and control of *Dreissena polymorpha* and other macrofouling bivalves in the Netherlands. In: Nalepa TF, Schloesser DW (eds.) Zebra mussels biology, impacts and control. Lewis, Boca Raton pp. 537–554

Jenner HA, Whitehouse JW, Taylor CJL, Khalanski M (1998) Cooling water management in European power stations: biology and control of fouling. Hydroecologie Appliquee 10(1–2):225

Johnson KI, Henager CH, Page TL, Hayes PF (1986) Engineering factors influencing *Corbicula* fouling in Nuclear Service Water Systems. In: Proceedings of the EPRI symposium on condenser macrofouling control technologies: the state of the art. Electric Power Research Institute, Palo Alto, CA

Kawabe A, Treplin FW (1986) Control of macrofouling in Japan, existing and experimental methods. Res Rep 23:1–42

Khalanski M (1993) Testing of five methods for the control of zebra mussels in cooling circuits of power plants located on the Moselle River. Proceedings: third international Zebra mussel conference, Toronto, Canada

Khalanski M, Bordet F (1980) Effects of chlorination on marine mussels. In: Jolly RL, Brungs WA, Cumming RB (eds.) Water chlorination chemistry environmental impacts and health effects, vol 3. Ann Arbor Science, Ann Arbor, Michigan, pp. 557–567

Kilgour BW, Mackie GL (1993) Colonization of different construction materials by the Zebra Mussel *Dreissena polymorpha* (Bivalvia: Dreissenidae), In: Nalepa TF, Schloesser DW (eds) Zebra mussels biology, impacts and control. Lewis, Boca Raton, pp. 167–173

Kovalak WP, Longton GD, Smithee RD (1993) Infestation of power plant water systems by the zebra mussel (*Dreissena polymorpha* Pallas). In: Nalepa TF, Schloesser DW (eds) Zebra mussels: biology, impacts and control. Lewis, Boca Raton, pp. 359–380
Kukulka DJ, Devgun M (2007) Fluid temperature and velocity effect on fouling. Appl Thermal Eng 27:2732–2744
Leitch EG (1993) Evaluation of coatings for "Zebra Mussels". Agenda and abstracts of third international Zebra mussel conference, Toronto, Canada, 1993
Lewis BG (1983) Development of a preliminary model to predict time in days for 100% kill of mussels subject to continuous chlorination. CERL note no TPRD/L/2469/N83, Central Electricity Research Laboratories, Leatherhead
Lewis D, Van Benschoten JE, Jensen (1993) A study to determine effective ozone dose at various temperatures for inactivation of adult Zebra mussels. Unpublished report for Ontario Hydro
Leglize L, Ollivier F (1981) Mise au point bi-bliographique sur la biologie et l'ecologie de *Dreissena polymorpha* PALAS (1771) repartition geographique en France et dans les pays limitrophes. EDF/DER Report HE/31–81.37
Ludensky M (2003) Control and monitoring of biofilms in industrial applications. Int Biodeter Biodegr 51:255–263
Lutey RW, King VM, Gleghorn M (1989) Mechanisms of action of dimethylamides as a penetrant/dispersant in cooling water systems. International water conference, paper no IWC-89–33
Martinez SS, Gallegos AA, Martinez E (2004) Electrolytically generated silver and copper ions to treat cooling water: an environmentally friendly novel alternative. Int J Hydrogen Energy 29:921–932
McLean RI (1973) Chlorine and temperature stress on selected estuarine invertebrates. Am Zool J Water Pollut Cont Fed 45:837–841
McMahon RF, Ussery TA, Miller AC (1993) Thermal tolerance in Zebra mussels (*Dreissena polymorpha*) relative to rate of temperature increase and acclimation temperature. In: Agenda and abstracts of the third international Zebra mussel conference, Toronto, Canada, 1993. pp 97–118
Meade RJ, Robertson LR, Taylor NR, LaZomby JG (1997) Method for preventing microbial deposits in the papermaking process with ethylene oxide/propylene oxide copolymers. US Patent No 5: 624, 575
Meesters KPH, Groenestijn JWV, Gerritse J (2003) Biofouling reduction in recirculating cooling systems through biofiltration of process water. Water Res 37:525–532
Melo LF, Bott TR (1997) Biofouling in water systems. Exp Therm Fluid Sci 14:375–381
Meyer B (2003) Approaches to prevention, removal and killing of biofilms. Int Biodeter Biodegr 51:249–253
Montanat M, Merle G, Migeon B (1980) Essai d'utilisation du dioxide de chlore dans la boucle TERA installee a Montereau. Resultats des tests de toxicite. EDF DER Report HE/3180.044
Mott IEC, Bott TR (1991) The adhesion of biofilms to selected materials of construction for heat exchangers. Proceedings 9th international heat transfer conference, Jerusalem. 5:21–26
Mott IEC, Santos R, Bott TR (1994) The effect of surface on the development of biofilms. Proceedings 3rd meeting European adhesion of micro-organisms to surfaces, Chatenay-Malabry. C21:1–2
Muller-Steinhagen H (1993) Fouling: the ultimate challenge for heat exchanger design. In: Lee JS, Chung SH, Kim KY (eds.) Transport phenomenon in thermal engineering, vol 2. Begell House, pp 811–823
Murthy PS, Venkatesan R, Nair KVK, Inbakandan D, Syed Jehan S, Magesh Peter D, Ravindran M (2005) Evaluation of sodium hypochlorite for fouling control in plate heat exchanger for seawater application. Int Biodeter Biodegr 55:161–170
Murthy PS, Venkatesan R, Nair KVK, Subramoniam T (2008) Larval settlement and surfaces: implications in development of antifouling strategies. Springer Ser Biofilms. doi: 10.1007/7142_2008_17
Neitzel DA, Johnson KI, Page TL, Young JS, Dalling PM (1984) Correlation of bivalve biological characterestics and service water system design. Bivalve fouling of nuclear power plant service

water systems, vol 1. NUREG/CR-4070, PNL-5300, vol 1.US Nuclear Regulatory Commission, Washington DC

Nuzzo RG, Fusco FA, Allara DL (1987) Spontaneously organized molecular assemblies 3. Preparation and properties of solution adsorbed monolayers of organic disulfides on gold surfaces. J Am Chem Soc 109:2358–2368.

Oldman J, Clearwater S, Hickey C, Macaskill B, Grange K, Handley S, Ray D (2004) Marsden B coal fired power station: effects of cooling water abstraction and discharge. NIWA Client Report HAM2004–108

Ormerod K, Lund Y (1995) The influence of disinfection processes on biofilm formation in water distribution systems. Water Res 29:1013–1021

Petrucci G, Rosellini M (2005) Chlorine dioxide in seawater for fouling control and cost of disinfection in potable waterworks. Desalination 182:283–291

Poleman HJG, Jenner HJ (2002) Pulse chlorination the best available technique in macrofouling mitigation using chlorine. Power Plant Chem 4:93–97

Prince EL, Muir AVG, Thomas WM, Stollard RJ, Sampson M, Lewis JA (2002) An evaluation of the efficacy of aqualox for microbiological control of industrial cooling tower systems. J Hosp Infect 52:243–249

Rabas TJ, Panchal CB, Sasscer DS, Schaefer R (1993) Comparison of river water fouling rates for spirally indented and plain tubes. Heat Trans Eng 14(4):58–73

Rajagopal S (1997) The ecology of tropical marine mussels and their control in industrial cooling water systems. Ph.D. thesis, University of Nijmegen, Nijmegen, The Netherlands

Rajagopal S, Van Der Velde G, Jenner HA (1994) Biology and control of brackish water mussel, *Mytilopsis leucophaeta* in the Velsen and Hemweg power stations, The Netherlands. Part II: control measures. Report no 163871-KES/WBR 94–3128, KEMA Environmental Research, pp. 45

Rajagopal S, Venugopalan VP, Azariah J, Nair KVK (1995) Response of the green mussel Perna viridis (L) to heat treatment in relation to power plant biofouling control. Biofouling 8:313–330

Rajagopal S, Nair KVK, Azariah J, Van der Velde G, Jenner HA (1996) Chlorination and mussel control in the cooling conduits of a tropical power station. Mar Environ Res 41(2):201–221

Rajagopal S, Nair KVK, Van der Velde G, Jenner HA (1997) Response of mussel *Brachidontes striatulus* to chlorination: an experimental study. Aquat Toxicol 39:135–149

Rajagopal S, Van der Velde G, Van der Gaag M, Jenner HA (2003) How effective is intermittent chlorination to control adult mussel fouling in cooling water systems. Water Res 37:329–338

Rajagopal S, Venugopalan VP, van der Velde G, Jenner HA (2006) Mussel colonization of a high flow artificial benthic habitat: byssogenesis holds the key. Mar Environ Res 62:98–115

Ramsey TGG, Tackett JH, Morris DW (1988) Effect of low-level continuous chlorination on *Corbicula fluminea.*Environ Toxicol Chem 7:855–856

Reid DC, Bott TR, Miller R (1992) Biofouling in stirred tank reactors effect of surface finish. In: Melo LF, Bott TR, Fletcher M, Capdeville B (eds.) Biofilms science and technology. Kluwer, Dordrecht, pp. 521–526

Rosmaninho R, Rizzo G, Muller-Steinhagen H, Melo LF (2005) Antifouling stainless steel based surfaces for milk heating processes. In: Muller-Steinhagen H, Malayeri MR, Watkinson AP (eds.) Proceedings of 6th international conference on heat exchanger fouling and cleaning challenges and opportunities, Kloster Irsee, Germany5–10 June 2005. ECI Symp Ser RP2:97–102

Roy D, Wong PK, Engelbrecht RS, Chian ES (1981) Mechanism of enteroviral inactivation by ozone. Appl Environ Microbiol 41:718–723

Sarunac N, Guida V, Weisman R, Zhang Z (1994) Integration of macrobiofouling control methods into an effective macrobiofouling control strategy. Proceedings condenser technology conference EPRI Report TR-1034753.39–3.68

Sasikumar N, Nair KVK, Azariah J (1992) Response of barnacles to chlorine and heat treatment: an experimental study for power plant biofouling control. Biofouling 6:69–79

Shin HJ, Wang YH, Sukenik CN (1999) Pyrolysis of self-assembled organic monolayers on oxide substrates. J Mater Res 14:2116–2123

Sheikholeslami R (2000) Composite fouling of heat transfer equipment in aqueous media a review. Heat Transfer Eng 21:34–42

Smeltzer MS (2008) Biofilms and aseptic loosening. Springer Ser Biofilms. doi: 10.1007/7142_2008_1

Sohn J, Amy G, Cho J, Lee Y, Yoon Y (2004) Disinfectant decay and disinfection by-products formation model development: chlorination and ozonation by-products. Water Res 38:2461–2478

Strauss SD (1989) New methods, chemical improve control of biological fouling. Power 51–52

Strauss SD, Puckorius PR (1984) Cooling water treatment for the control of scaling, fouling and corrosion. Power S1–S24

Syrett BC, Coit RL (1983) Causes and prevention of power plant condenser tube failures. Mater Perform 4:44–50

Thompson IS, Richardson CA, Seed R, Walker G (2000) Quantification of Mussel (*Mytilus edulis*) growth from power station cooling waters in response to chlorination procedures. Biofouling 16(1):1–15

Travade F, Khalanski M (1986) Controle de la salissure biologique des circuits de centrals thermiques cotieres parles moules (*Mytilus edulis*). Etude d'optimisation de la chlorination a Gravelines Haliotis 15:265–273

Turner HJ, Reynolds DM, Redfiled AC (1948) Chlorine and sodium pentachlorphenate as fouling preventatives in sea water conduits. Ind Eng Chem 40:450–453

Tuthill AH (1985) Sedimentation in condensers and heat exchangers. Causes Effects Power Eng 89:46–49

US Environmental Protection Agency (USEPA) (1994) National primary drinking water regulations: disinfectants and disinfection. USEPA, Washington, DC

US Environmental Protection Agency (USEPA) (2002) National recommended water quality criteria 2002: EPA-822-R-02–047. USEPA, Office of Water, Washington DC

Van Benschoten UJE, Jensen JN, Lewis D, Brady TJ (1993) Chemical oxidants for controlling Zebra mussels (*Dreissena polymorpha*): a synthesis of recent laboratory and filed studies. In: Nalepa TF, Schloesser DW (eds.) Zebra mussels biology, impacts and control. Lewis, London, pp. 599–619

Venugopalan VP, Nair KVK (1990) Effects of a biofouling community on cooling water characterestics of a coastal power plant. Can J Mar Sci 8:233–245

Vieira MJ, Oliveira R, Melo LF, Pinheiro MM, Van der Mei HC (1992) Biocolloids and biosurfaces. J Dispersion Sci Tech 13(4):437–445

Vreenegoor SM, Block RM, Rhoderick JC, Gullans SR (1977) The effects of chlorination on the osmoregulatory ability of the blue crab *Callinectes sapidus* Assoc. Southeastern Biol Bull 24–93

Walsh SE, Maillard JY, Russell AD, Catrenich CE, Charbonneau DL, Bartolo RG (2003) Development of bacterial resistance to several biocides and effects on antibiotic susceptibility. J Hosp Infect 55:98–107

Whitehouse JW (1985) Marine fouling and power stations. A collaborative research report by CEGB, EDF, ENEL and KEMA, pp. 1–48

Whittaker C, Ridgeway HR, Olson BH (1984) Evaluation of cleaning strategies for removal of biofilms from reverse osmosis membranes. Appl Environ Microbiol 48:395–403

Wiancko PM, Claudi R (1994) The Ontario Hydro final strategy for Zebra mussel control. Fourth international Zebra mussel conference, Madison, Wisonsin, USA, pp. 225

Wright JB, Michalopoulos DL (1996) Method and composition for enhancing biocidal activity. European Patent No EP0741109A2

Yang CF, Xu DQ, Shen ZQ (1994) A theoretical analysis and experimental study of the induction period of calcium carbonate scaling. J Chem Ind Eng (China) 45:199–205

Yang QF, Ding J, Shen ZQ (2000a) Investigation on fouling behaviors of low energy surface and fouling fractal characterestics. Chem Eng Sci 55:797–805

Yang QF, Ding J, Shen ZQ (2000b) Investigation of calcium carbonate scaling on ELP surface. J Chem Eng Jpn 33:591–596

Young IC, Rong L, William JM, Lewis F (2000) Study of scale removal methods in a double pipe heat exchanger. Heat Transfer Eng 21:50–57

Inhibition and Induction of Marine Biofouling by Biofilms

S. Dobretsov

Abstract Microbial biofilms, predominantly composed of bacteria and diatoms, affect the settlement of invertebrate larvae and algal spores by production of bioactive compounds that can inhibit or induce settlement of biofoulers. In this review I summarize the studies on the inductive and inhibitive properties of biofilms. Particular attention has been given to antifouling and inductive compounds from marine microorganisms and quorum sensing signaling in prokaryotes and eukaryotes. Additionally, future research directions in the field of marine microbial biofouling are highlighted.

1 Introduction: Biofilms and Biofouling

Any natural and man-made substrates in the marine environment are quickly subject to biofouling, which is due to different species of micro- and macroorganisms (Railkin 2004). The process of biofouling has three main stages: adsorption of dissolved organic molecules, colonization by prokaryotes and eukaryotes, and subsequent recruitment of invertebrate larvae and algal spores (Maki and Mitchell 2002). These stages can overlap, be successional, or occur in parallel.

Aggregates of microorganisms adhered to each other and/or to surfaces with a distinctive architecture can be referred as biofilms (Maki and Mitchell 2002). In marine environments biofilms mainly consist of numerous species of bacteria and diatoms (Railkin 2004) incorporated into a matrix of extracellular polymers (EPS) composed of high molecular weight polysaccharides (Donlan 2002). Other unicellular organisms, like flagellates, yeasts, sarcodines and ciliates, contribute less than 1% to the total number of cells in biofilms (Railkin 2004).

S. Dobretsov
Marine Science and Fisheries Department, Agriculture and Marine Sciences College, Sultan Qaboos University, Al-Khod 49, PO Box 123, Sultanate of Oman
Benthic Ecology, IFM-GEOMAR, Kiel University, Düsternbrooker Weg 20, 24105, Kiel, Germany
e-mail: sergey_dobretsov@yahoo.com, sergey@squ.edu.om

Springer Series on Biofilms, doi: 10.1007/7142_2008_10

Biofilm development is a multistep process dependent on the properties of the substratum and environment and on the composition of the biofilm. At the same time, all biofilms in their development pass through several stages, such as attachment of cells and slime production, biofilm growth and, finally, biofilm sloughing and cell detachment (Lewandowski 2000).

Every biofilm is unique and heterogeneous in space and time, ranging from single layer of bacterial cells to multilayer biofilms containing numerous species of bacteria, diatoms, *Archaea*, and *Eucarya* (Donlan 2002). Changes in environmental conditions, such as water turbulence, temperature, salinity, light regime, and amount of nutrients, immediately change the composition of biofilms (Wieczorek and Todd 1998; Lau et al. 2005) and the production of chemical compounds (Miao et al. 2006).

Bacteria in biofilms control their growth and densities by a regulatory mechanism named quorum sensing (QS). It consists of production and release of low molecular weight signal molecules that activate or de-activate target bacterial genes responsible for cell division and adhesion. Quorum sensing signals can play a role in interactions between bacteria and higher organisms, such as the squid *E. scolopes* (Ruby and Lee 1998) and the alga *Ulva* (*Enteromorpha*) sp. (Tait et al. 2005). Some marine organisms have the capacity to interfere with bacterial QS signals in order to control biofilm formation on their surface (Zhang and Dong 2004).

Biofilms can have a substantial impact on biofouling communities by mediation of protist's colonization, the settlement of invertebrate larvae and macroalgal spores (Qian 1999; Egan et al. 2002; Huang and Hadfield 2003). On the whole, the physical

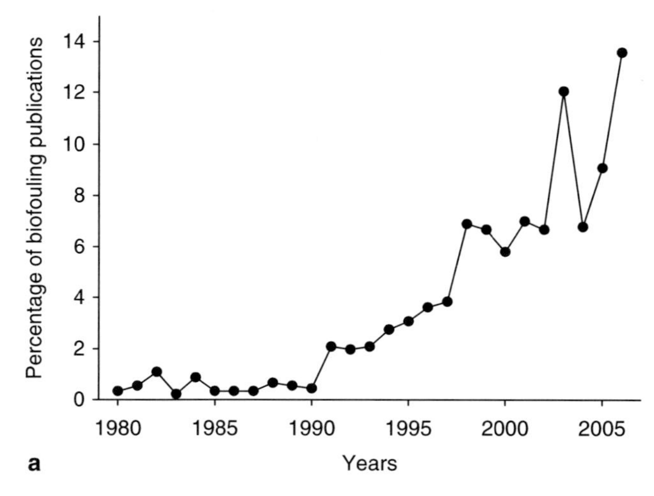

Fig. 1 Marine biofouling-related publication trends in the scientific literature. To access the frequency of marine biofouling-related papers, we ran a search on the Web of Science (Science Citation Index) for the period 1980–2006. **a** Rate of marine biofouling publications. **b** Main topics in marine biofouling studies. **c** Publications about the main marine biofouling organisms. Our search terms were "marine" plus specific terms presented in this figure. Because some articles considered multiple topics, the bars in **b** and **c** add up to more than 100%

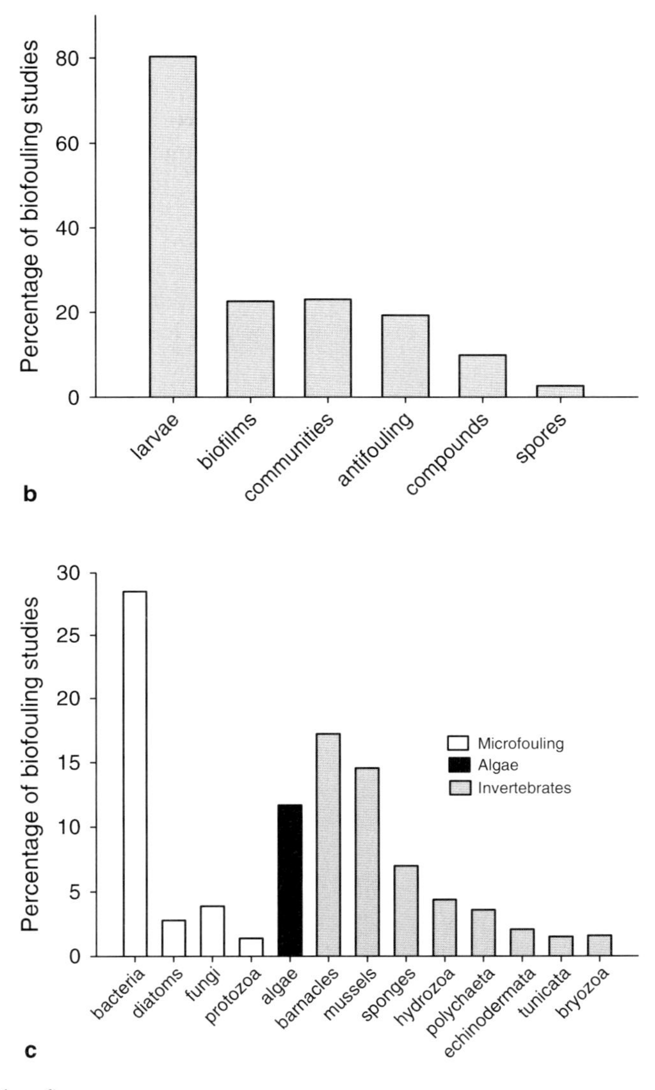

Fig. 1 (continued)

and chemical properties of biofilms, their composition, and accumulated chemical compounds inside the EPS matrix affect formation of biofouling communities (Maki and Mitchell 2002; Railkin 2004; Thiyagarajan et al. 2006).

Since 1972, the literature on biofouling has grown dramatically (Fig. 1a). This can be explained by several reasons. Biofouling is a serious problem for marine industries, aquaculture, and navies around the world (Railkin 2004; Yebra et al. 2004). The most effective methods of biofouling control are based on the application of highly toxic substances, like tributyl tin (TBT) and copper (Yebra et al. 2004). The recent ban of TBT and other tin-containing substances in antifouling paints stimulated biofouling studies and increased the necessity to find "environmentally friendly" non-toxic defensive compounds against biofouling.

According to the Web of Science, most biofouling publications deal with single species of invertebrate larvae, while investigations of biofilms and biofouling communities received less attention from researchers (Fig. 1b). Among biofilm studies, bacteria-related publications dominate (Fig. 1c).

In this review, I shall mainly cover different aspects of microbe–larval interactions. A number of reviews over the past years have presented information in part and in full about the antifouling compounds from cyanobacteria (Dahms et al. 2006), bacteria (Maki and Mitchell 2002; Dobretsov et al. 2006), and marine organisms (Railkin 2004, Fusetani et al. 2006), therefore these aspects will not be covered here and the reader is directed to these publications. I here review the interactions between microbes and macrofoulers with special emphasis on:

1. Induction and inhibition of larval settlement by biofilms
2. Chemical quorum sensing signaling between prokaryotes and eukaryotes
3. Directions for the future investigations

2 Effect of Biofilms on Larval Settlement

Biofilms are known to be important for the settlement of marine invertebrate larvae and spores of macroalgae (see reviews Wieczorek and Todd 1998; Maki and Mitchell 2002; Railkin 2004; Dobretsov et al. 2006). Marine biofilms can enhance (Kirchman et al. 1982; Patel et al. 2003; Qian et al. 2003; Huang and Hadfield 2003), inhibit (Maki and Mitchell 2002; Holmström et al. 2002; Egan et al. 2001; Dobretsov and Qian 2002, 2004) or have no effect on settlement of marine invertebrate larvae and macroalgal spores (Wieczorek and Todd 1998).

Different species of algae and invertebrates respond to biofilms differently. Larvae of specialist species may settle only on biofilms with a specific microbial composition, while generalists can settle on any kind of biofilm. The settlement response of the generalist larvae of *Hydroides elegans* depends on the density of bacteria in biofilms but has no relationship with bacterial community composition (Qian 1999; Unabia and Hadfield 1999; Lau et al. 2005; Nedved and Hadfield 2008). In contrast, larvae of *Balanus amphitrite* and *B. improvisus* differentiated the composition of intertidal and subtidal biofilms in experiments where there was a choice of sensing different settling cues (Qian et al. 2003; Thiyagarajan et al. 2005). Analogously, larvae of the bryozoan *Bugula neritina* in all cases attached on the subtidal biofilms when offered a choice of biofilms from different tidal regions (Dobretsov and Qian 2006). Since microbial community composition in biofilms varies substantially among tidal zones and substrata, the differential response of specialist larvae to biofilm composition may allow the larvae to evaluate substrata and then selected the suitable ones (Qian et al. 2003).

2.1 Induction of Larval Settlement by Biofilms

Numerous studies demonstrated an induction of larval settlement by single species of bacteria or diatoms (Table 1, see reviews: Wieczorek and Todd 1998; Maki and Mitchell 2002; Railkin 2004). Additionally, multispecies biofilms promote settlement and metamorphosis of polychaetes, hydroids, bryozoans, mollusks, tunicates, and barnacles (Table 1). The data presented in this table, clearly show that most of publications were focused on species with high economic significance, either because they can cause a significant biofouling problem (e.g., barnacles and polychaetes) or because they are commercially exploited (e.g., mollusks). The suitability of biofouling species for bioassays is another criterion for the selection of study objects. Most species that have been used can be reared and grown in the laboratory (e.g., barnacles, mollusks, and polychaetes) or can release lecitotrophic larvae (e.g., some bryozoa), which do not require intensive culturing and larval feeding (Wieczorek and Todd 1998).

Up to now only a limited number of biofouling field studies have been performed (Table 1). In most cases, investigators have studied the effect of monospecies microbial films on larval settlement and metamorphosis in laboratory experiments. These studies give only limited information about the role of biofilms in the field and cannot adequately predict the settlement of larvae and spores in response to mixed-species biofilms (Lau et al. 2002), in which most bacterial strains cannot be cultivated (Dobretsov et al. 2006).

The effect of diatoms on settlement and metamorphosis of invertebrate larvae has not been well documented (Table 1). Most of the diatoms isolated from Hong Kong biofilms induced larval settlement of the tube worm *Hydroides elegans* in laboratory experiments (Harder et al. 2002a). Further investigation demonstrated that carbohydrates but not proteins from EPS of the diatoms *Achnanthes* sp. and *Nitzschia constricta* induced larval settlement of *H. elegans* (Lam et al. 2005). Larval attachment of the bryozoan *Bugula neritina* correlated with the density of diatoms in biofilms (Kitamura and Hirayama 1987). No evidence that zoo- or phytoflagellates can induce or inhibit larval settlement was detected, despite the facts that flagellates are the third largest group in marine biofilms and that the densities of flagellates correlate with those of mollusks (Railkin 2004).

Water-soluble metabolites (Fitt et al. 1989; Rodriguez 1993; Dobretsov and Qian 2002), surface-associated signals (Kirchman et al. 1982; Szewzyk et al. 1991; Lam et al. 2005), and volatile molecules (Harder et al. 2002b) produced by bacteria and diatoms induce larval settlement. Compared to the large number of chemical compounds that have been isolated and identified from marine organisms, only a few inductive compounds from microorganisms have been identified. These compounds include lipids (Schmahl 1985), oligopeptides (Neumann 1979), glycoconjugates (Kirchman et al. 1982; Szewzyk et al. 1991), amino acids (Fitt et al. 1989), alkanes, alkenes, and hydroxyketones (Harder et al. 2002b). Therefore, isolation and identification of inductive compounds from microorganisms should be an important target of future investigations.

Table 1 Induction of larval settlement by microbial biofilms (bf)

Species of larvae (Phylum/Class)	Place of experiment	Inductive biofilms	References
Cnidaria			
Aurelia aurita	Laboratory	Monospecies bf of Micrococcaceae	Schmahl (1985)
Cassiopea andromeda	Laboratory	bf of the bacterium *Vibrio* sp.	Hofmann and Brand (1978); Neumann (1979)
Cyanea capitata	Laboratory	Multispecies bf	Brewer (1976)
Hydractinia echinata	Laboratory	bf of the bacterium *Alteromonas epejiana*	Leitz and Wagner (1993)
Obelia loveni	Laboratory	Multispecies bf	Dobretsov (1999)
Clava multicornis	Laboratory	Multispecies bf	Orlov (1996)
Acropora millepora	Laboratory	bf of the bacterium *Pseudoalteromonas*	Negri et al. (2001)
Heteroxenia fuscenscens	Laboratory	bf of the bacteria	Henning et al. (1991)
Acropora microphthalma	Laboratory	Multispecies bf	Webster et al. (2004)
Mollusca			
Mytilus galloprovincialis	Laboratory	Multispecies bf	Bao et al. (2007)
Mytilus edulis	Laboratory, field	Multispecies bf	Bayne (1964)
Pinctada maxima	Laboratory	Multispecies bf; monospecies bf of bacteria	Zhao et al. (2003)
Chlamys islandica	Laboratory	Multispecies bf	Harvey et al. (1995)
Crassostrea gigas, C. virginica	Laboratory	bf of the bacterium *Alteromonas colwelliana*	Fitt et al. (1989)
Saccostrea commercialis	Laboratory	Monospecies bf of bacteria	Anderson (1996)
Haliotis discus, H. rufescens, H. laevigata	Laboratory	Multispecies bf; monospecies bf of bacteria and diatoms	Morse et al. (1984); Roberts (2001)
Ostrea edulis	Laboratory	Multispecies bf	Knight-Jones (1951)
Placopecten magellanicus	Laboratory	Multispecies bf	Parsons et al. (1993)
Concholepas concholepas	Laboratory	Monospecies bf of bacteria	Rodriguez et al. (1993)
Bryozoa			
Bugula neritina	Laboratory, field	Multispecies bf; monospecies bf of bacteria and diatoms	Mihm et al. (1981); Kitamura and Hirayama (1987); Maki et al. (1989); Dahms et al. (2004); Dobretsov and Qian (2006)

(continued)

Table 1 (continued)

Species of larvae (Phylum/Class)	Place of experiment	Inductive biofilms	References
B. simplex, B. stolonifera, B. turrita	Laboratory	Multispecies bf	Brancato and Woollacott (1982)
Annelida			
Hydroides elegans	Laboratory, field	Multispecies bf; monospecies bf of bacteria and diatoms	Hadfield et al. (1994); Lau et al. (2002); Harder et al. (2002a); Lam et al. (2005)
Pomatoceros lamarkii	Laboratory	Multispecies bf	Hamer et al. (2001)
Janua brasiliensis	Laboratory	bf of the bacterium *Halomonas marina*	Kirchman et al. (1982)
Spirorbis borealis	Laboratory	Multispecies bf	Williams (1964)
S. corrallinae, S. tridentatus	Laboratory	Multispecies bf	De Silva (1962)
S. spirorbis	Laboratory, field	Multispecies bf	Wieczorek and Todd (1998)
Arthropoda/Cirripedia			
Balanus amphitrite	Laboratory, field	Multispecies bf; monospecies bf of bacteria and diatoms	Maki et al. (1988); Khandeparker et al. (2002); Qian et al. (2003); Patil and Anil (2005)
Balanus trigonus	Laboratory	Multispecies bf	Lau et al. (2005); Thiyagarajan et al. (2006)
Balanus cariosus, B. glandula	Field	Multispecies bf	Strathmann et al. (1981)
Notomegabalanus algicola	Field	Multispecies bf	Hentschel and Cook (1990)
Semibalanus balanoides	Laboratory, field	Multispecies bf	Le Tourneux and Bourget (1988); Thompson et al. (1998)
Echinodermata/Echinoidea			
Heliocidaris erythrogramma	Laboratory, field	Multispecies bf and the bacterium *Pseudoalteromonas luteoviolacea*	Huggett et al. (2006)
Acanthaster planci	Laboratory	Bacterial monospecies bf, multispecies bf	Johnson and Sutton (1994)
Arbacia punctulata, Lytechinus pictus	Laboratory	Multispecies bf	Cameron and Hinegardner (1974)
Strongylocentrotus droebachiensis	Laboratory	Multispecies bf	Pearce and Scheibling (1991)
S. purpuratus	Laboratory	Multispecies bf	Amador-Cano et al. (2006)

(continued)

Table 1 (continued)

Species of larvae (Phylum/Class)	Place of experiment	Inductive biofilms	References
Chordata/Ascidiacea			
Ciona intestinalis	Laboratory	bf of the bacterium *Pseudoalteromonas* sp.	Szewzyk et al. (1991)
		Multispecies bf	Wieczorek and Todd (1998)

It remains unknown how biofilm-derived cues trigger settlement and metamorphosis of invertebrate larvae and algal spores. The neuronal and genetic bases of the signal transduction pathways involved in this process have been studied only rarely. Previous studies suggested that specific larval receptors may be involved in the settlement process and that larval genes are differentially expressed during larval metamorphosis (Qian 1999). Lectins on the surface of the larvae of the polychaete *Janua brasiliensis* (Kirchman et al. 1982), *B. amphitrite* (Khandeparker et al. 2002), *Bugula* spp. (Maki et al. 1989), the hydrozoan *Obelia loveni* (Railkin 2004), and the green alga *Dunaliella* sp. (Mitchell and Kirchman 1984) play an important role in the recognition of marine biofilms. For example, incubation of *J. brasiliensis* larvae in solutions of D-glucose inhibited their settlement and metamorphosis (Kirchman et al. 1982). The attachment of the larvae was also suppressed if bacterial films in the experiment were pretreated with a solution of the lectin concanavalin A. This allowed the authors to formulate the hypothesis that lectins similar to concanavalin A are located on the surface of the larvae and interact by the "lock-and-key" scheme with bacterial polysaccharides and glycoproteins, which leads to the settlement of larvae and their metamorphosis (Kirchman et al. 1982). Beside the lectin receptors, G-protein-coupled receptors and two signal transduction systems (adenylate cyclase/cylic AMP and phosphatidyl-inositol/diacylglycerol/protein kinase C) play an important role in regulating metamorphosis of barnacles, mollusks and, possibly, polychaetes (Baxter and Morse 1992; Clare et al. 1995). Finding of larval receptors and identification of microbial cues involved in larval and spore settlement will help us understand the signal transduction pathways and facilitate the search for and development of new antifouling technologies.

2.2 Inhibition of Larval Settlement by Biofilms

Generally, the amount of inhibitive and inductive isolates in marine biofilms is approximately equal (Lau et al. 2002; Dobretsov and Qian 2004). Initially, it had been proposed that bacteria belonging mostly to the genus *Pseudoalteromonas* inhibit larval and spore settlement (Egan et al. 2001; Holmström et al. 2002). Later, it was shown that a wide range of bacterial taxa can inhibit larval settlement (Burgess et al. 2003; Dobretsov et al. 2006; Table 2). For instance, the marine bacteria

Table 2 Antialgal and antilarval compounds isolated from marine biofilms

Microbial strain	Antifouling compound	Effective concentrations ($\mu g\ mL^{-1}$)	Mode of action	Effective against	Reference
Bacterium *Alteromonas* sp.	Ubiquinone	12.5–25.0	NT	Barnacle *Balanus amphitrite* (LE)	Kon-ya et al. (1995)
Bacterium *Acinetobacter* sp.	6-Bromoindole-3-carbaldehyde	10	NT	Barnacle *B. amphitrite*	Olguin-Uribe et al. (1997)
Bacteria *Halomonas (Deleya) marina*, *Vibrio campbelli*	Polysaccharides	?	NT	Barnacle *B. amphitrite* (LE)	Maki et al. (1988)
Bacterium *Vibrio alginolyticus*	Heat stable, polar polysaccharide(s) >200 kDa	1,333–0.013	NT	Barnacle *B. amphitrite*, polychaete *Hydroides elegans*, bryozoan *Bugula neritina* (LE)	Dobretsov and Qian (2002); Harder et al. (2004)
Bacteria *Vibrio* sp. and an unidentified α-*Proteobacterium*	Heat stable, polar polysaccharides >100 kDa	?	NT	Polychaete *H. elegans*, bryozoan *B. neritina* (LE)	Dobretsov and Qian (2004)
Bacterium *Streptomyces fungicidicus*	Diketopiperazines	25–100	NT	Barnacle *B. amphitrite* (LE)	Li et al. (2006)
Bacterium *Pseudoalteromonas issachenkonii*	Proteolytic enzymes	0.001	NT	Bryozoan *B. neritina* (LE), barnacle *B. amphitrite*, byozoans *B. neritina*, *Schizoporella* sp. (FE)	Dobretsov et al. (2007b)
Bacterium *Shewanella oneidensis*	2-Hydroxymyristic acid	10	R	Alga *Ulva pertusa* spores (LE), *Ulva* sp. (FE)	Bhattarai et al. (2007)

(continued)

Table 2 (continued)

Microbial strain	Antifouling compound	Effective concentrations (μg mL^{-1})	Mode of action	Effective against	Reference
Bacterium *Shewanella oneidensis*	*cis*-9-Oleic acid	100	R	Alga *Ulva pertusa* (LE), *Delisia fimbriata*, *Sargassum* sp., *Ulva pertusa*, *B. amphitrite*, *Mytilus* sp., *Spirorbis borealis* (FE)	Bhattarai et al. (2007)
Cyanobacterium *Phormidium tenue*	Galactosyl diacylglycerol	?	T	Antialgal (LE)	Murakami et al. (1991)
Cyanobacterium *Nodularia harveyana*	Norharmalane	0.5–18.0	T	Alga *Nostoc insulare* (LE)	Volk (2006)
Bacterium *Pseudoalteromonas tunicata*	Heat-sensitive, polar compound 3–10 kDa	?	?	Algae *Polysiphonia* sp., *Ulva lactuea* (LE)	Egan et al. (2001)
Cyanobacterium *Gomphosphaeria aponina*	Aponin	?	T	Alga *Gymnodinium breve* (LE)	McCoy et al. (1979)
Fungus *Ampelomyces* sp.	3-Chloro-2,5-dihydroxybenzyl alcohol	0.67–3.81	NT	Tubeworm *Hydroides elegans*, barnacle *B. amphitrite* (LE)	Kwong et al. (2006)

T toxic; *NT* non-toxic; *R* repellent; *?* no information; *LE* laboratory experiment; *FE* field experiment

Halomonas (*Deleya) marina* (Maki et al. 1988) and bacteria belonging to the genera *Vibrio, Alteromonas, Flavobacterium, Micrococcus, Rhodovulum*, and *Pseudomonas* (Mary et al. 1993; Lau et al. 2003) inhibited larval attachment of the barnacle *B. amphitrite*. There is no any predictive relationship between the phylogenetic affiliation of bacteria and their antifouling activity (Lau et al. 2002; Patel et al. 2003; Dobretsov and Qian 2004).

Epibiotic bacteria associated with marine organisms have been proposed as an important source of antifouling compounds since they may help to protect their hosts from biofouling (Dobretsov and Qian 2002; Holmström et al. 2002). The bacterium *P. tunicata* is one of the first isolates that produced a range of antilarval, antialgal, antifungal, and antibacterial compounds (Holmström et al. 2002). Forty two bacterial isolates from different marine organisms produced antibacterial compounds, and one strain (*Pseudomonas* sp.) inhibited settlement of *B. amphitrite* larvae and *Ulva lactuca* spores (Burgess et al. 2003). Recent studies confirmed the antifouling activity of epibiotic bacterial strains isolated from marine sponges, corals, and algae (Holmström et al. 2002; Dobretsov and Qian 2002; Dobretsov and Qian 2004). These examples demonstrate that epibiotic bacteria associated with marine organisms can be an important source of antifouling compounds.

Even though biofilms and their compounds are key factors in the establishment of biofouling communities (Qian et al. 2003), only a few antifouling compounds have been isolated from marine bacteria so far (Dobretsov et al. 2006; Fusetani et al. 2006; Table 2). The first identified antilarval compound from marine bacteria – ubiquinone – was isolated from *Alteromonas* sp. (Kon-ya et al. 1995). The mode of action of this compound has not been discovered but the authors showed that ubiquinone inhibited larval settlement of the barnacle *B. amphitrite* in a non-toxic way (Table 2). In another study, 6-bromoindole-3-carbaldehyde isolated from the γ-Proteobacteria *Acinetobacter* sp. showed antifouling activity against larvae of the barnacle *B. amphitrite* (Olguin-Uribe et al. 1997). Several strains of cyanobacteria produce cytotoxic compounds that affect algal growth and survival (Volk 2006). Marine fungi have been shown to produce antifouling compounds as well (Kwong et al. 2006; Table 2). Antilarval and antialgal compounds isolated from marine microbes include lipids, polysaccharides, fatty acids, piperazines, and proteins (Table 2). Recently, a proteolytic enzyme from the deep-sea bacterium *Pseudoalteromonas issachenkonii* has been isolated (Dobretsov et al. 2007a). This enzyme inhibited larval settlement of the bryozoan *Bugula neritina* at a concentration of 1 ng mL^{-1} and was non-toxic. This concentration is much lower than that of other known antifouling compounds (Table 2, Dobretsov et al. 2006; Fusetani et al. 2006). This suggests that microbial enzymes may be a good alternative for toxic antifouling compounds.

So far, it has been postulated that antifouling compounds produced by marine organisms are mostly non-polar, poorly water-soluble secondary metabolites, which are effective at low concentrations (Steinberg et al. 2001). However, the data presented in Table 2 clearly show that microorganisms may produce both antifouling water-soluble and non-water-soluble compounds. Therefore, these compounds need to be investigated in the future.

Most of the antifouling compounds from marine microorganisms have been tested only under laboratory conditions (Table 2). Stability and performance of antifouling compounds in the field may be different from laboratory conditions (Dobretsov et al. 2006; Dahms et al. 2006). Until now, only a few antifouling compounds from microorganisms have been incorporated into non-toxic paint matrixes and tested in field experiments (e.g., Burgess et al. 2003; Dobretsov et al. 2007a; Bhattarai et al. 2007). In future experiments it is necessary to test all potent antifouling compounds from microbes both in the laboratory and in the field.

3 Quorum Sensing in Biofouling Communities

There are a number of different quorum sensing (QS) signaling systems employed by bacteria, but overall the mechanism of QS remains consistent through all prokaryotes (Whitehead et al. 2001). Generally, a small chemical compound ("autoinducer" or "signal") is produced by bacteria and then transported or diffused outside the cell. So far, *N*-acetyl-l-homoserine lactones, furanosylborate, cyclic thiolactone, hydroxy-palmitic acid, methyl dodecenoic acid, and farnesoic acid have been identified as QS signals (Parsek and Greenberg 2000). When the bacterial density in biofilms is high enough, these molecules reach a threshold concentration and begin to bind to a receptor protein. This process promotes the transcription of a number of genes, which regulates cell division and controls biofilm formation and composition (Parsek and Greenberg 2000).

Based on the properties of bacterial signal receptors, QS signaling can be grouped into two categories (Fig. 2). Most Gram-negative bacteria represent one category where *N*-acetyl-L-homoserine lactone (AHL) signal molecules, LuxR-type signal receptor, and LuxR-type I synthase are the major components (Dong et al. 2002). In contrast, cell-to-cell signaling in most Gram-positive bacteria occurs via a phosphorylation–dephosphorylation mechanism that is mediated by a two-component QS system. Here, oligopeptides signals are transported outside the cell and detected by a membrane-bound sensor (Novick 2003), which affects a response regulator by a phosphorelay (Zhang and Dong 2004).

Information concerning the presence of AHLs and other QS molecules in the marine environment is scarce (Dobretsov et al. 2007b). In the light organ of the sepiolid squid *Euprymna scolopes*, light emissions are regulated by AHLs of the marine symbiotic bacterium *Vibrio fisheri* (Ruby and Lee 1998). Recently, the presence of QS signals was demonstrated in marine snow (Gram et al. 2002). The production of AHLs by bacteria associated with marine sponges was reported by Taylor et al. (2004). The production of QS signals by tropical marine 2-, 4-, and 6-day-old subtidal biofilms has recently been investigated (Huang et al. 2007). A QS inducer, *N*-dodecanoyl-L-homoserine lactone, was detected in 6-day-old biofilms at a concentration of 3.36 mM L^{-1} by GC-MS. These findings suggest that QS signals might be produced in situ, but more investigations are needed.

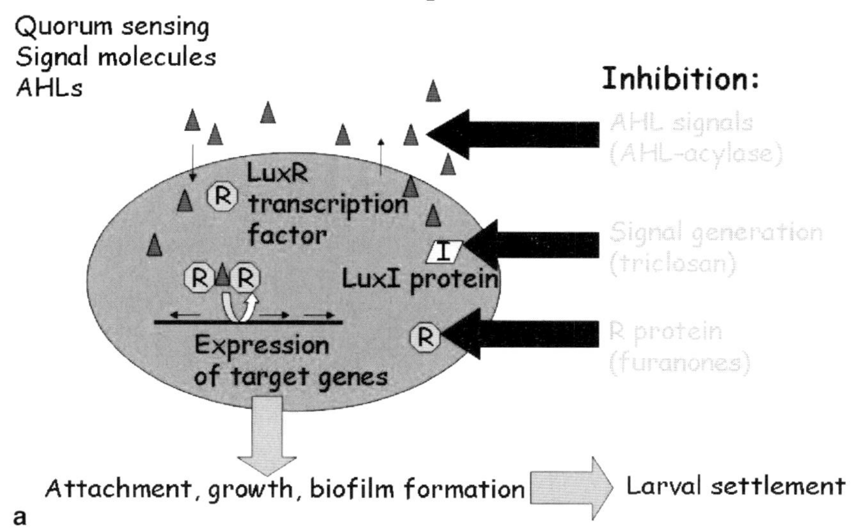

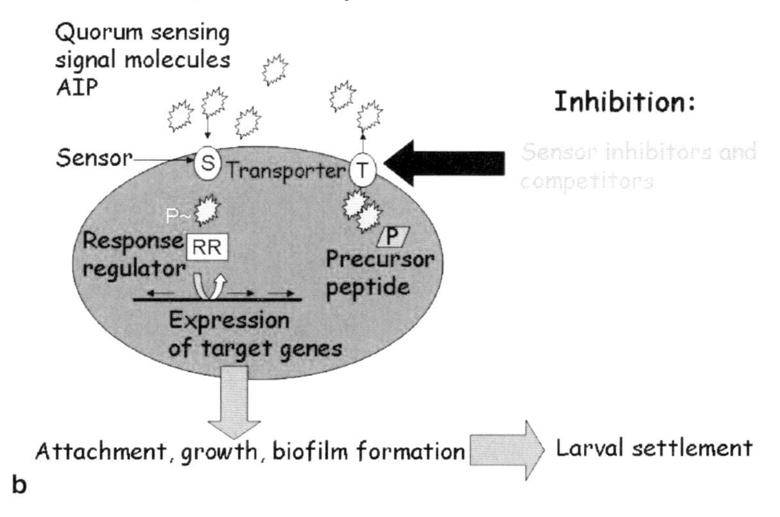

Fig. 2 General scheme of quorum sensing (QS) in **a** Gram-negative and **b** Gram-positive bacterial cells and its inhibition by chemical compounds. *AHL N*-Acetyl-l-homoserine lactone, *AIP* autoinducing peptide, *R* R-protein, *P* phosphate

Any reagent that prevents accumulation or recognition between QS signals and receptor proteins might block QS-dependent gene expression (Fig. 2; Zhang and Dong 2004). This inhibits bacterial attachment and disrupts biofilm formation. For example, bacterial AHL signal degradation enzymes (AHL-lactonase and AHL-acylase; Fig. 2) inhibit QS (Zhang and Dong 2004). Triclosan – a potent inhibitor of the enoyl-ACP reductase that is involved in AHL biosynthesis – reduces AHL production and blocks bacterial QS (Dobretsov et al. 2007b). Halogenated furanones, produced by the red alga *Delisea pulchra* (Givskov et al. 1996), by *Streptomyces* spp. (Cho et al. 2001) and by the magnolia *Hortonia* sp. (Bauer and Robinson 2002) result in acceleration of LuxR synthase degradation and lead to QS inhibition (Fig. 2a). There are several known mechanisms that interfere with the two-component systems of Gram-positive bacteria (Fig. 2b). For example, several phenolic inhibitors, such as closantel, cause structural alterations to the receptor kinase and inhibit QS (Stephenson et al. 2000). Additionally, a truncated autoinducing peptide (AIP) lacking the N-terminal tail shows a wide QS inhibitory activity towards all the four AIP-specific groups of *Streptococcus aureus* (Zhang and Dong 2004). These examples show that QS of Gram-positive and Gram-negative bacteria can be effectively blocked by QS inhibitors.

Because biofilms enhance settlement of invertebrate larvae and algal spores (see Table 1; Wieczorek and Todd 1998; Maki and Mitchell 2002; Railkin 2004), QS blockers can control larval settlement indirectly by regulating the microbial community structure of biofilms and the density of bacteria, which in turn affects larval behavior (Dobretsov et al. 2007b; Fig. 2). This result demonstrates the possibility of using QS inhibitors for control of both micro- and macrofouling communities.

4 Conclusions and Future Perspectives

The data presented in this review clearly demonstrate that marine biofilms are a key factor for the settlement of macrofoulers. Up to now, inhibitive and inductive metabolites have been isolated predominantly from bacteria and diatoms (Tables 1 and 2). Only one inhibitive compound from a marine fungus has been isolated and identified (Kwong et al. 2006). At the same time, biofilms consist of numerous species of bacteria, diatoms, flagellates, fungi, sarcodines, and ciliates (Railkin 2004), which may produce both inhibitive and inductive compounds. Therefore, less investigated groups of microorganisms, like marine microalgae, flagellates, fungi, sarcodines, and ciliates, may have a high antifouling potential.

Natural biofilms have been shown to be complex and dynamic communities; interactions within their components play an important role in the production of chemical compounds. For example, the proportion of inductive, non-inductive, and inhibitive strains of microorganisms in biofilms determines larval settlement (Lau et al. 2002; Dahms et al. 2004). Nevertheless, most investigators have been dealing with single species of microorganisms belonging to particular taxa (Tables 1, 2), while the combined effect of different microbial taxa on larval settlement might be different and should be explored in future studies.

Because only a limited amount of marine organisms can be cultivated in the laboratory, it is necessary to investigate the performance of inductive or antifouling compounds under natural conditions on a variety of species. Up to now, only a limited number of field investigations of larval responses towards biofilms have been carried out (Table 1), and only in a few cases have microbial antifouling compounds been tested under natural conditions (i.e., Burgess et al. 2003; Bhattarai et al. 2007; Dobretsov et al. 2007a). Thus, little is known about the antifouling performance of biocides in nature, and in future experiments antifouling compounds should be tested under field conditions.

The interaction between marine microbes and larvae is another area where more information is needed. In several invertebrate species the larval receptors (Hadfield et al. 2000; Jeffery 2002), the signal transduction pathways (Morse 1990; Carpizo-Ituarte and Hadfield 2003; Amador-Cano et al. 2006), potential genes (Seaver et al. 2005; Frobius and Seaver 2006), and expressed proteins (Sanders et al. 2005; Gallus et al. 2005) involved in larval settlement and metamorphosis have been characterized. In future studies, it will be necessary not only to identify inductive and inhibitive compounds from microorganisms but also to identify the genes responsible for production of these compounds, as well as the larval receptors involved in their detection. A better understanding of the signal transduction pathways of settlement processes will allow us to improve bioassay systems and find new antifouling and inductive compounds.

Under different culture conditions marine microorganisms can produce different chemical compounds. It has been shown that bacteria were inhibitive to the larvae of *B. amphitrite* at salinities of 35 and 45 ‰ but were inductive at 15 and 25‰ (Khandeparker et al. 2002). Additionally, inductiveness of biofilms and production of compounds varied at different temperatures (Miao et al. 2006). These results indicate that the performance of antifouling compounds and their production by microorganisms should be investigated under field conditions.

In the coming decades, the marine environment will be subject to profound abiotic changes, such as elevated water temperature, changes in salinity, decrease of pH, and elevated ultraviolet radiation. These changes will affect not only the survival of propagules but also their recruitment, which is controlled by the quality and quantity of settlement cues produced by marine microorganisms. Additionally, climate changes can modify the composition of microbial biofilms and their metabolites, which in turn, can change propagule settlement. Therefore, it would be interesting to investigate and predict possible effects of climate changes on microbial biofilms, macrofouling communities, and their interactions.

Many marine organisms can control epibiosis on their surface by production of chemical compounds (Harder 2008; Dobretsov and Qian 2002). Can we "learn from nature" and manipulate biofilm properties in order to increase their antifouling or inductive properties? Several recent attempts have been made, which include the immobilization of live bacterial cells (Holmström et al. 2000), deterrence of microbes (Mitchell and Kirchman 1984, Railkin 2004), and inhibition of bacterial QS signals (Dobretsov 2007b). In "living paints" the bacteria that release antifouling compounds are immobilized in polymers and maintained alive. Such coatings would have an indefinite lifespan in comparison to traditional antifouling coatings, which fail when they exhaust their reservoir of biocides. Theoretically, such technology is well within the realms of possibility, since the immobilization of bacteria in artificial matrices is established (Holmström et al. 2000).

Overall, our data suggest that microorganisms are an important source of biologically active metabolites for the antifouling industry and aquaculture. Additional screening of different microbial taxa will result in the isolation of novel and potent biofouling compounds. Future studies should include genetic, molecular, biochemical, and microbiological multidisciplinary approaches for the investigation of microbe–larva interactions. Additionally, the future development of antifouling coatings with microbial compounds requires a successful collaboration between academic and industrial researchers (Rittschof).

Acknowledgemnts We thank Dr. F. Weinberger for his constructive comments on the manuscript. The author's studies were supported by an Alexander von Humboldt Fellowship.

References

Amador-Cano G, Carpizo-Ituarte E, Cristino Jorge D (2006) Role of protein kinase C, G-protein coupled receptors, and calcium flux during metamorphosis of the sea urchin *Strongylocentrotus purpuratus*. Biol Bull 210:121–131

Anderson MJ (1996) A chemical cue induces settlement of Sydney rock oysters, *Saccostrea commercialis*, in the laboratory and in the field. Biol Bull 190:350–358

Bao W-Y, Satuito CG, Yang J-L, Kitamura H (2007) Larval settlement and metamorphosis of the mussel *Mytilus galloprovincialis* in response to biofilms. Mar Biol 150:565–574

Bauer WD, Robinson JB (2002) Disruption of bacterial quorum sensing by other organisms. Curr Opin Biotechnol 13:234–237

Baxter GT, Morse DE (1992) Cilia from abalone larvae contain a receptor-dependent G-protein transduction system similar to that in mammals. Biol Bull 183:147–154

Bayne BL (1964) Primary and secondary settlement in *Mytilus edulis* L. (Mollusca). J Anim Ecol 33:513–523

Bhattarai HD, Ganti VS, Paudel B, Lee YK, Lee HK, Hong Y-K, Shin HW (2007) Isolation of antifouling compounds from the marine bacterium, *Shewanella oneidensis* SCH0402. World J Microbiol Biotechnol 23:243–249

Brancato MS, Woollacott RM (1982) Effect of microbial films on settlement of bryozoan larvae (*Bugula simplex, B. stolonifera, B. turrita*). Mar Biol 71:51–56

Brewer RH (1976) Larval settling behaviour in *Cyanea capillata* (Cnidaria: Scyphozoa). Biol Bull 150:183–199

Burgess JG, Boyd KG, Armstrong E, Jiang Z, Yan L, Berggren M, May U, Pisacane T, Granmo A, Adams DR (2003) The development of a marine natural product-based antifouling paint. Biofouling 19:197–205

Cameron RA, Hinegardner RT (1974) Initiation of metamorphosis in laboratory cultured sea urchins. Biol Bull 146:335–342

Carpizo-Ituarte EJ, Hadfield MG (2003) Transcription and translation inhibitors permit metamorphosis up to radiole formation in the serpulid polychaete *Hydroides elegans* Haswell. Biol Bull 204:114–125

Cho KW, Lee HS, Rho JR, Kim TS, Mo SJ, Shin J (2001) New lactone-containing metabolites from a marine-derived bacterium of the genus *Streptomyces*. J Nat Prod 64:664–667

Clare AS, Thomas RF, Rittschof D (1995) Evidence for the involvement of cyclic-AMP in the modulation of barnacle settlement. J Exp Biol 198:655–664

Dahms H-U, Dobretsov S, Qian P-Y (2004) The effect of bacterial and diatom biofilms on the settlement of the bryozoan *Bugula neritina*. J Exp Mar Biol Ecol 313:191–209
Dahms H-U, Xu Y, Pfeiffer C (2006) Antifouling potential of cyanobacteria: a mini-review. Biofouling 22:317–327
De Silva PHDP (1962) Experiments on the choice of substrate by Spirorbis larvae (Serpulidae). J Exp Biol 39:483–490
Dobretsov SV (1999) Macroalgae and microbial film determine substrate selection in planulae of *Gonothyraea loveni* (Allman 1859) (Cnidaria: Hydrozoa). Zoosystematica Rossica 1:109–117
Dobretsov S, Qian P-Y (2002) Effect of bacteria associated with the green alga *Ulva reticulata* on marine micro- and macrofouling. Biofouling 18:217–228
Dobretsov S, Qian P-Y (2004) The role of epibotic bacteria from the surface of the soft coral *Dendronephthya* sp. in the inhibition of larval settlement. J Exp Mar Biol Ecol 299:35–50
Dobretsov S, Qian P-Y (2006) Facilitation and inhibition of larval attachment of the bryozoan *Bugula neritina* in association with mono-species and multi-species biofilms. J Exp Mar Biol Ecol 333:263–264
Dobretsov S, Dahms H-U, Qian P-Y (2006) Inhibition of biofouling by marine microorganisms and their metabolites. Biofouling 22:43–54
Dobretsov S, Xiong H, Xu Y, Levin LA, Qian PY (2007a) Novel antifoulants: inhibition of larval attachment by proteases. Mar Biotech 9:388–397
Dobretsov S, Dahms H-U, Huang YL, Wahl M, Qian PY (2007b) The effect of quorum sensing blockers on the formation of marine microbial communities and larval attachment. FEMS Microbiol Ecol 60:177–188
Dong YH, Gusti AR, Zhang Q, Xu JL, Zhang LH (2002) Identification of quorum-quenching *N*-acyl homoserine lactonases from Bacillus species. Appl Environ Microbiol 68:1754–1759
Donlan RM (2002) Biofilms: microbial life on surfaces. Emerg Infect Dis 8:881–890
Egan S, James S, Holmström C, Kjelleberg S (2001) Inhibition of algal spore germination by the marine bacterium *Pseudoalteromonas tunicata*. FEMS Microbiol Ecol 35:67–73
Egan S, James S, Kjelleberg S (2002) Identification and characterization of a putative transcriptional regulator controlling the expression of fouling inhibitors in *Pseudoalteromonas tunicatae*. Appl Environ Microbiol 68:372–378
Fitt WK, Labare MP, Fuqua WC, Walch M, Coon SL, Bonar DB, Colwell RR, Weiner RM (1989) Factors influencing bacterial production of inducers of settlement behaviour of larvae of the oyster *Crassostrea gigas*. Microb Ecol 17:287–298
Frobius AC, Seaver EC (2006) *Capitella* sp I homeobrain-like, the first lophotrochozoan member of a novel paired-like homeobox gene family. Gene Expr Patterns 6:985–991
Fusetani N, Clare AS, Mueller WEG (eds.) (2006) Antifouling compounds. Progress in molecular and subcellular biology, vol 42. Springer, Berlin Heidelberg New York
Gallus L, Ramoino P, Faimali M, Piazza V, Maura G, Marcoli M, Ferrando S, Girosi L, Tagliafierro G (2005) Presence and distribution of serotonin immunoreactivity in the cyprids of the barnacle *Balanus amphitrite*. Eur J Histochem 49:341–347
Givskov M, deNys R, Manefield M, Gram L, Mayimilien R, Eberl L, Molin S, Steinberg PD, Kjelleberg S (1996) Eukaryotic interference with homoserine lactone-mediated prokaryotic signalling. J Bacteriol 178:6618–6622
Gram L, Grossart HP, Schlingloff A, Kiőrboe T (2002) Possible quorum sensing in marine snow bacteria: production of acylated homoserine lactones by *Roseobacter* strains isolated from marine snow. Appl Environ Microbiol 68:4111–4116
Hadfield MG, Unabia CC, Smith CM, Michael TM (1994) Settlement preferences of ubiquitous fouler *Hydroides elegans*. In: Thompson M-F, Sarojini R, Nagabhushanam R (eds.) Recent developments in biofouling control. Balkema, Rotterdam, pp. 65–74
Hadfield MG, Meleshkevitch EA, Boudko DY (2000) The apical sensory organ of a gastropod veliger is a receptor for settlement cues. Biol Bull 198:67–76
Hamer JP, Walker G, Latchford JW (2001) Settlement of *Pomatoceros lamarkii* (Serpulidae) larvae on biofilmed surfaces and the effect of aerial drying. J Exp Mar Biol Ecol 260:113–132

Harder T, Lam C, Qian P-Y (2002a) Induction of larval settlement in the polychaete *Hydroides elegans* by marine biofilms: an investigation of monospecific diatom films as settlement cues. Mar Ecol Prog Ser 229:105–112

Harder T, Lau SCK, Dahms H-U, Qian PY (2002b) Isolation of bacterial metabolites as natural inducers for larval settlement in the marine polychaete *Hydroides elegans* (Haswell). J Chem Ecol 28:2029–2043

Harder T, Dobretsov S, Qian PY (2004) Waterborne polar macromolecules act as algal antifoulants in the seaweed *Ulva reticulata*. Mar Ecol Prog Ser 274:133–141

Harder T (2008) Marine epibiosis: concepts, ecological consequences and host defence. Springer Series on Biofilms, doi:7142_2008_16

Harvey M, Miron G, Bourget E (1995) Resettlement of Iceland scallop (*Chlamys islandica*) spat on dead hydroids: response to chemical cues from the protein-chitinous perisarc and associated microbial film. J Shellfish Res 14:383–388

Henning G, Benayahu Y, Hofmann DK (1991) Natural substrates, marine bacteria and a phorbol ester induce metamorphosis of the soft coral *Heteroxenia fuscenscens* (Anthozoa: Octocorallina). Verh Dtsch Zool Ges 84:486–487

Hentschel JR, Cook PA (1990) The development of marine fouling community in relation to the primary film of microorganisms. Biofouling 2:1–11

Hofmann DK, Brand U (1978) Induction of metamorphosis in the symbiotic scyphozoan *Cassiopea andromeda*: role of marine bacteria and of biochemicals. Symbiosis 4:99–116

Holmström C, Steinberg P, Christov V, Christie G, Kjelleberg S (2000) Bacteria immobilized in gels: improved methodologies for antifouling and biocontrol applications. Biofouling 15:109–117

Holmström C, Egan S, Franks A, McCloy S, Kjelleberg S (2002) Antifouling activities expressed by marine surface associated *Pseudoalteromonas* species. FEMS Microbiol Ecol 41:47–58

Huang S, Hadfield MG (2003) Composition and density of bacterial biofilms affect metamorphosis of the polychaete *Hydroides elegans*. Mar Ecol Prog Ser 260:161–172

Huang Y-L, Dobretsov S, Ki J-S, Yang L-H, Qian P-Y (2007) Presence of acyl-homoserine lactones in subtidal biofilms and the implication in inducing larval settlement of the polychaete *Hydroides elegans*. Microb Ecol 54:384–392

Huggett MJ, Williamson JE, de Nys R, Kjelleberg S, Steinberg PD (2006) Larval settlement of the common Australian sea urchin *Heliocidaris erythrogramma* in response to bacteria from the surface of coralline algae. Oecologia 149:604–619

Jeffery CJ (2002) New settlers and recruits do not enhance settlement of a gregarious intertidal barnacles in New South Wales. J Exp Mar Biol Ecol 275:131–146

Johnson CR, Sutton DC (1994) Bacteria on the surface of crustose coralline algae induce metamorphosis of the crown of thorns starfish *Acanthaster planci*. Mar Biol 120:305–310

Khandeparker L, Anil AC, Raghukumar S (2002) Factors regulating the production of different inducers in *Pseudomonas aeruginosa* with reference to larval metamorphosis in *Balanus amphitrite*. Aquat Microb Ecol 28:37–54

Kirchman D, Graham D, Reish D, Mitchell R (1982) Lectins may mediate in the settlement and metamorphosis of *Janua (Dexiospira) brasiliensis* Grube (Polychaeta: Spirorbidae). Mar Biol Lett 3:201–222

Kitamura H, Hirayama K (1987) Effect of primary films on the settlement of larvae of a bryozoan *Bugula neritina*. Bull Jpn Soc Sci Fish 53:1377–1381

Knight-Jones EW (1951) Gregariousness and some other aspects of the settling behaviour in Spirorbis. J Mar Biol Ass UK 30:201–222

Kon-ya K, Shimidzu N, Otaki N, Yokoyama A, Adachi K, Miki W (1995) Inhibitory effect of bacterial ubiquinones on the settling of barnacle, *Balanus amphitrite*. Experientia 51:153–155

Kwong TFN, Li M, Li X, Qian PY (2006) Novel antifouling and antimicrobial compound from a marine-derived fungus *Ampelomyces* sp. Mar Biotech 8:634–640

Lam C, Harder T, Qian P-Y (2005) Induction of larval settlement in the polychaete *Hydroides elegans* by extracellular polymers of benthic diatoms. Mar Ecol Prog Ser 286:145–154.

Lau SCK, Mak KK, Chen F, Qian P-Y (2002) Bioactivity of bacterial strains from marine biofilms in Hong Kong waters for the induction of larval settlement in the marine polychaete *Hydroides elegans*. Mar Ecol Prog Ser 226:301–310
Lau SCK, Thiyagarajan V, Qian P-Y (2003) The bioactivity of bacterial isolates in Hong Kong waters for the inhibition of barnacle (*Balanus amphitrite* Darwin) settlement. J Exp Mar Biol Ecol 282:43–60
Lau SCK, Thiyagarajan V, Cheung SCK, Qian P-Y (2005) Roles of bacterial community composition in biofilms as a mediator for larval settlement of three marine invertebrates. Aquat Microb Ecol 38:41–51
Le Tourneux F, Bourget E (1988) Importance of physical and biological settlement cues used at different scales by the larvae of *Semibalanus balanoides*. Mar Biol 97:57–66
Leitz T, Wagner T (1993) The marine bacterium *Alteromonas espejiana* induces metamorphosis of the hydroid *Hydractinia echinata*. Mar Biol 115:173–178
Lewandowski Z (2000) Structure and function of biofilms. In: Evans LV (ed.). Biofilms: recent advances in their study and control. Harwood Academic, Amsterdam, pp. 1–17
Li X, Dobretsov S, Xu Y, Xiao X, Hung OS, Qian PY (2006) Antifouling diketopiperazines produced by a deep-sea bacterium, *Streptomyces fungicidicus*. Biofouling 22:201–208.
Maki JS, Mitchell R (2002) Biofouling in the marine environment. In: Bitton G (ed.) Encyclopedia of environmental microbiology. Wiley, New York, pp. 610–619
Maki JS, Rittschof D, Costlow JD, Mitchell R (1988) Inhibition of attachment of larval barnacles, *Balanus amphitrite*, by bacterial surface films. Mar Biol 97:199–206
Maki JS, Rittschof D, Schidt AR, Snyder AG, Mitchell R (1989) Factors controlling settlement of bryozoan larvae: a comparison of bacterial films and unfilmed surfaces. Biol Bull 177:295–302
Mary A, Mary V, Rittschof D, Nagabhushanam R (1993) Bacterial-barnacle interaction: potential of using juncellins and antibiotics to alter structure of bacterial communities. J Chem Ecol 19:2155–2167
McCoy FL, David Jr I, Eng-Wilmont L, Martin DF (1979) Isolation and partial purification of a red tide *(Gymnodinium breve)* Cytolytic factor(s) from cultures of *Gomphosphaeria aponina*. J Agric Food Chem 27:69–79
Miao L, Kwong TFN, Qian PY (2006) Effect of culture conditions on mycelial growth, antibacterial activity, and metabolite profiles of the marine-derived fungus *Arthrinium c.f. saccharicola*. Appl Microb Biotechnol 72:1063–1073
Mihm JW, Banta WC, Loeb GI (1981) Effects of absorbed organic and primary biofilms on bryozoan settlement. J Exp Mar Biol Ecol 54:167–179
Mitchell R, Kirchman D (1984) The microbial ecology of marine surfaces. In: Costlow JD, Tipper RC (eds.) Marine biodeterioration: an interdisciplinary study. Naval Institute Press, Anapolis, MA, pp. 49–56
Morse DE (1990) Recent progress in larval settlement and metamorphosis: closing gaps between molecular biology and ecology. Bull Mar Sci 46:465–483
Morse ANC, Froyd CA, Morse DE (1984) Molecules from cyanobacteria and red algae that induce larval settlement and metamorphosis in the mollusk *Haliotis rufescens*. Mar Biol 81:293–298
Murakami N, Morimoto T, Imamura H, Ueda T, Nagai S, Sakakibara J, Yamada N (1991) Studies on glycolipids: III. glycoglycolipids from an axenically cultures cyanobacterium, *Phormidium tenue*. Chem Pharm Bull 39:2277–2281
Nedved BT, Hadfield MG (2008) *Hydroides elegans* (Annelida: Polychaeta): a model for biofouling research. Springer Series on Biofilms, doi:7142_2008_15
Negri AP, Webster NS, Hill RT, Heyward AJ (2001) Metamorphosis of broadcast spawning corals in response to bacteria isolated from crustose algae. Mar Ecol Prog Ser 223:121–131
Neumann R (1979) Bacterial induction of settlement and metamorphosis in the planula larvae of *Cassiopea andromeda* (Cnidaria: Schiphozoa, Rhizostomeae). Mar Ecol Prog Ser 1:21–28
Novick RP (2003) Autoinduction and signal transduction in the regulation of staphylococcal virulence. Mol Microbiol 48:1429–1449
Olguin-Uribe G, Abou-Mansour E, Boulander A, Debard H, Francisco C, Combaut G (1997) 6-Bromoindole-3-carbaldehyde from an *Acinetobacter* sp. bacterium associated with the ascidian *Stomoza murrayi*. J Chem Ecol 23:2507–2521

Orlov D (1996) Observations on the settling behaviour of planulae of *Clava multicornis* Forskal (Hydroidea, Athecata). Scientia Marina 60:121–128

Parsek MR, Greenberg EP (2000) Acyl-homoserine lactone quorum sensing in gram-negative bacteria: a signaling mechanism involved in associations with higher organisms. Proc Natl Acad Sci U S A 97:8789–8793

Parsons GJ, Dadswell MJ, Roff JC (1993) Influence of biofilm on settlement of sea scallop *Placopecten magellanicus* (Gmelin, 1791), in Passamaquoddy Bay, New Brunswick, Canada. J Shellfish Res 12:279–283

Patel P, Callow ME, Joint I, Callow JA (2003) Specificity in larval settlement modifying response of bacterial biofilms towards zoospores of the marine alga *Enteromorpha*. Environ Microbiol 5:338–349

Patil JS, Anil AC (2005) Influence of diatom exopolymers and biofilms on metamorphosis in the barnacle *Balanus amphitrite*. Mar Ecol Prog Ser 301:231–245

Pearce CM, Scheibling RE (1991) Effect of macroalgae, microbial films, and conspecifics on the induction of metamorphosis of the green sea urchin *Strongylocentrotus droebachiensis*. J Exp Mar Biol Ecol 147:147–162

Qian P-Y (1999) Larval settlement of polychaetes. Hydrobiologia 402:239–253

Qian P-Y, Thiyagarajan V, Lau SCK, Cheung SCK (2003) Bacterial community profile in biofilm and attachment of the acorn barnacle *Balanus amphitrite*. Aquat Microb Ecol 33:225–237

Railkin AI (2004) Marine biofouling: colonization processes and defenses. CRC, Boca Raton, FL

Roberts R (2001) A review of settlement cues for larval abalone (*Haliotis* spp.). J Shellfish Res 20:571–586

Rodriguez SR, Ojeda FP, Inestrosa NC (1993) Settlement of benthic marine invertebrates. Mar Ecol Prog Ser 97:193–207

Ruby EG, Lee KH (1998) The *Vibrio fischeri Euprymna scolopes* light organ association: current ecological paradigms. Appl Environ Microbiol 64:805–812

Sanders MB, Billinghurst Z, Depledge MH, Clare AS (2005) Larval development and vitellin-like protein expression in *Palaemon elegans* larvae following xeno-oestrogen exposure. Integrat Compar Biol 45:51–60

Seaver EC, Thamm K, Hill SD (2005) Growth patterns during segmentation in the two polychaete annelids, *Capitella* sp I and *Hydroides elegans*: comparisons at distinct life history stages. Evol Develop 7:312–326

Schmahl G (1985) Bacterial induced stolon settlement in the scyphopolyp of *Aurelia aurita* (Cnidaria, Scyphozoa). Helgolander Meeresuntersuchungen 39:33–42

Steinberg PD, De Nys R, Kjelleberg S (2001) Chemical mediation of surface colonization. In: McCkintock JB, Baker JB (eds.), Marine chemical ecology. CRC, Boca Raton, FL, pp. 355–387

Stephenson K, Yamaguchi Y, Hoch JA (2000) The mechanism of action of inhibitors of bacterial two-component signal transduction systems. J Biol Chem 275:38900–38904

Strathmann RR, Branscomb ES, Vedder K (1981) Fatal errors in set as a cost of dispersal and the influence of internal flora on set of barnacles. Oecologia 48:13–18

Szewzyk U, Holmstrom C, Wrangstadh M, Samuelsson MO, Maki JS, Kjelleberg S (1991) Relevance of the exopolysaccharide of marine *Pseudomonas* sp. strain S9 for the attachment of *Ciona intestinalis* larvae. Mar Ecol Prog Ser 75:259–265

Tait K, Joint I, Daykin M, Milton DL, Williams P, Camara M (2005) Disruption of quorum sensing in seawater abolishes attraction of zoospores of the green alga *Ulva* to bacterial biofilms. Environ Microbiol 7:229–240

Taylor MW, Schupp PJ, Baillie HJ, Charlton TS, de Nys R, Kjelleberg S, Steinberg PD (2004) Evidence for acyl homoserine lactone signal production in bacteria associated with marine sponges. Appl Environ Microbiol 70:4387–4389

Thiyagarajan V, Hung OS, Chiu JMY, Wu RSS, Qian P-Y (2005) Growth and survival of juvenile barnacle *Balanus amphitrite*: interactive effects of cyprid energy reserve and habitat. Mar Ecol Prog Ser 299:229–237

Thiyagarajan V, Lau SCK, Cheung SCK, Qian PY (2006) Cypris habitat selection facilitated by microbial films influences the vertical distribution of subtidal barnacle *Balanus trigonus*. Microb Ecol 51:431–440

Thompson RC, Norton TA, Hawkins SJ (1998) The influence of epilithic microbial films on the settlement of *Semibalanus balanoldes* cyprids – a comparison between laboratory and field experiments. Hydrobiologia 376:203–216

Unabia CRC, Hadfield MG (1999) Role of bacteria in larval settlement and metamorphosis of the polychaete *Hydroides elegans*. Mar Biol 133:55–64

Volk RB (2006) Antialgal activity of several cyanobacterial exometabolites. J Appl Phycol 18:145–151

Webster NS, Smith LD, Heyward AJ, Watts JEM, Webb RI, Blackall LL, Negri AP (2004) Metamorphosis of a scleratinian coral in response to microbial biofilms. Appl Envir Microb 70:1213–1221

Wieczorek SK, Todd CD (1998) Inhibition and facilitation of the settlement of epifaunal marine invertebrate larvae by microbial biofilm cues. Biofouling 12:81–93

Williams GB (1964) The effect of extracts of Fucus serratus in promoting the settlement of larvae of Spirorbis borealis (Polychaetea). J Mar Biol Ass UK 44:397–414

Whitehead NA, Barnard AM, Slater H, Simpson NJ, Salmond GP (2001) Quorum sensing in Gram-negative bacteria. FEMS Microbiol Rev 25:365–404

Yebra DM, Kiil S, Dam-Johansen K (2004) Antifouling technology – past, present and future steps towards efficient and environmentally friendly antifouling coatings. Prog Org Coatings 50:75–104

Zhang LH, Dong YH (2004) Quorum sensing and signal interference: diverse implications. Mol Microbiol 53:1563–1571

Zhao B, Zhang S, Qian P-Y (2003) Larval settlement of the silver- or goldlip pearl oyster *Pinctada maxima* (Jameson) in response to natural biofilms and chemical cues. Aquaculture 220:883–901

A Triangle Model: Environmental Changes Affect Biofilms that Affect Larval Settlement

P.Y. Qian (✉) and H.-U. Dahms

Abstract Biofilms are ubiquitous – covering every exposed surface in marine environment and thus playing a key role in mediating biotic interactions and biogeochemical activities occurring on the surfaces. For the propogates of marine organisms, biofilm attributes serve as inhibitive or inductive cues for the attachment of settling larvae and algal spores of potential colonizers. Microbes in biofilms are not only the sources of chemical cues but also consumers of chemical cues. As microbes in biofilm are very sensitive to changes in ambient environment, the production of chemical cues by the microbes will change in response to spatio-temporal variation of microbial density, community structure, topography, dynamics, and the microbial physiological conditions in biofilms. These lead to changes in physical and chemical biofilm properties and in the bioactivity of biofilm for attachment of marine propogates. While there have been a number of reviews on the effect of biofilms on settlement of marine invertebrate larvae and algal spores, the effects of environmental changes on microbial community structure dynamics and bioactivities of biofilms remain much unexplored. Recent advances in molecular fingerprinting techniques have made it possible to precisely study the linkage between environmentally driven changes in biofilms and larval settlement. We are now gaining a better picture of the triangle relationship between environmental variables, biofilm dynamics and bioactivity, and the behavior of settling larvae or spores of marine organisms. Here, we would like to formally introduce a triangle model to provide a conceptual framework for interactions between environmentally induced biofilm changes that in turn affect the settlement of dispersal propogates.

P.Y. Qian
Department of Biology and Coastal Marine Lab Hong Kong, University of Science and Technology, Clearwater Bay, Kowloon, Hong Kong
e-mail: boqianpy@ust.hk

Springer Series on Biofilms, doi: 10.1007/7142_2008_19

1 Introduction

Biofilms are surface-attached microbial communities consisting of multiple layers of cells embedded in hydrated matrices (Kierek-Pearson and Karatan 2005). They are ubiquitous on aquatic substrata. The physical, chemical, and biological events leading to the establishment of attached communities of microorganisms as biofilms have been thoroughly reviewed (Characklis and Marshall 1990; Caldwell et al. 1993; Palmer and White 1997). Biofilms have a wide array of interactions with other organisms: with interspecific interactions to even transphyletic microbiota (archaea, eubacteria, fungi, unicellular eucaryotes; see Paerl and Pinckney 1996) as well as with macroalgae and invertebrates (Hadfield and Paul 2001).

Several studies show that biofilms can either foster (Kirchman et al. 1982; Callow and Callow 2000; Qian et al. 2003; Hung et al. 2005a, b; Lau et al. 2005) or reduce settlement of marine invertebrate larvae (Holmstrøm et al. 1992; Dobretsov and Qian 2004; Dahms and Qian 2005), or have no effect (Todd and Keough 1994; Lau et al. 2003a). Microbes forming biofilms (e.g., bacteria, diatoms, fungi) attract other organisms by providing resources (e.g., food, habitat, protection), by benefitting invertebrate larvae in neutralizing and marking adverse substratum properties (Costerton et al. 1995), or by signaling habitats with different fitness expectations (Gordon 1999). Biofilms may also provide important integrated historical environmental information (Harder et al. 2002a). The physical, chemical, and biotic properties of biofilms (Qian et al. 2003), their composition, and the dynamics of resulting bioactivities, therefore, have a share in structuring macrobenthic communities (Underwood and Fairweather 1989).

Larvae were shown to choose when and where to settle to some extent (Pawlik 1992; Qian et al. 2003). They can deferentiate between biofilms of different density (e.g., Maki et al. 1988; Neal and Yule 1994), community composition (e.g., Holmstrøm et al. 1992; Lau et al. 2002; Patel et al. 2003; Qian et al. 2003; Lau et al. 2005), age (e.g., Szewzyk et al. 1991; Keough and Raimondi 1996), metabolic activity (e.g., Wieczorek and Todd 1998; Holmstrøm and Kjelleberg 1999; Hung et al. 2005a, b; Lau et al. 2005), a wide variety of microbial products ranging from small-molecule metabolites to high molecular weight extracellular polymers (Harder et al. 2002a; Lau et al. 2003a; Lam et al. 2005a, b), or different habitats (Keough and Raimondi 1996; Thiyagarajan et al. 2005; Dobretsov et al. 2006). While some invertebrate species only settle on biofilms with viable bacterial cells (Lau et al. 2003b), others can settle on biofilms of non-viable cells (Hung et al. 2005a). Saying this, it has to be kept in mind that larvae respond differentially to biofilms also as a function of their own (larval) metabolic activity, density, and composition (Holmstrøm and Kjelleberg 1999). In addition, macroorganisms show different taxonic, ontogenetic, and physiological patterns for settlement (Wahl 1989).

Whereas colonization cues, particularly with respect to invertebrate larval settlement and metamorphosis, have been studied intensively (see reviews by Chia 1989; Pawlik 1992; Qian 1999; Rodriguez et al. 1993), the role of biofilms in larval settlement have just begun to be explored (Wieczorek and Todd 1998; Holmstrøm and Kjelleberg 2000; Qian et al. 2003). This holds particularly for aspects of biofilm dynamics, both in the process of biofilm development and at a mature stage. Biofilms change phenotypically and genetically according to internal (e.g., age, bioactive compounds, competition) and external environmental conditions (Todd and Keough 1994; Cooksey and Wigglesworth-Cooksey 1995; Wimpenny 2000).

In this review, we pursue the hypothesis that environmentally caused variations of biofilms are critical for settlement processes and the structure of resulting communities. We put forward a conceptual triangle model to describe how the environment, microbes in biofilm, and the settling propogates of marine organisms can interact. The model also describes how these interactions change structure, composition, and bioactivities of biofilms and, in turn, affect possible distribution patterns of marine macroorganisms. Environmental factors could be biotic (e.g., grazing, competition, physical disturbance) or abiotic, but we will focus on the latter. We particularly attempt in this review to discuss, with respect to larval settlement, the unavoidable phenomena of biofilm development in assay wells and the effects of biofilms on compound uptake and compound production.

2 Triangle Model

A triangle model is introduced to provide a conceptual framework for interactions between environmentally caused changes in biofilm bioactivity that in turn affect the settlement of dispersal stage (Fig. 1). Biofilm cues that affect potential colonizers

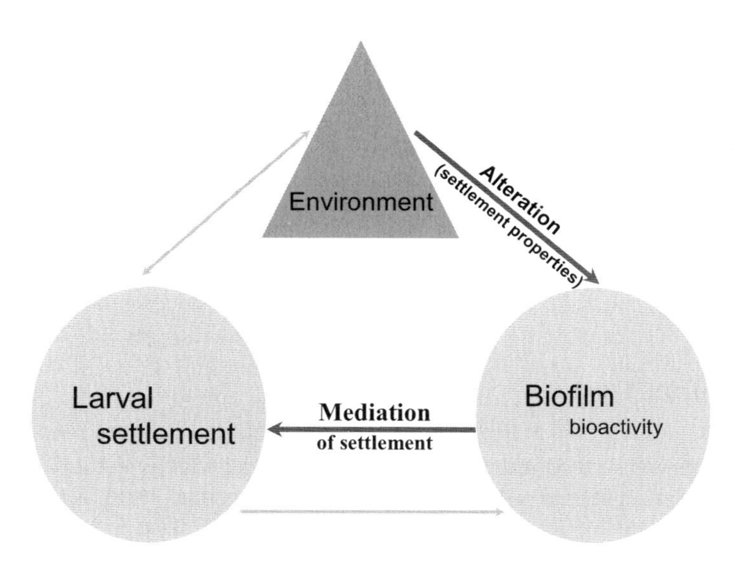

Fig. 1 Elements of a triangle model that demarcate environmental effects on biofilm that affect larval settlement of marine substrata

differentially by providing either inhibiting or attracting colonization cues can be caused by environmentally caused dynamic changes in biofilm properties, such as in the abundance and composition of microorganisms in biofilms, and in the physical structure, composition, and physiology of biofilms.

3 Biofilm Properties

Biofilm properties that are relevant for the mediation of larval settlementand are prone to environmental changes include age, taxon diversity, density, quorum sensing, extracellular polymers (EPS), and other metabolites (see reviews by Dobretsov et al. 2006, Qian et al. 2007). Any environmental factor that affects such biofilm properties in turn would also affect larval settlement.

3.1 Biofilm Taxon Diversity

Several studies indicate that settlement responses are determined by bacterial community composition rather than by cell density or biomass (Dahms et al. 2004; Lau et al. 2005; Dobretsov and Qian 2006). So far, no correlation could be shown between the phylogenetic affiliation of bacteria and their effects on larval or spore settlement (Lau et al. 2002; Patel et al. 2003; Dobretsov and Qian 2004) although more species in the genera *Peudoaltermonos, Altermonos, Vibrio, Streptomyces* and *Actenomyces* appear to be bioactive (see review by Dobretsov et al. 2006).

3.2 Biofilm Density

Larval settlement of *Hydroides elegans* differ only slightly among biofilms developed at different salinities, but not among those developed at different temperatures (Lau et al. 2005). This settlement response was moderately correlated with bacterial density but had no relationship with bacterial community composition of the biofilm.

3.3 Biofilm Age

Maki et al. (1988) showed that the effect of bacteria from natural biofilms that inhibit barnacle attachment depend on biofilm age. Laboratory experiments of Lau et al. (2003a) demonstrated that inductive bacterial strains were more active in their stationary phase than in their log phase. Since biofilm age is generally correlated with biofilm thickness or bacterial density, the latter will provide a good indicator of substratum stability.

3.3.1 Quorum Sensing

The capacity for intercellular communication has major effects on the formation and community structure of biofilms (Dobretsov et al. 2007, Huang et al. 2007b). Gram-negative bacteria may communicate between each other through the use of quorum sensing signals, such as *N*-acetyl-L-homoserine lactones (AHL; see Parsek et al. 1999; Huang et al. 2007a). These small molecules act as extracellular signals that activate transcription and modulate physiological processes when accumulated in the presence of increased cell densities (Parsek and Greenberg 2005).

3.4 Extracellular Polymers (EPS)

Microbial cells in biofilms are enmeshed in an extensive matrix of extracellular polysaccharides (Decho 1990). Its extensive mucoid network facilitates the attachment of bacteria (Stevenson and Peterson 1989) and other microbes as well as the settlement of invertebrate larvae and algal spores (Holmstrøm and Kjelleberg 1999, Lau et al. 2003a). The EPS production of diatoms is largely affected by environmental parameters (Wolfstein and Stal 2002) such as the provision of nutrients and light. Differences (i.e., molecular weight and monomer composition) in EPS obtained from diatoms grown under different environmental conditions (temperature and salinity) are reflected in the larval settlement response (Lam et al. 2005b), confirming the ability of larvae to distinguish between biofilms of varying composition, physiological condition, and growth phase (Wieczorek and Todd 1998).

3.5 Biofilm Chemical Compound Diversity

Marine microbes are a potent source of bioactive compounds. But, so far, only a limited number of marine microbes have been screened and only a few antisettlement compounds have been isolated and identified (Dobretsov et al. 2006). This holds particularly for microbes enmeshed in biofilms. Recently, five antifouling diketopiperazines were isolated from the deep sea bacterium *Streptomyces fungicidicus* (Li et al. 2006) and more bioactive compounds were isolated from other microbes (Yang et al. 2006; Xu et al. 2007).

The synthesis of bioactive metabolites by microbes changes with environmental conditions (Kjelleberg et al. 1993, Miao et al. 2006; Yang et al. 2007). For example, bacterial strains of the same species can produce different compounds under different cultural or environmental conditions, and provide a variable mount of bioactive compounds (Armstrong et al. 2001). In entire multispecies biofilms, changes in the type and amount of compound production are likely to occur as well (Cooksey and Wigglesworth-Cooksey 1995; Qian et al. 2007).

4 Environmental Effects on Biofilm Development and Variability

Biofilm community alterations can be caused by abiotic factors (e.g., depth, illumination, exposure time, tidal height, flow regime, physical disturbance, latitude, season, water chemistry, nutrient supply; see Characklis and Marshall 1990) or by biotic factors (e.g., availability and physiological condition of colonizing species, competition and cooperation among species, biological disturbance; see Clare et al. 1992; Fenchel 1998; Dahms and Qian 2005). Spatial variability of biofilms provides neocolonization possibilities that also affect the vertical distribution of microbes (Caldwell et al. 1993; Costerton et al. 1995). Heterogeneity with vertical depth is of particular relevance for the interpretation of successional events following disturbances that create open spaces within natural aged communities, where space is commonly limited (e.g., Butler and Chesson 1990). The conceptual frame of "patch-dynamics" in metapopulations (Wright et al. 2004) suggests that disturbances provide neocolonization opportunities at any developmental stage of a community in a mosaic fashion (Butler and Chesson 1990). Such a stochastic concept is much more applicable to the process of biofilm formation and subsequent overgrowth by macroorganisms than any scenario of gradual succession (see Henschel and Cook 1990). Colonization *in situ* is a dynamic process, where chance effects rather than deterministic processes become prevailant. Some phases in the colonization process may be accelerated or slowed down, occur reversely or simultaneously (Palmer and White 1997).

Biofilm community succession can be affected by a number of ecological, biological, and physiological events initiated by primary colonizers, as well as by surface modifications, which determine the types/species of microbes to be recruited as secondary colonizers (Costerton et al. 1995). Synergistic and/or competitive interactions among colonizers, together with the arrival of new recruits and/or loss of previous colonists, continuously shape the biofilm community (Wimpenny 2000). As the thickness of the biofilm increases, sharp vertical gradients as well as horizontal patches of pH, dissolved oxygen, and metabolic byproducts usually develop within biofilms.

Colonization events are thus suggested to be ruled predominantly by the following factors:

1. Qualitative and quantitative aspects of colonizers (i.e., which and how many can approach and attach to a surface) that themselves provide particular biotic functions
2. Physical and chemical conditions of the seawater/substratum interface

Biotic parameters also include microbes with good adhesion properties, which would have a selective advantage even under turbulent conditions (Beech et al. 2000). This ability would be enhanced by the secretion of EPS that is resistant to high fluid shear and chemical agents (Stewart 2002). Microbes may retain the ability to detach from biofilms when conditions become unfavorable (Maki 1999).

5 Abiotic Environmental Factors that Structure Biofilms and their Bioactivity

The relationship between habitat and biofilm community is tight so that the structure, composition, and/or physiology of a biofilm community will effectively reflect key environmental factors at substratum interfaces. Environmental gradients and changes in abiotic factors that affect the bioactivity of biofilms include: depth, illumination, exposure time, latitude, season, water chemistry, nutrient supply, and substratum characteristics.

5.1 *Light*

Photosynthesis in diatoms and cyanobacteria can result in extra amounts of EPS exudates (Wolfstein and Stal 2002), with effects that are mentioned above. Hung et al. (2005a, b) studied whether either UV-A or UV-B radiation can indirectly affect larval attachment of barnacles by altering the settlement bioactivity of biofilms. Both UV-A and UV-B caused a decrease in the percentage of respiring bacterial cells in microbial films and this effect increased with an increase of UV energy. At the same energy level, UV-B caused a greater decrease in respiring bacterial cell densities than UV-A (Hung et al. 2005a, b). However, despite strong UV radiation, the bioactivity of biofilm that mediates cyprid settlement remained unchanged, indicating that increased UV radiation may not significantly affect the barnacle recruitment by means of affecting the inductive larval attachment cues of microbial films (Hung et al. 2005a). In contrast, larval settlementof *Hydroides elegans* decreased with increased UV radiation, indicating that enhanced UV radiation may have a significant effect on the larval settlement of *H. elegans* by affecting a biofilm's inductive cues (Hung et al. 2005b). These findings suggest that the effects of light on biofilm bioactivity will depend on the larval/spore's response to biofilm properties.

5.2 *Flow*

Under turbulent conditions, bacteria that are capable of rapid adhesion have an advantage for settling and growing. Adhesion is enhanced by the secretion of EPS that is resistant to high fluid shear (Ophir and Gutnick 1994). This way, flow can structure microbial communities that are shown to affect settlement differentially. However, there is no study hitherto of hydrodynamic effects on larval settlement (Qian et al. 1999, 2000) that has considered the effects of flow on biofilms.

5.3 Temperature and Salinity

Lau et al. (2005) studied temperature and salinity effects on the density and total biomass of bacterial communities that in turn affected the settlement of barnacles and a tubeworm. Larval settlement of *Balanus amphitrite* and *B. trigonus* was induced by biofilms developed at high temperatures (23°C and 30°C), but was unaffected (*B. amphitrite*) or inhibited (*B. trigonus*) by those developed at low temperature (16°C). The settlement response of these barnacles did not correlate with the biomass or the bacterial density of the biofilms, but did coincide with marked differences in bacterial community compositions at different temperatures. Chiu et al. (2006) found variations in microbial community structure of microbial films as determinants in the control of larval metamorphosis. Microbial films that developed at higher temperatures (23°C and 30°C) induced higher rates of larval metamorphosis than biofilms developed at lower temperatures (16°C). However, no significant conclusion on the interactive effect between temperature and salinity and larval settlement could be drawn.

5.4 Nutrients

The chemical composition (e.g., nutrient load) of ambient waters strongly determines the number, diversity, and metabolic states of planktonic bacteria, as well as their tendency to adhere to surfaces (Schneider and Marshall 1994). Until now there has been no study available that links nutrients, biofilms, and colonization *in situ*. In the laboratory, Huang et al. (2007b) demonstrated that nutrient availability and de novo protein synthesis mediated biofilm formation of *Pseudoalteromonas spongiae* under static and starving conditions, which in turn affected the inductiveness of biofilms for the larval settlementof *H. elegans*. The effects of organic substances in the form of amino acids on biofilm bioactivity were studied by Jin and Qian (2004, 2005). They found that aspartic acid and glutamic acid significantly increased bacterial abundance, modified the bacterial community structures of biofilms, and elevated the inductive effect of biofilms. Alanine and asparagine increased, while isoleucine decreased, the bioactivity of biofilms by changing their bacterial species composition, but not the bacterial density. Leucine, threonine, and valine did not alter bacterial community structures or bioactivities of the biofilm in that study. In a recent study, Hung et al. (2007) found that biofilm developed at the same intertidal height of different habitats with contrasting environmental conditions showed remarkable differences in bioactivity for barnacle larval settlement, suggesting that the nutrient condition at different habitats is the key factor governing the different bioactivities of those biofilms.

5.5 Intertidal Versus Subtidal Biofilms

In experiments with *Bugula neritina*, larvae preferentially attached to subtidal biofilms (Dobretsov and Qian 2006). In the latter study, subtidal biofilms, diatom density, EPS thickness, and biofilm age, but not bacterial density, correlated positively with enhanced larval attachment. Minchinton and Scheibling (1993) showed that the diatom *Achnantes parvula* guided the attachment of the barnacle *Semibalanus balanoides* on high intertidal regions. In a study by Qian et al. (2003), cyprids of *B. amphitrite* preferred intertidal biofilms (i.e., 6-day old) over unfilmed surfaces for attachment. Cyprids also preferred biofilms of mid-intertidal height over high-intertidal or subtidal heights. There was no correlation between attachment and any of the three biofilm attributes (i.e., biomass, abundance of bacteria and diatoms). Qian et al. (2003) therefore concluded that changes in bacterial community profiles in the biofilm affected the attractiveness of the biofilm to barnacle larvae. In our previous study, we found that mid-intertidal biofilms induced cyprid settlement of *B. amphitrite*, while subtidal biofilms from the same site did not induce cyprid settlement of *B. trigonus* (Thiyagarajan et al. 2005).

5.6 Seasonality

One possible cause of temporal variation of immigration responses to microbial communities could be temporal differences in the "sensitivity" of biota to microbial cues (i.e., a form of intrinsic variability). There is substantial evidence for behavioral differences on the population level, as well as of cohorts or generations. Seasonally distinct behavioral responses at immigration would appear to be adaptive where these are related to variations in selection pressure, such as the seasonality of competitors for resources or predators (Raimondi and Keough 1990). The composition, quantity, and metabolic characteristics of biofilm communities change seasonally in the field (e.g., Anderson 1995) and it seems reasonable that immigrating organisms respond to these temporal alterations.

Wieczorek et al. (1996) showed marked seasonal variations in the effects of biofilm cues on the larval settlement of certain marine invertebrate groups and taxa to hard substrata under natural conditions, which are not to be explained by larval availability alone. Also, a reversal of the biofilm effect on larval settlement response with season, from inhibitory to facilitatory, was noted for certain species.

This may also hold for biochemical compounds (Cooksey and Wigglesworth-Cooksey 1995; Qian et al. 2007), but has not been characterized as yet – neither under laboratory nor under field conditions. It is necessary to monitor biofilm cue production and release rate under natural conditions where the cells occur in heterogeneous consortia within biofilms (Paerl and Pinckney 1996).

6 Conclusions

It has to be emphasized that relevant settlement signals have not reasonably been characterized as yet, neither for facilitative nor for inhibitory biofilms. Also, responses of invading biota to monospecific biofilms in the laboratory may not reflect responses to complex microbial communities in the field. This emphasizes the need for long-term assessments of biofilm effects on settlement, under field conditions, if appropriate conclusions are to be drawn about species-specific larval responses to biofilms with the consequence of community alterations.

Besides knowing settlement-mediating chemical signals that are produced by biofilms, the producers of inhibitive or inductive chemical cues need to be identified. In addition, any synergistic effects in heterogeneous consortia within multispecies biofilms need to be studied. As emphasized in this review, we particularly need to investigate the environmental conditions that modify entire multispecies biofilm bioactivity, since the settlement-mediating bioactivity of biofilms varies with environmental conditions in the laboratory as under natural conditions. The investigation of dynamic biofilm bioactivity is complicated by several biofilm properties, such as cell density, biofilm thickness, structural alterations, microbial taxon diversity, bioactive compound diversity, and the differential production of compounds. New methods need to be developed that allow one to genetically identify and measure microbial abundances and diversity, and to analyze such functions as the production and storage of toxic, deterring, attracting or biocommunicative compounds that mediate the colonization of invertebrate larval settlers in the marine environment.

Acknowledgements This contribution is supported by a RGC grant (HKUST6402/05M) and a COMAR grant (COMRRDA06/07.SC01 and the CAS/SAFEA International Partnership Program for Creative Research Team) to P-Y Qian.

References

Anderson MJ (1995) Variations in biofilms colonizing artificial surfaces: seasonal effects and effects of grazers. *J Mar Biol Assoc UK* 75:705–714

Armstrong E, Yan L, Boy KG, Wright PC, Burgess JG (2001) The symbiotic role of marine microbes on living surfaces. *Hydrobiologia* 461:37–40

Beech IB, Gubner R, Zinkevich V, Hanjangsit L, Avci R (2000) Characterisation of conditioning layers formed by exopolymeric substances of *Pseudomonas* NCIMB 2021 on surfaces of AISI 316 stainless steel. *Biofouling* 16(2–4):93–104

Butler AJ, Chesson PL (1990) Ecology of sessile animals on subtidal hard substrata: the need to measure variation. *Aust J Ecol* 15:521–531

Caldwell DE, Korber DR, Lawrence JR (1993) Analysis of biofilm formation using 2D vs 3D imaging. *J Appl Bacteriol Symp Suppl* 74:52–66

Callow ME, Callow JA (2000) Substratum location and zoospore behavior in the fouling alga *Enteromorpha*. *Biofouling* 15:49–56

Characklis WG, Marshall KC (1990) Biofilms: a basis for an interdisciplinary approach. In: Characklis WG, Marshall KC (eds.) Biofilms. Wiley, New York, pp. 3–15

Chia FS (1989) Differential larval settlement of benthic marine invertebrates. In: Ryland JS, Tyler PA (eds.) Reproduction, genetics and distributions of marine organisms. Olsen and Olsen, Fredensborg, Denmark, pp. 3–12

Chiu JMY, Thiyagarajan V, Tsoi MMY, Qian P-Y (2006) Qualitative and quantitative changes in marine biofilms as a function of temperature and salinity in summer and winter. *Biofilms* 1–13

Clare AS, Rittschof D, Gerhart DJ, Maki JS (1992) Molecular approaches to nontoxic antifouling. *Invert Reprod Develop* 22(1–3):67–76

Cooksey KE, Wigglesworth-Cooksey B (1995) Adhesion of bacteria and diatoms to surfaces in the sea: a review. *Aquat Microbial Ecol* 9:87–96

Costerton JW, Lewandowski Z, de Berr D, Caldwell D, Kroner D, James G (1995) Microbial Biofilms. *Annu Rev Microbiol* 49:711–745

Dahms HU, Qian PY (2005) Exposure of biofilms to meiofaunal copepods affects the larval settlement of *Hydroides elegans* (Polychaeta). *Mar Ecol Prog Ser* 297:203–214

Dahms HU, Dobretsov S, Qian PY (2004) The effect of bacterial and diatom biofilms on the settlement of the bryozoan *Bugula neritina*. *J Exp Mar Biol Ecol* 313:191–209

Decho AW (1990) Microbial exopolymer secretions in ocean environments: their role(s) in food webs and marine processes. *Oceanogr Mar Biol Ann Rev* 28:73–153

Dobretsov S, Qian PY (2004) The role of epibotic bacteria from the surface of the soft coral *Dendronephthya* sp. in the inhibition of larval settlement. *J Exp Mar Biol Ecol* 299:35–50

Dobretsov S, Qian PY (2006) Facilitation and inhibition of larval attachment of the bryozoan *Bugula neritina* in association with mono-species and multi-species biofilms. *J Exp Mar Biol Ecol* 333:263–264

Dobretsov S, Dahms HU, Qian PY (2006) Inhibition of biofouling by marine microorganisms and their metabolites. *Biofouling* 22:43–45

Dobretsov S, Dahms HU, Huang YL, Wahl M, Qian PY (2007) The effect of quorum sensing blockers on the formation of marine microbial communities and larval attachment. *FEMS Microbiol Ecol* 60:177–188

Fenchel T (1998) Formation of laminated cyanobacterial mats in the absence of benthic fauna. *Aquat Microbiol Ecol* 14:235–240

Gordon JW (1999) Genetic enhancement in humans. *Science* 283 (5410):2023

Hadfield MG, Paul VJ (2001) Natural chemical cues for settlement and metamorphosis of marine-invertebrate larvae: 431–461. In: McClintock JB, Baker BJ (eds.) Marine chemical ecology. CRC, Boca Raton, pp. 1–610

Harder T, Lam C, Qian PY (2002a) Induction of larval settlement in the polychaete *Hydroides elegans* by marine biofilms: an investigation of monospecific diatom films as settlement cues. *Mar Ecol Prog Ser* 229:105–112

Henschel JR, Cook PA (1990) The development of a marine fouling community in relation to the primary film of microorganisms. *Biofouling* 2:1–11

Holmstrøm C, Kjelleberg S (1999) Marine *Pseudoalteromonas* species are associated with higher organisms and produce active extracellular compounds. *FEMS Microbiol Ecol* 30:285–293

Holmstrøm C, Kjelleberg S (2000) Bacterial interactions with marine fouling organisms. In: Evans LV (ed.) Biofilms: recent advances in their study. Harwood Academic, Amsterdam, pp. 101–117

Holmstrøm C, Rittschof D, Kjelleberg S (1992) Inhibition of settlement by larvae of *Balanus amphitrite* and *Cliona intestinalis* by a surface-colonizing marine bacterium. *Appl Environ Microb* 58:2111–2115

Huang YL, Dobretsov S, Ki JS, Yang LH, Qian PY (2007a) Presence of acyl-homoserine lactone in subtidal biofilm and the implication in larval behavioral response in the polychaete *Hydroides elegans*. *Microbial Ecol* 54:384–392

Huang YL, Dobretsov SV, Xiong HR, Qian PY (2007b) Effect of biofilm formation by *Pseudoalteromonas spongiae* on induction of larval settlement of the polychaete *Hydroides elegans*. *Appl Environ Microbiol* 73:6284–6288

Hung OS, Gosselin LA, Thiyagarajan V, Wu RSS, Qian PY (2005a) Do effects of ultraviolet radiation on microbial films have indirect effects on larval attachment of the barnacle *Balanus amphitrite*? *J Exp Mar Biol Ecol* 323:16–26
Hung OS, Thiyagarajan V, Wu RSS, Qian PY (2005b) Effect of ultraviolet radiation on biofilms and subsequent larval settlement of *Hydroides elegans*. *Mar Ecol Prog Ser* 304:155–166
Hung OS, Thiyagarajan V, Zhang R, Wu RSS, Qian PY (2007) Attachment of *Balanus amphitrite* larvae to biofilms originated from contrasting environments. *Mar Ecol Prog Ser* 333:229–242
Jin T, Qian PY (2004) Effect of mono-amino acids on larval metamorphosis of the polychaete *Hydroides elegans*. *Mar Ecol Prog Ser* 267:223–232
Jin T, Qian PY (2005) Amino acid exposure modulates the bioactivity of biofilms for larval settlement of *Hydroides elegans* by altering bacterial community components. *Mar Ecol Prog Ser* 297:169–179
Keough MJ, Raimondi PT (1996) Responses of settling invertebrate larvae to bioorganic films: effects of large-scale variation in films. *J Exp Mar Biol Ecol* 207:59–78
Kierek-Pearson K, Karatan E (2005) Biofilm development in bacteria. *Adv Appl Microbiol* 57:79–111
Kirchman D, Graham D, Reish D, Mitchell R (1982) Lectins may mediate in the settlement and metamorphosis of *Janua (Dexiospira) brasiliensis* Grube (Polychaeta: Spirorbidae). *Mar Biol Lett* 3:201–222
Kjelleberg S, Albertson N, Flardh K, Holmquist L, Jouper-Jaan A, Marouga R, Ostling J, Svenblad B, Weichart D (1993) How do non-differentiating bacteria adapt to starvation? *Antonie Leeuwenhoek* 63:333–341
Lam C, Harder T, Qian PY (2005a) Induction of larval settlement in the polychaete *Hydroides elegans* by extracellular polymers of benthic diatoms. *Mar Ecol Prog Ser* 286:145–154
Lam C, Harder T, Qian PY (2005b) Growth conditions of benthic diatoms affect quality and quantity of extracellular polymeric larval settlement cues. *Mar Ecol Prog Ser* 294:109–116
Lau SCK, Mak KKW, Chen F, Qian PY (2002) Bioactivity of bacterial strains from marine biofilms in Hong Kong waters for the induction of larval settlement in the marine polychaete *Hydroides elegans*. *Mar Ecol Prog Ser* 226:301–310
Lau SCK, Harder T, Qian PY (2003a) Induction of larval settlement in the serpulid polychaete *Hydroides elegans* (Haswell): role of bacterial extracellular polymers. *Biofouling* 19:197–204
Lau SCK, Thiyagarajan V, Qian PY (2003b) The bioactivity of bacterial isolates in Hong Kong waters for the inhibition of barnacle (*Balanus amphitrite* Darwin) settlement. *J Exp Mar Biol Ecol* 282:43–60
Lau SCK, Thiyagarajan V, Cheung SCK, Qian PY (2005) Roles of bacterial community composition in biofilms as a mediator for larval settlement of three marine invertebrates. *Aquat Microb Ecol* 38:41–51
Li X, Dobretsov S, Xu Y, Xiao X, Qian PY (2006) Antifouling diketopiperazines produced by a deep-sea bacterium *Streptomyces fungicidicus*. *Biofouling* 22:201–208
Maki JS, Rittschof D, Costlow JD, Mitchell R (1988) Inhibition of attachment of larval barnacles, *Balanus amphitrite*, by bacterial surface films. *Mar Biol* 97:199–206
Maki JS (1999) The influence of marine microbes on biofouling. In: Fingerman M, Nagabhushanam R, Thompson F (eds.) Recent advances in marine biotechnology, vol 3. Biofilms, bioadhesion, corrosion, and biofouling. Science, Enfield, NH, pp. 147–171
Miao L, Kwong TFN, Qian PY (2006) Effect of culture conditions on the mycelial growth, antibacterial activity and metabolite profiles of the marine-derived fungus *Arthrinium* cf *saccharicola*. *Appl Microbiol Biotechnol* 72:1063–1073
Minchinton TE, Scheibling RE (1993) Variations in sampling procedure and frequency affect estimates of recruitment of barnacle. *Mar Ecol Prog Ser* 99:83–88
Neal AL, Yule AB (1994) The tenacity of *Elminius modestus* and *Balanus perforatus* cyprids to bacterial films grown under different shear regimes. *J Mar Biol Assoc UK* 74:251–257
Ophir T, Gutnick DL (1994) A role for exopolysaccharides in the protection of microorganisms from dessication. *Appl Environ Microbiol* 60:740–745

Paerl HW, Pinckney JL (1996) A mini-review of microbial consortia: their roles in aquatic production and biogeochemical cycling. *Microb Ecol* 31:225–247

Palmer RJ, White DC (1997) Developmental biology of biofilms: implications for treatment and control. *Trends Microbiol* 5:435–439

Parsek MR, Greenberg EP (2005) Sociomicrobiology: the connections between *quorum sensing* and biofilms. *Trends Microbiol* 13:27–33

Parsek MR, Val DL, Hanzelka BL, Cronan JEJr, Greenberg EP (1999) Acyl homoserine-lactone quorum-sensing signal generation. *Proc Natl Acad Sci U S A* 96(8):4360–4365

Patel P, Callow ME, Joint I, Callow JA (2003) Specificity in larval settlement modifying response of bacterial biofilms towards zoospores of the marine alga *Enteromorpha*. *Environ Microb* 5:338–349

Pawlik JR (1992) Chemical ecology of the settlement of benthic marine invertebrates. *Oceanogr Mar Biol Ann Rev* 30:273–335

Qian PY (1999) Larval settlement of polychaetes. *Hydrobiologia* 402:239–253

Qian PY, Rittschoff D, Sreedhar B, Chia FS (1999) Macrofouling in unidirectional flow: miniature pipes as experimental models for studying the effects of hydrodynamics on invertebrate larval settlement. *Mar Ecol Prog Ser* 191:141–151

Qian PY, Rittschof D, Sreedhar B (2000) Macrofouling in unidirectional flow: miniature pipes as experimental models for studying the interaction of flow and surface characteristics on the attachment of barnacle, bryozoan and polychaete larvae. *Mar Ecol Prog Ser* 207:109–121

Qian PY, Thiyagarajan V, Lau SCK, Cheung S (2003) Relationship between community and the attachment of acorn barnacle *Balanus amphitrite* Darwin. *Aquat Microbiol Ecol* 33:225–237

Qian PY, Lau SCK, Dahms HU, Dobretsov S, Harder T (2007) Marine biofilms as mediators of colonization by marine macroorganisms implications for antifouling and aquaculture. *Mar Biotech* 9:399–410

Raimondi PT, Keough MJ (1990) Behavioural variability in marine larvae. *Aust J Ecol* 15:427–437

Rodriguez SR, Ojeda FP, Inestrosa NC (1993) Settlement of benthic marine invertebrates. *Mar Ecol Prog Ser* 97:193–207

Schneider RP, Marshall KC (1994) Retention of the Gram negative marine bacterium SW8 on surfaces – effects of microbial physiology, substratum nature and conditioning films. *Colloids Surf B: Biointerfaces* 2:387–396

Stevenson RJ, Peterson CG (1989) Variation in benthic diatom (Bacillariophyceae) immigration with habitat characteristics and cell morphology. *J Phycol* 25:120–129

Stewart PS (2002) Mechanisms of antibiotic resistance in bacterial biofilms. *Int J Med Microbiol* 292:107–113

Szewzyk U, Holmstroem C, Wrangstadh M, Samuelsson MO, Maki JS, Kjelleberg S (1991) Relevance of the exopolysaccharide of marine *Pseudomonas* sp. strain S9 for the attachment of *Ciona intestinalis* larvae. *Mar Ecol Prog Ser* 75:259–265

Thiyagarajan V, Hung OS, Chiu JMY, Wu RSS, Qian PY (2005) Growth and survival of juvenile barnacle *Balanus amphitrite*: interactive effects of cyprid energy reserve and habitat. *Mar Ecol Prog Ser* 299:229–237

Todd CD, Keough MJ (1994) Larval settlement in hard substratum epifaunal assemblages: a manipulative field study of the effects of substratum filming and the presence of incumbents. *J Exp Mar Biol Ecol* 181:159–187

Underwood AJ, Fairweather PG (1989) Supply side ecology and benthic marine ecology. *Trends Ecol Evol* 4:16–20

Wahl M (1989) Marine epibiosis. I. Fouling and antifouling: some basic aspects. *Mar Ecol Prog Ser* 58:175–189

Wieczorek SK, Todd CD (1998) Inhibition and facilitation of settlement of epifaunal marine invertebrate larvae by microbial biofilm cues. *Biofouling* 12:81–118

Wieczorek SK, Murray AWA, Todd CD (1996) Seasonal variation in the effects of hard substratum biofilming on settlement of marine invertebrate larvae. *Biofouling* 10:309–330

Wimpenny J (2000) An overview of biofilms as functional communities. In: Allison D, Gibert P, Lappin-Scott H, Wilson M (eds.) Community structure and co-operation in biofilms. Cambridge University Press, Cambridge, UK, pp. 1–24

Wolfstein K, Stal LJ (2002) Production of extracellular polymeric substances (EPS) by benthic diatoms: effect of irradiance and temperature. *Mar Ecol Prog Ser* 236:13–22

Wright JP, Gurney WSC, Jones CG (2004) Patch dynamics in a landscape modified by ecosystem engineers. *Oikos* 105:336–348

Xu Y, Miao L, Li XC, Xiao X, Qian PY (2007) Antibacterial and antilarval activity of deep-sea bacteria from sediments of the West Pacific Ocean. *Biofouling* 23:131–137

Yang LH, Li XC, Yan SC, Sun HZ, Qian PY (2006) Antifouling properties of 10b-formamidodalihinol-A and kalihinol A isolated from the marine sponge *Acanthella cavernosa*. *Biofouling* 22:23–32

Yang LH, Xiong HR, Lee OO, Qi SH, Qian PY (2007) Effect of agitation on violacein production in *Pseudoalteromonas luteoviolacea* isolated from a marine sponge. *Lett Appl Microbiol* 44:625–630

Index

Printing: Krips bv, Meppel, The Netherlands
Binding: Stürtz, Würzburg, Germany